Recovery of Byproducts from Acid Mine Drainage Treatment

Scrivener Publishing
100 Cummings Center, Suite 541J
Beverly, MA 01915-6106

Publishers at Scrivener
Martin Scrivener (martin@scrivenerpublishing.com)
Phillip Carmical (pcarmical@scrivenerpublishing.com)

Recovery of Byproducts from Acid Mine Drainage Treatment

Edited by

Elvis Fosso-Kankeu,
Christian Wolkersdorfer and Jo Burgess

Scrivener Publishing

WILEY

This edition first published 2020 by John Wiley & Sons, Inc., 111 River Street, Hoboken, NJ 07030, USA and Scrivener Publishing LLC, 100 Cummings Center, Suite 541J, Beverly, MA 01915, USA

For more information about Scrivener publications please visit www.scrivenerpublishing.com.

Wiley Global Headquarters
111 River Street, Hoboken, NJ 07030, USA

For details of our global editorial offices, customer services, and more information about Wiley products visit us at www.wiley.com.

Library of Congress Cataloging-in-Publication Data

ISBN 978-1-119-62007-5

Cover image: Pixabay.Com
Cover design by Russell Richardson

Set in size of 11pt and Minion Pro by Manila Typesetting Company, Makati, Philippines

Printed in the USA

Contents

Preface

Mining still plays a considerable role in supporting the economy of various countries around the world. However, its activities can have negative effects on both the ecosphere and the anthroposphere, when the generation and deposition of solid and liquid residues is not managed properly. Poor management of such residues, especially when the mining company just focuses on treating the pollution, might overlook that prevention and reduction of waste results in a smaller footprint and in reducing their liability. As these mining residues can be considered resources, they could even contribute to defraying their management costs.

Mining residues result from the exploration, mining, and processing of the rocks and ores from mines and quarries. They mainly consist of unaltered natural material from crushing, ore processing, or enrichment and might contain chemicals or inorganic and organic additives in the tailings material. These solid wastes are usually deposited as tailings dams and waste rock dumps. Globally, these residues constitute the largest, single source of waste. Tailings dams in landfill sites can contribute to environmental pollution through various ways: contamination of streams by acid mine drainage (AMD) or acid rock drainage (ARD), contamination of streams due to surface run-off from the impoundment area, air and water contamination due to wind erosion of dried-out tailings, and physical and aesthetic modifications of the environment.

AMD and ARD are acidic mining-impacted waters with a pH below 5.6. It is usually highly mineralized with elevated concentrations of potentially toxic elements. They result from the oxidation of pyrite, marcasite, or pyrrhotite when exposed to oxygen and water. Major sources of AMD include drainage from underground mine shafts, runoff and discharges from open pits, and ARD emerges from mine waste dumps, tailings, and ore stockpiles. Yet, it should be mentioned that, by far, not all mine water is acidic—there are many mine sites which discharge mine water of a quality that is not harmful to the environment.

According to regulations, adherence to good practices requires that three main approaches must be considered for mining residue management: rehabilitation, recovery, and reuse. This minimizes the negative effects on the environment, ensuring optimum use of natural resources and reducing waste. Therefore, the recovery of residual values from tailings is an important option that can be optimized through the use of sustainable, affordable, and effective technologies that minimize waste by-products. Furthermore, proper design, construction, and performance monitoring of tailings cover systems is required to ensure control of contaminant and acidity loading. This could be better achieved through using suitable models, allowing more accurate predictions even in different climatic regions.

Mine water formation is considered a serious environmental threat worldwide and is mainly caused by mining activities. Recommended approaches for the treatment of AMD/

ARD require treatment beyond the neutralization and desalination processes to reduce waste on site as much as possible. Often, the sludge produced during neutralization and desalination is enriched with various metals. Therefore, it has the potential to be recycled or used to create products with commercial value. This requires an integrated and interdisciplinary approach to drive the green and circular economy.

This book presents 11 chapters focusing on the management of tailing dams, prediction of acidic mine water formation, established treatment methods that have successfully converted acidic mine water to usable water, and approaches to using acidic mine water as a resource for relevant commodities.

Our aim was to show the broad range of research conducted about the valorization and treatment of acidic mine water. It also shows the variability in expressing what is important for an individual author or research group, providing the reader insights into the thinking about acidic mine water and solving issues around it, and we kept some repetition to allow using an individual chapter by itself. The chapters also show that it is still a long way to go before we can solve the mine water issue in its full scale. A lot of work has been done since the first papers on high density sludge were published in the 1970s, but these chapters show that there is still a lot of research and development needed to find the silver bullet of mine water treatment. Therefore, the chapters also outline the shortcomings of various technologies or the problems which still persist with the valorization of mine water or mining residues.

Although this edition does not extensively cover geochemical predictions, we however acknowledge that they could be considered to improve future mine water treatment and recovery of valuable solutes.

The editors and the publisher are grateful to the reviewers who have contributed to improving the quality of the book through their constructive comments. The editors also thank the publisher for including this book in their series.

This book will be of interest to scientists and engineers in academia or industry, working on projects related to the management of solid and liquid wastes from mining activities. We hope that you, our reader, will gain some value from the book and the many references therein.

Elvis Fosso-Kankeu, Christian Wolkersdorfer and Jo Burgess
June 2020

Part 1

PREDICTION AND PREVENTION OF AMD FORMATION

1

Management of Metalliferous Solid Waste and its Potential to Contaminate Groundwater: A Case Study of O'Kiep, Namaqualand South Africa

Innocentia G. Erdogan[1,2,3], Elvis Fosso-Kankeu[1*], Seteno K.O. Ntwampe[2,3], Frans B. Waanders[1] and Nils Hoth[4]

[1]Water Pollution Monitoring and Remediation Initiatives Research Group (WPMRIRG) in the CoE of C-Based Fuels School of Chemical and Minerals Engineering, North-West University Potchefstroom, South Africa
[2]Bioresource Engineering Research Group (BioERG) and Chemical Engineering Department, Cape Peninsula University of Technology, Cape Town, South Africa
[3]Faculty of Engineering, Chemical Engineering Department, Cape Peninsula University of Technology, Cape Town, South Africa
[4]Institute of Mining and Special Civil Engineering, Technische Universität Bergakademie Freiberg, Saxony, Germany

Abstract
South Africa is an important mining region, and nearly every province has remnants of active and closed metalliferous mining sites. Metalliferous mines produce a large quantity of metalliferous solid waste (MSW) because the ore is minutely portioned in comparison to the mined material. In the metalliferous industry, production of Cu, Pb, and Zn culminates in the degradation of the immediate environment with the primary effects of the MSW being contamination of groundwater by dissolved, potentially toxic elements (PTE), and *via* generation of acid rock drainage (ARD), the acidification of nearby soil, an underlying condition for PTE mobilization. O'Kiep, in particular, is characterized by remnants of a closed metalliferous mine (CMM). This mine ceased production in 2004; albeit, producing MSW which were deposited in the vicinity of the community of O'Kiep. Currently, there are minimal solutions to the challenges associated with ARD from MSW with groundwater deterioration in O'Kiep being a public health concern as the groundwater in this area is severely contaminated by PTEs, attributed to leachate formation and weathering of MSW including demonstrable ARD formation as confirmed by static and humidity cell tests which indicate environmental contamination from the MSW associated with mining activities. Due to its biological and chemical composition, ARD formation in O'Kiep needs to be effectively managed to minimize the negative effects of the MSW in the vicinity of the community.

Keywords: Acid rock drainage, closed metalliferous mines, metalliferous solid waste, potentially toxic elements, groundwater contamination

**Corresponding author*: Elvis.FossoKankeu@nwu.ac.za [ORCID: 0000-0002-7710-4401]
[ORCID: Erdogan, 0000-0001-8754-0277; Ntwampe, 0000-0001-7516-6249; Waanders, 0000-0002-8431-6189]

Elvis Fosso-Kankeu, Christian Wolkersdorfer and Jo Burgess (eds.) Recovery of Byproducts from Acid Mine Drainage Treatment, (3–22) © 2020 Scrivener Publishing LLC

List of Abbreviations

ARD	Acid rock drainage
CMM	Closed metalliferous mine
MSW	Metalliferous solid waste
MT	Metalliferous tailings
PTE	Potentially toxic element
SMW	Stockpiled metalliferous waste
SOM	Stockpiled overburden material
TSF	Tailings storage facility

1.1 Introduction

Mining and mineral processing industries have been the critical focus of research in many countries due to increasingly stringent regulatory conditions as sustainability concerns increase and the effect of global warming and environmental degradation in general [1]. Several non-metalliferous and metalliferous mines globally are situated in arid and semi-arid regions and are operated through both surface and underground mining methods [2]. They have been an important industry in many countries worldwide [3] including South Africa, where metalliferous deposits containing Au, Pt, Zn, and Cu have played and some continue to play a vital role [4].

The economic benefits of metalliferous mines have often surpassed the major challenges posed to the environment in which they are operated as they provide employment to vulnerable communities; albeit, enormous quantities of metalliferous solid waste (MSW) are generated. A number of these challenges include the dust, groundwater, and surface water contamination and an incapability to use the land for developmental purposes during post-mining operations [5]. The closed metalliferous mines (CMMs) and MSW generated pose a serious health risk to ecosystems and neighboring communities due to contaminant dispersion, particularly in the form of potentially toxic elements (PTEs) with special emphasis on Sb, Al, Cr, Cu, Fe, Pb, Mn, Ni, Ag, Zn, and sulfide-bearing minerals [6]. Additionally, MSW are reported to be sources of PTE containing leachates, as oxidation of (di-)sulfides leads to the formation of acid rock drainage (ARD), a process microbially mediated [7], and the creation of secondary contaminants such as PTE *via* dissolution of MSW [8]. Changing climatic patterns, especially in arid regions, further play a contributory role by facilitating the dispersion and mobility of PTEs contributing to minimal vegetation growth [9], which appears to be the case in O'Kiep, Namaqualand, South Africa.

O'Kiep is a former copper mining area with ore mineralization being dominated by Cu-rich sulfidic ores, i.e., bornite and chalcopyrite, which are the most abundant copper-bearing minerals in the area. Some of these constituents undergo oxidation. The CMM which ceased production in 2004 produced a large number of metalliferous tailings (MTs) with approximately 5.8 Mt of material. MSW from the CMMs were deposited in the vicinity of the community of O'Kiep [10, 11].

The potential formation of ARD poses a threat to the environment resulting in the contamination of surface as well as groundwater reservoirs, which became evident in the hydrochemical study of open pit groundwater of O'Kiep [12]. Therefore, such environmental contamination and associated challenges attributed to problems caused by mining activities need to be addressed, as highlighted in the Constitution of the Republic of South Africa, 1996 (Act 108 of 1996) [13]. Generally, South Africa has numerous challenges concerning ARD formation due to previous and current mining practices [14].

O'Kiep is affected by MSW which is a potential cause of diseases in humans [15]. These challenges can be escalated when contaminated water reaches the aquifer, the only valuable groundwater resource especially in areas that have no peripheral rivers. This is also the case for O'Kiep, with the lower Orange River being the only perennial river located approximately 150 km from the town. It is currently a major source of piped potable water for agricultural, domestic, and industrial use [12]. As the lower Orange River is running dry due to climatic changes and minimal rainfall, this community will soon have to rely on groundwater.

Because minimal research has been conducted on environment health and the local community of O'Kiep, this chapter discusses these effects of CMM and MSW and the potential formation of ARD.

1.2 CMMs: Overview and Challenges

The number of CMMs globally is in the hundreds of thousands, and all are capable of generating ARD, increasing PTE solubility; thus mobility, culminating in the degradation of the environment [16, 17]. For example, CMM in Romania known to produce Cu and U is a potential danger to the health of the communities of Anina, Ciudanovița, Lavrion, Lișava, Moldova Nouă, and Năvodari as well as to the surrounding environment [16, 18, 19]. Similarly, a CMM located in Malaysia yielded 2.47 Mt of concentrate comprising of approximately 600 Mt of Cu, 45 t of Au, and 294 t of Ag, produced approximately 250 Mt of stockpiled overburden material (SOM) and MTs of approximately 150 Mt [20]. Another CMM located in Bulgaria producing Cu concentrates created environmental risks for the region, especially for the local aquifer, whereby rainfall fills the pit, forming a water body containing high concentrations of PTE. It was determined that the PTEs were highly mobilized during the wet season, with many reported cases of contamination by PTEs close to the mine [21]. Similarly, a CMM in Alaska with MSW comprising of Cu, Fe, Pb, and Zn sulfide minerals which were exposed to biochemical weathering culminated in CMM and remnant MSW extending into the coastal zone of Prince William Sound with field investigations revealing that the oxidation of sulfidic MSW at these sites will generate ARD with subsequent transportation of PTEs into the marine environment [22].

Thousands of CMMs with MSW containing PTEs from Au, Cu, Pb, and Zn mining in the USA have been reported [23]. These mines once reflected the historical development of the American continents, yet they represent a possible threat to human health with local environments being the repository for the contaminants they generated [24]. CMMs often contain unmined SOM, stockpiled metalliferous waste (SMW) and MTs that weather and leach to the surrounding environment. Several of these CMMs are located on or adjacent to public land. To mitigate such effects, governments needs to provide a wide range of scientific expertise to assist and collaborate with the local municipalities whereby environmental strategies to mitigate adverse environmental effects can culminate in effective remediation of CMMs. In Cyprus, CMMs caused severe off-site ecological challenges and health hazards for the local residents [25]. Groundwater sources were affected by the leaching of PTEs, surface water was contaminated due water erosion, and harmful dust comprising of PTEs spread because of wind erosion. In addition to the ecological hazards associated with the CMMs, several of these locations are aesthetically unpleasant and remain a financial liability to investors and the community in general, due to the downgrading of nearby areas, non-development, and therefore, loss of income. These factors are essential for countries such as Cyprus, whereby tourism is a noteworthy source of revenue for local residents [26]. Similarly, the O'Kiep region is known for its seasonal flowers; however, these flowers, plants, and wild animals have been diminishing with the environment being damaged due to mining activities [27].

1.3 Metalliferous Solid Waste

MSW is the high-volume material estimated at thousand million tons generated annually that originates from the excavation and further chemical and physical processing of a varied range of non-metalliferous and metalliferous minerals by both surface and underground methods [2]. These mines are generally essential for profitability of low-grade ore bodies, and consequently, volumes of MSW are generated. It is produced during the process of beneficiation, extraction, and mineral processing and can be divided into: 1) coarse-grained waste rock material generated during mining, 2) SOM, 3) SMW that are usually stored in heaps, and 4) fine-grained MTs, usually stored in hydraulic-fill structures. Additionally, the oxidation of MSW can generate leachates and gaseous by-products during MSW oxidation [28, 29]. Approximately, one-half of the MSW generated is SMW and one-third is MT, with 61% of the MSW originating from copper-bearing waste.

The SOMs, SMW, and MTs do contain PTEs that are sources of environmental contamination from mining activity. They might pose human health hazards and agricultural produce contamination through contaminated surface or groundwater usage for irrigation purposes, culminating in the uptake of PTEs by vegetation in which animal graze thus further bioaccumulating in the food chains [30]. Both SOMs and SMW generally have reduced water holding capacity, low organic matter content; albeit, with elevated levels of PTEs [31]. Interaction of pyrite-rich SMW and groundwater bodies at CMMs is an environmental concern globally [32]. When the SMW is rich in sulfidic constituents, potentiality of ARD formation and the release of dissolved metals under aerobic conditions can ensue [33].

1.3.1 Stockpiled Overburden Materials

The SOMs are site-specific and differ from one mine to another due to different geological settings and characteristics of the ore being mined. In mining, SOM lies above an area that lends itself to economic exploitation, and SOMs are distinct from MTs, as SOMs are typically not contaminated with high concentrations of PTEs. The mining industry has to handle and dispose of the overburden material [34]. SOMs may also be used to restore and rehabilitate an exhausted mining site to its original condition [35]. Though attempts have been made to use the overburden material in O'Kiep, these SOMs lie above a partially rehabilitated CMM and have consequently been contaminated over time (Figure 1.1).

Figure 1.1 Contaminated overburden material from a partially closed metalliferous mine in O'Kiep.

1.3.2 Stockpiled Metalliferous Waste

Metalliferous mining operations of some ores produce large quantities of waste known as MT and SMW [36]. SMW are made of coarse-grained rock, crushed from larger rocks to fine particles which are heterogeneous and are stored nearby the mining site [37]. The structure of the SMW has a serious effect on the groundwater because of the oxidation of metal sulfides present and because it often occupies large areas, which might result in environmental concerns [38]. Occasionally, SMWs are stored within a tailings storage facility (TSF) to prevent ARD formation, as most SMW contain PTEs, sulfides, and radioactive minerals [39]. SMW can also be a resource of metals and minerals or have other applications at the mine site, such as backfilling of underground mines or for capping of TSF [37]. The chemical and physical characteristics of SMW vary depending on the geological setting, geochemistry, and mineralogy of the ore being mined, and the type of process used to beneficiate that ore [40]. In the case of O'Kiep, the SMW is mixed with slag (Figure 1.2),

Figure 1.2 Stockpiled metalliferous wastes and contaminated soil.

Figure 1.3 Open pit in O'Kiep.

with a depth of approximately 8 m as the mine generated an estimated 5 Mt of slag. During the mine's life, attempts to recover Cu from the slag by flotation were unsuccessful resulting in slags or SMW heaps. Rehabilitation of a mine site can be possible to limit the disposal of the slag, with reinvestigations into the possibility of reprocessing of slag as a Cu resource using advance methods [41]. Also, in Japan, MTs containing pyrite (FeS_2) and substantial quantities of copper and zinc were reported in a study by Tabelin *et al.* [42]. For O'Kiep, MSW severely contaminated nearby soils [43] and acidified the soils, thus increasing the mobility of PTEs [44], to areas in the vicinity of the houses (Figure 1.2). These SMW have become problematic since it constitutes a source of ARD formation and groundwater contamination of the open pit [12] (Figure 1.3).

Similar findings were reported by Chileshe *et al.* [45] which revealed that the SMW in Zambia were highly contaminated with PTEs, which could pose hazards to human health and contaminate groundwater. Namaqualand (South Africa) is well known for its widespread distribution of small mines and has an extensive mining history; currently, it has been listed as a highly contaminated region with major environmental damage [41, 46]. Most SMW in the region was characterized by Cr, Pb, and P which are high-risk PTEs for the local ecosystem and inhabitants of the area. Overall, the wind-blown dust from the SMW and TSF has also been considered a health problem in O'Kiep and could result in respiratory diseases (Figure 1.4). However, rehabilitation of the SMW and TSF remains a part of the challenges for the local population, which must be earmarked for future rehabilitation; albeit, such an undertaking has not materialized yet.

1.3.3 Metalliferous Tailings

South Africa holds approximately 400 TSF covering an estimated 400 km^2 [47]. In Witwatersrand basin alone, 270 TSFs can be found and most of them are unlined and produce SMW leachate. According to Gauteng's Department of Agriculture and Rural Development, these mine residues cover an area of 321 km^2. Biotoxicity of MTs is primarily due to its low pH leachate and the high concentrations of PTEs contained within the waste (Figure 1.5).

Additionally, MTs produced by low grade, high tonnage operations are progressively increasing their environmental footprint across Southern Africa [48] (Table 1.1). MTs often lead to high bulk densities and low infiltration rates and have low organic matter [49].

Figure 1.4 Aerial view of the mine tailings storage facility in O'Kiep, Namaqualand Region South Africa (made with CapeFarmMapper, https://gis.elsenburg.com/apps/cfm), [51].

Similar to SMW, the PTEs in the TSF can pose a severe threat to the environment and human health [50]. MTs are often fine-grained and silt-sized particles with high PTE concentrations. Upon oxidation, MT with high pyritic contents and PTEs are associated with reduced plant productivity, environmental contamination, and human health concerns [36]. It is well known that a high concentration of PTEs could lead to numerous clinical outcomes of the population who live on these contaminated sites (Figure 1.3). MTs are usually placed in impoundment areas exposed to precipitation and water runoff, which can

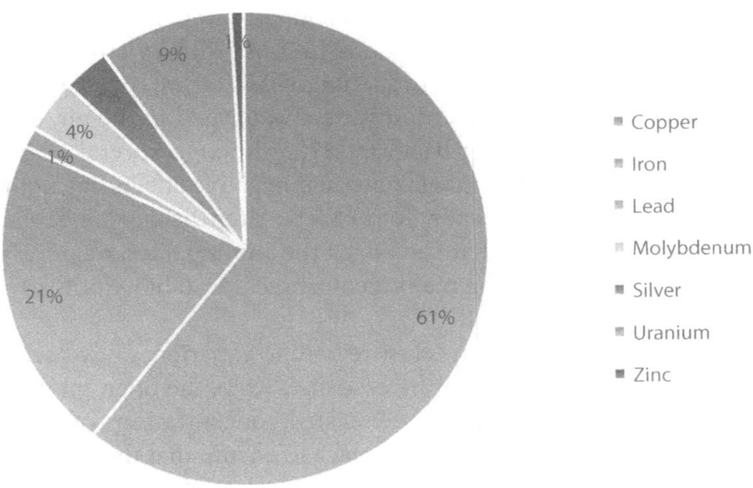

Figure 1.5 Metalliferous solid waste generation [55].

Table 1.1 Metalliferous solid waste generation.

Minerals mined	Stockpiled metalliferous waste (Mt)	Metalliferous tailings (Mt)	Potential risk	References
Antinomy	–	420–525	–	[52]
Copper	0.2–124	0.1–791	Massive release of PTEs and Groundwater acidification	[51, 53]
Iron	102	75		[54]
Lead	2	9	PTEs in groundwater contamination	[54]
Silver	20	6		[54]
Zinc	1	6–50,000	–	[52, 54]

allow PTEs to be leached [39]. Some MTs such as those in Limpopo province (South Africa) were investigated and reported to have an average pH of 8.0 indicating an alkaline medium; albeit, with the concentration of PTEs, i.e., As, Cd, and Cr, which strongly bind to the smaller fractions in mining waste. Results show a high concentration and easily extractable Cu in the MTs, which indicated bioavailability thus environmental and human exposure risks [5].

1.4 Environmental and Social Impact of CMMs and MSW

As has been shown above, CMMs are potential sources of environmental contamination and may pose a health risk to local populations [56]. A primary environmental concern associated with active and CMMs is the production of ARD from the oxidation of disposed sulfidic waste material when exposed to moisture and oxygen in the air, especially in the presence of iron and sulfur-oxidizing bacteria [57]. Although several CMMs and MSW exist in South Africa, information on their potential negative effects has not been consolidated in the literature reviewed especially for O'Kiep. After the closure of the mine, MSW was abandoned, resulting in degraded soils near the mining sites, i.e., areas which were pristine, a phenomenon often causing major geotechnical instability of the waste resulting in further PTEs mobility [58].

Additionally, communities after mine closure tend to experience adverse environmental and other social impact phenomena. In the case of O'Kiep, the open pit groundwater has been characterized as having a low pH and elevated concentrations of PTEs by Erdogan *et al.* [12]. Additionally, measures were not in place to ensure that the CMM in O'Kiep was rehabilitated effectively. This resulted in the negative effects of CMM and MSW not being addressed thus a lack of accountability.

Generally, unrehabilitated CMMs can generate ARD, increasing PTEs solubility, and the degradation of the environment, including local water sources. CMMs can also result in vast vegetation reduction attributed to changes in the soil quality and structure [59]. On the other hand, the presence of carbonates and other caustic type gangue minerals can lead to reactions to reduce the ARD formation potential; albeit, this can still result in poor-quality drainage with high concentrations of sulfate and PTEs [37]. One of the challenging issues associated with soil contamination is PTEs mobility which is initiated by MSW weathering by wind action which is highly likely in O'Kiep since the MSW has resulted in soil degradation [43] and possible groundwater contamination [12]. The possible effects of the MSW near local communities are the high concentrations of PTEs and low soil organic matter availability due to the arid nature of the area which makes PTEs mobility easier.

Currently, mining industries are consistently under intense scrutiny from the general public and governments due to some improper disposal methods of MSW after mine closure. Contaminant containing MSW in South Africa is currently a challenge of higher concern both for the mining industry and the government. This includes several TSF that have been abandoned and never been rehabilitated. Subsequent analysis by Wapwera *et al.* [60] showed that there are radioactive substances in CMM operations which can exceed the international guidelines; therefore, they can result in observable consequences on the health of the local population, who reside in the vicinity of such mines. Generally, Cu mining waste make-up the largest percentage of metal mining and processing wastes generated globally. There is a broad range of naturally occurring radioactive materials that have been concentrated or exposed in MSW, and as a result, animal and human exposure becomes unavoidable.

Mining and extraction of Cu by surface or underground methods can concentrate and expose radionuclides in the SMW and MTs [61]. Nkosi *et al.* [62] reported that there is increasing evidence that environmental factors such as air contamination by volatile constituents from SMW resulted in an increased risk of chronic respiratory symptoms of individuals in communities located near SMW in South Africa. Additionally, long-term aerosols containing Cu constituents can irritate human respiratory mucous membranes, causing headaches, dizziness, and nausea [63]. Similarly, drinking water containing excessive quantities of Cu can lead to nausea, vomiting, diarrhoea, and if taken intentionally, can cause liver and kidney damage including death. Since the highest chance of ingesting excess Cu is *via* drinking water in the form of groundwater, it is essential to monitor Cu release to aquifers such as that of O'Kiep [17].

1.5 Soil Contamination

The CMM and MSW in O'Kiep has contaminated nearby soils (Figure 1.6) with elevated concentration of S, Cu, F, Ba, Mn, and Cl [10, 43, 44]. Lee *et al.* [64] comprehended spatial variation of PTEs in the soil to identify adequate measures to preventing pristine soil contamination at closed metalliferous mining areas, whereby the spatial distribution of Cu and Pb was monitored. This is mainly done to develop rehabilitation strategies to mitigate environmental contaminant mobility, i.e., PTEs distribution to contaminant free areas, or into other natural resources including groundwater.

Figure 1.6 Contaminated soils of O'Kiep.

1.6 Groundwater Contamination

Several factors are associated with MSW and CMMs that affect groundwater, with one being ARD seepage [38]. Furthermore, PTEs stored in MTs, soils, and MSW have become potential secondary groundwater contaminants [65]. Open pits are formed during surface mining operations and are thereafter filled with water, either through groundwater recharge or surface run-off. Generally, the success in closing open pit mines has varied tremendously [66] and the groundwater from these open pits end up being contaminated by ARD. CMMs can present challenges for environmental restoration due to the presence of a high concentration of sulfidic constituents in the MSW, with rainfall further exacerbated the problem as it would result in large volumes of contaminated groundwater recharges and surface run-off open pit water [20]. There are well-known examples of legacy sites requiring perpetual treatment of their water such as open pits, with O'Kiep being an example [12]; albeit, some other open pit water had achieved various beneficial traits for end-users such as agricultural use [67]. Static tests suggested that the soils of O'Kiep had a high acid-producing potential, which was confirmed by humidity cell tests [10], with the likelihood of weathering, of either rock, soil, and MSW under moisturous conditions leading to acidification of any stagnant water.

1.7 Atmospheric Contamination

Metalliferous mine dust associated with PTE containing aerosols released into the atmosphere through the dust and other gases from the MSW is one such environmental contamination challenge related to the cause of adverse health effects in humans and vegetation [41, 68, 70]. Despite occupational health improvements within the mining industry, the release of metalliferous dust into the atmosphere remains a human health challenge, especially in regions with poorly developed regulatory standards and whereby historic mining has left a legacy of exposed MSW. The global challenge associated with unrehabilitated MTs is blowing dust under desertification conditions attributed to changes in climatic patterns which could further be entrenched in arid regions [69]. Dust generation is a problem in drier climates and typically where the TFs are exposed.

1.8 Metalliferous Solid Waste Management

The MSW should be subjected to extensive testing such as mineralogy and geology characterization and hydrogeological modeling. The suite of tests performed should include static tests such as acid-base accounting [40] and kinetic testing, such as humidity cells tests. All these tests have been used in various scientific exercises and MSW management planning [71].

1.9 Rehabilitation and Restoration Strategies

Mining industries in many countries are required to follow environmental and rehabilitation standards to ensure the area mined is rehabilitated closely to its original state. This is due to cumulative quantities of MSW produced annually. Environmental and communal issues associated with the disposal of such waste on pristine land, mining companies are in search of alternative techniques of MSW disposal even for repurposing [72]. The restoration efforts of the CMM, however, remain mostly ineffective since vegetation that was planted could not withstand post-mining activities due to the severely deteriorated hydrogeochemical soil qualities. However, it is anticipated that enhanced knowledge on growing conditions and selection of suitable plant species could contribute to the improvement of a phytoremediation rehabilitation strategy for a low-cost and maintainable restoration program of the CCMs [26]. Furthermore, this will have a direct usefulness to the areas such as O'Kiep that are similarly affected by the existence of PTEs in the environment. This indicates that rehabilitation and treatment might be costly in the long term as further and localized environmental deterioration advances itself when rehabilitation is neglected.

As a result, the recent two decades have witnessed a global surge in research on post-mining landscape restoration, yielding a suite of techniques, including phytoremediation [31]. The rehabilitation schemes proposed are required for the restoration of ecological indicators of mine sites and to reduce human health exposure to minimize risks [73]. Such remediation strategies would thus reduce acid-producing constituents in MSW [37]. However, restoration of MSW and contaminated soil through vegetation growth is a favorable, eco-friendly, and cost-effective technology for long term directed site rehabilitation [74].

The rehabilitation of mine sites can be very complex due to a variety of PTEs and soil quality which might influence the remediation scheme [75]. Other studies propose microorganisms which can play a role in the biogeochemical phases of the soil in order to control PTEs behavior in contaminated environments. However, certain microorganisms can accelerate the oxidation and dissolution of (di-)sulfide minerals, leading to the formation of ARD [75]. Generally, different biological strategies have been explored for the elimination of PTEs and remediation systems even for ARD. These include the use of algae, fungi, bacteria, and wild yeast in passive treatment technologies [73]. According to Bruneel *et al.* [75], few studies reporting on CMMs have been conducted in Sub-Saharan African countries including South Africa. However, no remediation studies have been conducted in O'Kiep to address even the microbiological treatment of the open pit groundwater let alone the surrounding environment.

Mine waste storage areas need to be constructed and protected in such a way that their adverse effects on human health and the natural environment are minimized in the long-term [28]. Only a few studies have been reported in South Africa that explore the restoration, remediation, and rehabilitation practices areas after mining disturbance. According to Claveria *et al.* [76], high concentrations of As and Cu, and their cumulative behavior in nature could be considered as a reason enough for site remediation of metalliferous soils to reduce these PTEs influence on the mine site, using suitable post-mining rehabilitation schemes. The TSF in O'Kiep could be safely cordoned off, and an ARD canal could be erected around the TSF with a settling pond which can be interlinked to a passive treatment system. Furthermore, the place should be enclosed in order to prevent children and people of the nearby community from entering the site which is not the case currently. A similar remediation strategy was recommended by Kříbek *et al.* [77] in the study of environmental impacts and remediation measures in Zambia's Kabwes Pb-Zn smelter. MTs have been re-used and recycled as backfill post, extraction of minerals repurposed and used in the production of bricks [78].

Within the last decade, research into challenges of revegetation of TSF has expanded considerably especially in CMM; therefore, reducing environmental damage including landscape modification with the vegetation regrowing and the burden of contamination on the food chain being reduced will result in a reduced threat to human health. Where plants are to be used for restoration and rehabilitation of CMM's, the plants can be genetically adapted to metal-enriched soils [79], either through directed evolution and the use of advanced molecular biology techniques.

1.10 ARD Formation and Groundwater Contamination

South African government's wastewater management regulations are essential in contributing to the management of ARD in the mining industry. This challenge of ARD formation has featured substantially both in national and international scientific reports [80]. In South Africa, ARD became a matter of concern in 2002, when the West Rand Basin was flooded with 20 million liters of ARD [81]. In a recent ARD study, Naidoo [82], reported that ARD is also a concern in countries such as Australia, Canada, the United States of America, and Germany. South Africa's ARD contamination challenges were observed nationwide in: 1) Mpumalanga and KwaZulu-Natals' coal fields, 2) Witwatersrand goldfields, and 3) the O'Kiep copper fields [12], but natural attenuation increased the pH values to circumneutral in several areas. One main environmental and health concern was the initially vast production of ARD, in the city of Johannesburg, which the Department of Water Affairs estimating that at up to 92 million liters per day are produced. ARD mobilizes PTEs in the environment and contaminates surface water supplies [83]. ARD can represent a complex challenge due to the MSW oxidation state [84]. The TSF in O'Kiep also has a potential of generating ARD during the wet season (Figure 1.7), which can leach into soils with secondary potential groundwater contamination.

Furthermore, conventional wastewater treatment methods using chemical agents were determined not to be appropriate for all cases of ARD treatment. There are on-going research activities associated with ARD from Cu mining. Although O'Kiep was identified by the Inter-Ministerial Committee on mine water management in South Africa as a lower

Figure 1.7 Leachates from metalliferous storage facility in O'Kiep.

priority area, further assessment and monitoring is required at this stage as evidenced by leachates from the TSF.

1.11 Overview of Challenges Associated with CMMs

The chronology of challenges associated with CMMs is challenging (Figure 1.8). There is still a concern with both the detrimental health effects and environmental impacts of sub-optimal management of waste and increasing contamination in O'Kiep.

CMMs generate a large quantity of waste in the form of aerosols, liquids and solids. Numerous of these wastes are potentially hazardous to the environment and are dangerous to living organisms, including human beings. The MSW generated comprises of PTEs, and only 0.4% is discharged with ARD, and 0.007 % of mining waste takes the form of air emissions; referred to as aerosols [85]. Cases of inadequate mine waste management practice are amongst the most noticeable features of the world-wide CMMs. Mining waste in the form of SOM, SMW, MTs, from unrehabilitated CMMs under suitable circumstances can results in direct discharges into waterways that can result in long-term environmental and social consequences. There is enough evidence that improper disposal of these wastes may cause contamination of air, groundwater,

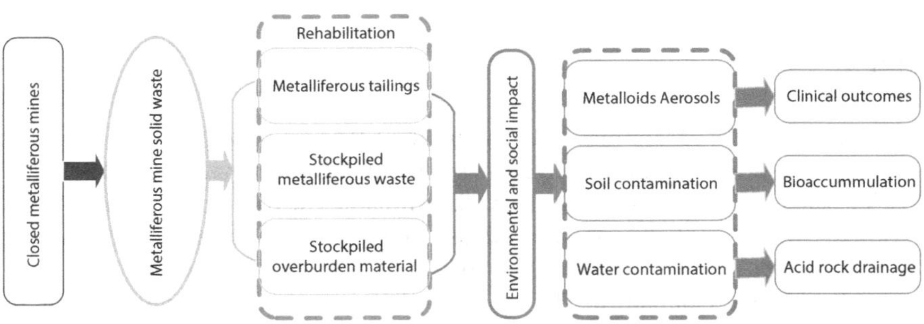

Figure 1.8 Overview of challenges associated with closed metalliferous mining.

surface water, and soils and bioaccumulation in the wildlife thus biomagnify in the food chain. There is a need for comprehensive long-term strategies for transforming the area of O'Kiep to move toward a zero-contaminant environmental footprint. As PTE's are persistent in the environment [85]. In line with international standards, local authorities and society are encouraged to seek rehabilitation strategies whereby MSW can be recycled and reduced. According to Adler *et al.* [86], negative externalities associated with CMMs were not internationalized; as a result, the mining industry failed to prepare for closure adequately and to dispose of MSW in a manner that is in line with current global best practice. In particular, cumulative harm to the population within the vicinity of the mine resulted in modified water tables during the rainy season, culminating in contaminated groundwater sources, ARD formation and aquifer instability. All these challenges must be addressed before they cause even more devastating socio-economic and environmental consequences.

1.12 Conclusion

The CMMs and MSW, in general, need environmental health and safety assessments. Additionally, MSW are potential sources of PTEs and were identified as a greater threat to the communities surrounding mining sites. Accumulation of high concentrations of PTEs in areas associated with Cu mining can cause human health risks and environmental challenges. It is essential to determine the firmness of MSW heaps as well as their potential failure during the wet season to ensure implementation of mitigation strategies to minimize contamination. To reduce the mining footprint and effects as well as to limit population exposure to harmful MSW, access to MSW may be advisable with its redistribution to environmentally acceptable sites while other treatment strategies plans are being developed to resolve the present contamination, a strategy which can provide an immediate relief in rehabilitation challenges in sites such as O'Kiep. Also, SMW treatment is needed to prevent additional oxidation of sulfide minerals by air and moisture contact which could further result in the release of PTEs into groundwater bodies and their deposition as aerosols in pristine areas further away from the mining site. Therefore, there is a need for appropriate remedial procedures to initially determine the properties of all waste generating metalliferous dust in order to also reduce its negative effects, with further environmental assessments and clinical outcomes being monitored on individuals living in the vicinity of the CMM and MSW being of paramount importance.

References

1. Farjana, S.H., Huda, N., Mahmud, M.P., Saidur, R., A Review on Impact of Mining and Mineral Processing Industries through Life Cycle Assessment. *J. Clean. Prod.*, 231, 120–1217, 2019.
2. Preite, V., Sailer, C., Syllwasschy, L., Bray, S., Ahmadi, H., Krämer, U., Yant, L., Convergent evolution in *Arabidopsis halleri* and *Arabidopsis arenosa* on calamine metalliferous soils. *Philos. Trans. of the R. Soc.*, 374, 1777, 20180243, 2019.
3. Palumbo-Roe, B. and Colman, T., The nature of waste associated with closed mines in England and Wales, Nottingham, UK, British Geological Survey, 82 pp. 2010. http//geolib.bgs.ac.uk.

4. Bell, F.G., De Bruyn, I.A., Stacey, T.R., Some examples of the impact of metalliferous mining on the environment: A South African perspective. *Bull. Eng. Geology Environ.*, 61, 1, 1–20, 2002.
5. Gitari, M.W., Akinyemi, S.A., Ramugondo, L., Matidza, M., Mhlongo, S.E., Geochemical fractionation of metals and metalloids in tailings and appraisal of environmental pollution in the abandoned Musina Copper Mine, South Africa. *Environ. Geochem. Health.*, 40, 6, 2421–2439, 2018.
6. Prasad, M.N.V. and Shih, K., *Environmental materials and waste: Resource recovery and pollution prevention*, Acad. Press, USA, 2016.
7. Komnitsas, K., Xenidis, A., Adam, K., Oxidation of pyrite and arsenopyrite in sulphidic spoils in Lavrion. *Miner Eng.*, 8, 12, 1443–1454, 1995.
8. Kwon, M.J., Yang, J.S., Lee, S., Lee, G., Ham, B., Boyanov, M.I., Kemner, K.M., O'Loughlin, E.J., Geochemical characteristics and microbial community composition in toxic metal-rich sediments contaminated with Au–Ag mine tailings. *J. Hazard. Mater.*, 296, 147–157, 2015.
9. Navarro, M.C., Pérez-Sirvent, C., Martínez-Sánchez, M.J., Vidal, J., Tovar, P.J., Bech, J., Abandoned mine sites as a source of contamination by heavy metals: A case study in a semi-arid zone. *J. Geochem. Explor.*, 96, 2, 183–193, 2008.
10. Erdogan, I.G., Fosso-Kankeu, E., Ntwampe, S.K., Waanders, F.B., Hoth, N., Rand, A., Acid rock drainage prediction of metalliferous soils from O'Kiep, Namaqualand, South Africa: A Humidity Cell Test assessment. *IMWA Conference – Mine Water: Technological and Ecological Challenges*, Perm, Russia, 15–19 July, 2019.
11. Clifford, T.N., Barton, E.S., The O'kiep Copper District, Namaqualand, South Africa: A review of the geology with emphasis on the petrogenesis of the cupriferous Koperberg Suite. *Miner. Deposita.*, 47, 8, 837–857, 2012.
12. Erdogan, I., Moncho, T., Fosso-Kankeu, E., Ntwampe, S., Waanders, F., Hoth, N., Rand, A., Fourie, B., Hydrochemical Characteristics of Open-Pit Groundwater from a Closed Metalliferous Mine in O'kiep, Namaqualand Region, South Africa, in: *9th Int'l Conference on Advances in Science, Engineering, Technology & Waste Management (ASETWM-17)*, Nov. 27–28, pp. 64–68, 2017.
13. Constitution of the Republic of South Africa, *(Act 108 of 1996). Republic of South Africa, Government Gazette, Pretoria, South Africa*, 1996, Available at: https://www.gov.za/sites/www.gov.za/files/images/a108-96.pdf (Accessed on 4 October 2019).
14. Naidoo, D., The national water and sanitation summit, The Star, South Africa, Department of Water and Sanitation, South Africa, 1–79, 2014.
15. Lee, H., Kabir, M.I., Kwon, P.S., Kim, J.M., Kim, J.G., Hyun, S.H., Rim, Y.T., Bae, M.S., Ryu, E.R., Jung, M.S., Contamination assessment of abandoned mines by integrated pollution index in the Han River watershed. *The Open Environ. Pollut. Toxicol. J.*, 1, 27–33, 2009.
16. Pehoiu, G., Radulescu, C., Murarescu, O., Dulama, I.D., Bucurica, I.A., Teodorescu, S., Stirbescu, R.M., Health Risk Assessment Associated with Abandoned Copper and Uranium Mine Tailings. *Bull. Eng. Geology Environ.*, 102, 4, 1–7, 2019.
17. Ippolito, J.A., Cui, L., Novak, J., Johnson, M.G., Biochar for Mine-land Reclamation, in: *Biochar from Biomass and Waste*, pp. 75–90, Elsevier, USA, 2019.
18. Komnitsas, K., Kontopoulos, A., Lazar, I., Cambridge, M., Risk assessment and proposed remedial actions in coastal tailings disposal sites in Romania. *Miner Eng*, 11, 12, 1179–1190, 1998.
19. Kontopoulos, A., Komnitsas, K., Xenidis, A., Papassiopi, N., Environmental characterisation of the sulphidic tailings in Lavrion. *Miner Eng.*, 8, 10, 1209–1219, 1995.
20. Van Der Ent, A. and Edraki, M., Environmental geochemistry of the abandoned Mamut copper mine (Sabah) Malaysia. *Environ. Geochem. Health.*, 40, 1, 189–207, 2018.
21. Nikolov, H. and Borisova, D., Long term monitoring of water basin of an abandoned copper open pit mine, in: *EGU General Assembly Conference Abstracts*, vol. 14, p. 10171, 2012,

Available at: http://meetingorganizer.copernicus.org/egu2012/egu2012-10171.pdf (Accessed on 4 October 2019).

22. Koski, R.A. and Munk, L., *Chemical data for rock, sediment, biological, precipitate, and water samples from abandoned copper mines in Prince William Sound, Alaska*, US Geological Survey Western Mineral Resources, USA, 2007, Available at: https://pubs.usgs.gov/of/2007/1359/of2007-1359.pdf (Accessed on 9 October 2019).
23. Venkateswarlu, K., Nirola, R., Kuppusamy, S., Thavamani, P., Naidu, R., Megharaj, M., Abandoned metalliferous mines: Ecological impacts and potential approaches for reclamation. *Rev. Environ. Sci. Biotechnol.*, 15, 2, 327–354, 2016.
24. Morganwalp, D.W. and Buxton, H.T., US Geological Survey Toxic Substances Hydrology Program, in: *Proceedings of the Technical Meeting*, Charleston, South Carolina, March 8-12, 1999, (no. 4018). US Department of the Interior, US Geological Survey. 1999.
25. Antivachis, D.N., Chatzitheodoridis, E., Skarpelis, N., Komnitsas, K., Secondary sulphate minerals in a Cyprus-type ore deposit, Apliki, Cyprus: Mineralogy and its implications regarding the chemistry of pit lake waters. *Mine Water Environ.*, 36, 2, 226–238, 2017.
26. Helsen, S., Rommens, T., De Ridder, A., Panayiotou, C., Colpaert, J., Remediation and rehabilitation of abandoned mining sites in Cyprus, in: *EGU General Assembly Conference Abstracts*, vol. 11, p. 6090, 2009, Available at: http://meetingorganizer.copernicus.org/EGU2009/EGU2009-6090.pdf (Accessed on 20 June 2019).
27. Cowling, R., *Namaqualand: a succulent desert*, Penguin Random House, South Africa, 2015, Available at: https://cvlesalfabegues.com/search/namaqualand-a-succulent-desert (Accessed on 15 June 2019).
28. Vallero, D.A. and Blight, G., Mine Waste: A Brief Overview of Origins, Quantities, and Methods of Storage, in: *Waste*, 2nd ed., T.M. Letcher and D.A. Vallero (Eds.), pp. 129–151, Academic Press, USA, 2019.
29. Szczepańska, J. and Twardowska, I., Mining waste, in: *Waste Management Series*, vol. 4, pp. 319–385, Elsevier, South Africa, 2004.
30. Chaturvedi, N., Dhal, N.K., Reddy, P.S.R., Comparative phytoremediation potential of *Calophyllum inophyllum* L., *Bixa orellana* L. and *Schleichera oleosa* (lour.) Oken on iron ore tailings. *Int. J. Mining, Reclam. Environ. Manag.*, 26, 2, 104–118, 2012.
31. Festin, E.S., Tigabu, M., Chileshe, M.N., Syampungani, S., Odén, P.C., Progresses in restoration of post-mining landscape in Africa. *J. Res.*, 30, 2, 381–396, 2019.
32. Galanopoulos, E., Skarpelis, N., Argyraki, A., Supergene alteration, environmental impact and laboratory scale acid water treatment of Cyprus-type ore deposits: Case study of Mathiatis and Sha abandoned mines. *Geochem.: Explor. Environ., Anal.*, 2018–070, 19, 4, 299–315, 2019.
33. Campbell, I.C. and Beardall, J., Ok Tedi copper mine, Papua New Guinea, stimulates algal growth in the Fly River. *Sustain. Water. Resour. Manag.*, 5, 2, 425–437, 2019.
34. Shanmuga, P.R. and Kumar, K.G., Geotechnical Aspects of Overburden Materials for Stowing in Underground Coal Mines. *Int. J. Eng. Sci. Res. Technol.*, 4, 03, 2015.
35. Kogel, J.E., Trivedi, N.C., Barker, J.M., Krukowski, S.T., Krukowski, *Industrial minerals & rocks: Commodities, markets, and uses*, SME, UK, 2006.
36. Karna, R.R. and Hettiarachchi, G.M., Subsurface Submergence of Mine Waste Materials as a Remediation Strategy to Reduce Metal Mobility: An Overview. *Curr. Pollut. Rep.*, 4, 1, 35–48, 2018.
37. Tayebi-Khorami, M., Edraki, M., Corder, G., Golev, A., Re-Thinking Mining Waste through an Integrative Approach Led by Circular Economy Aspirations. *Minerals*, 9, 5, 286, 2019.
38. Likus-Cieślik, J., Pietrzykowski, M., Szostak, M., Szulczewski, M., Spatial distribution and concentration of sulfur in relation to vegetation cover and soil properties on a reclaimed sulfur mine site (Southern Poland). *Environ. Monit. Assess.*, 189, 2, 87, 2017.

39. Beauford, R., *Rare earth element minerals and ores*, USA, 2012, *Available at:* http://rareearthelements.us/mineral_ores (Accessed on 15 May 2019).
40. Yucel, D.S. and Baba, A., Prediction of acid mine drainage generation potential of various lithologies using static tests: Etili coal mine (NW Turkey) as a case study. *Environ. Monit. Assess.*, 188, 8, 473, 2016.
41. Rozendaal, A., Rudnick, T.K., Heyn, R., Mesoproterozoic base metal sulphide deposits in the Namaqua Sector of the Namaqua-Natal Metamorphic Province, South Africa: A review. *South. Afr. J. Geology.*, 120, 1, 153–186, 2017.
42. Tabelin, C., Sasaki, A., Igarashi, T., Tomiyama, S., Villacorte-Tabelin, M., Ito, M., Hiroyoshi, N., Prediction of acid mine drainage formation and zinc migration in the tailings dam of a closed mine, and possible countermeasures, in: *MATEC Web of Conferences*, vol. 268, p. 06003, EDP Sciences. 2019. https://doi.org/10.1051/matecconf/201926806003.
43. Moncho, T., Erdogan, I., Emandien, M., Ntwampe, S., Fosso-Kankeu, E., Waanders, F., Rand, A., Fourie, B., Prediction of metals bioavailability in the soils of O'kiep, South Africa, in: *9th Int'l Conference on Advances in Science, Engineering, Technology & Waste Management (ASETWM-17)*, Parys, South Africa, Nov. 27-28, 2017, Available at: https://www.eares.org/siteadmin/upload/8974EAP1117023.pdf (Accessed on 1 September 2019).
44. Erdogan, I., Moncho, T., Fosso-Kankeu, E., Ntwampe, S., Waanders, F., Hoth, N., Rand, A., Potential toxic elements contamination of soils in O'Kiep, an arid region of Namaqualand, South Africa, in: presented at the *10th Int'l Conference on Advances in Science, Engineering, Technology & Healthcare (ASETH-18)*, Cape Town (South Africa), Nov. 19–20, 2018, 2018, Available at: https://www.eares.org/siteadmin/upload/1888EAP1118238.pdf (Accessed on 19 June 2019).
45. Chileshe, M.N., Syampungani, S., Festin, E.S., Tigabu, M., Daneshvar, A., Odén, P.C., Physicochemical characteristics and heavy metal concentrations of copper mine wastes in Zambia: implications for pollution risk and restoration. *J. Res.*, 1–11, 2019. https://doi.org/10.1007/s11676-019-00921-0
46. Hohne, S. and Hansen, R., Preliminary conceptual geo-environmental model of the abandoned copper mines of the Okiep Copper District, Namaqualand, Northern Cape, in: *Sustainable Development through Mining. Council for Geoscience, Council for Geoscience, Report Number 2008-070*, 2008.
47. Fosso-Kankeu, E., Waanders, F., Botes, W., Recovery of Base Metals from Mine Tailings Dumps collected in the Vicinity of Potchefstroom: Leaching assisted by Complexing Agent, in: *7th International Conference on Latest Trends in Engineering and Technology (ICLTET'2015)*, 2015, Available at: http://iieng.org/images/proceedings_pdf/2743E1115026.pdf (Accessed on 19 June 2019).
48. Amponsah-Dacosta, M. and Reid, D.L., Mineralogical characterization of selected South African mine tailings for the purpose of mineral carbonation, in: *An Interdiscip. Response. Min. Water Chall., Sui, Sun, Wang*, (Eds.), pp. 686–692, 2014.
49. Titshall, L.W., Hughes, J.C., Bester, H.C., Characterisation of alkaline tailings from a lead/zinc mine in South Africa and evaluation of their revegetation potential using five indigenous grass species. *South. Afr. J. Plant. Soil.*, 30, 2, 97–105, 2013.
50. García-Giménez, R. and Jiménez-Ballesta, R., Mine tailings influencing soil contamination by potentially toxic elements. *Environ. Earth Sci.*, 76, 1, 51, 2017.
51. Xtract, Resources, and Plc. 2015. Available at: https://xtractresources.com/okiep_glance_OFF.htm (Accessed on 19 June 2019).
52. Godfrey, L.K., Oelofse, S.H., Phiri, A., Nahman, A., Hall, J., Mineral waste: The required governance environment to enable re-use, pp. 1–55, 2007. http://researchspace.csir.co.za/dspace/handle/10204/3541

53. Earthworks, Oxfam America, Dirty Metals: Mining Communities and the Environment. Earthworks, Washington DC, 2004. Available at: https://www.miningresettlement.org/elibrary/dirty-metals-mining-communities-and-the-environment (Accessed on 15 August 2019).
54. Derkics, D., Report to congress: Wastes from the extraction and beneficiation of metallic ores, phosphate rock, asbestos, overburden from uranium mining, and oil shale, 1985. Final report. OSTI ID: 5487228. United States.
55. EPA, *Environmental Protection Agency-Office of Solid Waste: Wastes from the extraction and beneficiation of metallic ores, phosphate rock, asbestos, overburden from uranium mining, and oil shale*, U.S. Environmental Protection Agency Office of Solid Waste, Washington, D.C, 1985, Available at: https://www.epa.gov/sites/production/files/2016-03/documents/epa-530-sw-85-033.pdf (Accessed on 26 September 2019).
56. Oyourou, J.N., McCrindle, R., Combrinck, S., Fourie, C.J.S., Investigation of zinc and lead contamination of soil at the abandoned Edendale mine, Mamelodi (Pretoria, South Africa) using a field-portable spectrometer. *J. S. Afr. Inst. Min. Metall.*, 119, 1, 55–62, 2019.
57. Kotsiopoulos, A. and Harrison, S., Co-disposal of benign desulfurised tailings with sulfidic waste rock to mitigate ARD generation: Influence of flow and contact surface. *Miner Eng.*, 116, 62–71, 2018.
58. Mehta, N., Cocerva, T., Cipullo, S., Padoan, E., Dino, G.A., Ajmone-Marsan, F., Cox, S.F., Coulon, F., De Luca, D.A., Linking oral bioaccessibility and solid phase distribution of potentially toxic elements in extractive waste and soil from an abandoned mine site: Case study in Campello Monti, NW Italy. *Sci. Total Environ.*, 651, 2799–2810, 2019.
59. Zawadzki, J., Przeździecki, K., Miatkowski, Z., Determining the area of influence of depression cone in the vicinity of lignite mine by means of triangle method and LANDSAT TM/ETM+ satellite images. *J. Environ. Manag.*, 166, 605–614, 2016.
60. Wapwera, S.D., Egbu, C.O., Parsa, A.G., Ayanbinpe, G.M., Abandoned mines, homes for the people: Case study of Jos Tin-mining region. *Int. J. Hous. Mark. Anal.*, 8, 2, 239–264, 2015.
61. Vearrier, D., Curtis, J.A., Greenberg, M.I., Technologically enhanced naturally occurring radioactive materials. *Clin. Toxicol.*, 47, 1556–3650, 2009.
62. Nkosi, V., Wichmann, J., Voyi, K., Chronic respiratory disease among the elderly in South Africa: any association with proximity to mine dumps? *Environ. Health.*, 14, 1, 33, 2015.
63. Paredes, J.M., *Measuring the effects of mining on Perus public health: is the apurimac region prepared to assess heavy metal exposure?*, Duke University, 2016, Available at: https://dukespace.lib.duke.edu/dspace/handle/10161/11882 (Accessed on 23 September 2019).
64. Lee, H., Choi, Y., Suh, J., Lee, S.H., Mapping copper and lead concentrations at abandoned mine areas using element analysis data from ICP–AES and portable XRF instruments: A comparative study. *Int. J. Environ. Res. Public Health.*, 13, 4, 384, 2016.
65. Qin, W., Han, D., Song, X., Engesgaard, P., Effects of an abandoned Pb-Zn mine on a karstic groundwater reservoir. *J. Geochem. Explor.*, 200, 221–233, 2019.
66. Vandenberg, J., McCullough, C., Castendyk, D., Key issues in mine closure planning related to pit lakes, in: *Agreeing on solutions for more sustainable mine water management—Proceedings of the 10th ICARD & IMWA Annual Conference*, vol. 156, 2015.
67. Villain, L., Sundström, N., Perttu, N., Alakangas, L., Öhlander, B., Evaluation of the effectiveness of backfilling and sealing at an open-pit mine using ground penetrating radar and geoelectrical surveys, Kimheden, northern Sweden. *Environ. Earth Sci.*, 73, 8, 4495–4509, 2015.
68. Entwistle, J.A., Hursthouse, A.S., Reis, P.A.M., Stewart, A.G., Metalliferous Mine Dust: Human Health Impacts and the Potential Determinants of Disease in Mining Communities. *Curr. Poll. Rep.*, 1–17, 2019.
69. Hattingh, J. and Van Deventer, P.W., The Effect of the Chemical Properties of Tailings and Water Application on the Establishment of a Vegetative Cover on Gold Tailings Dams: Report to the

Water Research Commission, 899/1/04, Water Research Commission, Pretoria, 1770051104, pp. 1–162, 2004. Available at: http://www.wrc.org.za/wp-content/uploads/mdocs/899-1-04.pdf (Accessed on 10 September 2019).

70. Mark, R., *Closure of tailings facilities: Current practice review and guidelines for success, Doctoral Thesis*, Wits University, South Africa, 2006, Available at: http://wiredspace.wits.ac.za/handle/10539/1850 (Accessed on 24 September 2019).
71. European Commission, *Reference document on best available techniques for management of tailings and waste-rock in mining activities*, EU Commission, Belgium, 2009, Available at: https://eippcb.jrc.ec.europa.eu/reference/BREF/mmr_adopted_0109.pdf (Accessed on 27 September 2019).
72. Hudson-Edwards, K. and Dold, B., Mine waste characterization, management and remediation. *Minerals*, 5, 82–85, 2015.
73. Ali, A., Wahid, F., Guo, D., Bio-Processing of Mining Solid Waste and Resource Recovery. *Biol. Processing Solid. Waste*, 169, 165–186, 2019.
74. Petrisor, I.G., Dobrota, S., Komnitsas, K., Lazar, I., Kuperberg, J.M., Serban, M., Kuperberg, Serban, M., Artificial inoculation—perspectives in tailings phytostabilization. *Int. J. Phytoremediat.*, 6, 1, 1–15, 2004.
75. Bruneel, Odile, N., Mghazli, L., Sbabou, M., Hery, C., Casiot, A., Filali-Maltouf. Role of microorganisms in rehabilitation of mining sites, focus on Sub Saharan African countries. *J. Geochem. Explor.*, 205, 106327, 2019.
76. Claveria, R.J.R., Perez, T.R., Perez, R.E.C., Algo, J.L.C., Robles, P.Q., The identification of indigenous Cu and As metallophytes in the Lepanto Cu-Au Mine, Luzon, Philippines. *Environ. Monit. Assess.*, 191, 3, 185, 2019.
77. Křibek, B., Nyambe, I., Majer, V., Knésl, I., Mihaljevič, M., Ettler, V., Vaněk, A., Penížek, V., Sracek, O., Soil contamination near the Kabwe Pb-Zn smelter in Zambia: Environmental impacts and remediation measures proposal. *J. Geochem. Explor.*, 197, 159–173, 2019.
78. Kuhn, K. and Meima, J.A., Characterization and Economic Potential of Historic Tailings from Gravity Separation: Implications from a Mine Waste Dump (Pb-Ag) in the Harz Mountains Mining District, Germany. *Minerals*, 9, 5, 303, 2019.
79. Bini, C., Wahsha, M., Spiandorello, M., Land contamination and soil evolution in abandoned mine areas (Italy), in: *EGU General Assembly Conference Abstracts*, vol. 16, 2014, Available at: http://adsabs.harvard.edu/abs/2014EGUGA.16.2379B (Accessed on 26 September 2019).
80. Feris, L. and Kotzé, L.J., The regulation of acid mine drainage in South Africa: Law and governance perspectives. *PER: Potchefstroomse Elektroniese Regsblad*, 17, 5, 2105–2163, 2014.
81. DWAF, Department of Water Affairs. A Guideline for the Assessment, Planning and management of groundwater resources in South Africa, pp. 1–1, 4–2, 2008. http://www.dwa.gov.za/Documents/Other/Water%20Resources/GroundwaterPlanGuideMar08full.pdf
82. Naidoo, S., *Acid mine drainage in South Africa: Development actors, policy impact and broader implications*, Springer, Switzerland, 2017.
83. Olalde, M., *What's left in the wake of South Africa's abandoned gold mines*, GreenBiz Group Inc, USA, 2016, Available at: https://www.greenbiz.com/article/whats-left-wake-south-africas-abandoned-gold-mines. (Accessed on 29 September 2019).
84. Elghali, A., Benzaazoua, M., Bussière, B., Genty, T., Spatial mapping of acidity and geochemical properties of oxidized tailings within the former Fagle/Telbel mine site. *Minerals*, 9, 3, 180, 2019.
85. Makgae, M., Key areas in waste management: A South African perspective, in: *Integrated Waste Management-Volume II*, IntechOpen, UK, 2011.
86. Adler, R.A., Claassen, M., Godfrey, L., Turton, A.R., Water, mining, and waste: An historical and economic perspective on conflict management in South Africa. *Econ. Peace Secur. J.*, 2, 2, 32–41, 2007.

2

Mine Water Treatment and the Use of Artificial Intelligence in Acid Mine Drainage Prediction

Viswanath Ravi Kumar Vadapalli*, Emmanuel Sakala, Gloria Dube and Henk Coetzee

Council for Geoscience, Silverton, Pretoria, South Africa

Abstract
Acid Mine Drainage (AMD) emanating from coal, gold and copper mining has been widely reported with various negative environmental effects. Challenges associated with mine water can be experienced at a local and a regional scale. Such challenges include contamination of potable water and agricultural lands, and disrupted growth and reproduction of aquatic plants and animals. Therefore, it is critical to implement long term mine water management solutions including treatment of AMD. Treatment options can be broadly classified into passive and active treatment technologies. Both active and passive treatment technologies have their own advantages and disadvantages.

Prediction of AMD quality bears important consequences for long term management of water resources and therefore it is critical to improve the predictive capability of mine water using reliable and modern techniques. Artificial Intelligence (AI) is currently seeing a major interest in all spheres of life and interest from society in general. In this chapter, the authors highlight certain important aspects regarding AMD: generation, remediation, quality prediction using conventional and AI techniques and their limitations. A case study using a hybrid AI system to predict mine water quality is presented and discussed.

Keywords: Acid mine drainage, passive treatment, active treatment, artificial intelligence, hybrid AI system

List of Abbreviations

AI	Artificial Intelligence
ALD	Anoxic Limestone Drain
AMD	Acid Mine Drainage
ANFIS	Adaptive Fuzzy Interference System
ANN	Artificial Neural Network
BPNN	Back-Propagation Neural Network

**Corresponding author*: vvadapalli@geoscience.org.za [ORCID: 000-0002-9594-6717]
[ORCID: Sakala, 0000-0001-8698-5012; Dube, 000-0002-3040-2380; Coetzee, 0000-0003-3906-8224]

Elvis Fosso-Kankeu, Christian Wolkersdorfer and Jo Burgess (eds.) Recovery of Byproducts from Acid Mine Drainage Treatment, (23–48)

BWPCW	Brugspruit Water Pollution Control Works
DAS	Dispersed Alkaline System
DO	Dissolved Oxygen
DWAF	Department of Water Affairs and Forestry
ECL	Environmental Critical Level
EWRP	eMalahleni Water Reclamation Plant
GARD	Global Acid Rock Drainage
HDS	High Density Sludge
HiPRO	High Recovery Precipitating Reverse Osmosis
LDS	Low Density Sludge
LSTM	Long Short-Term Memory
MEND	Mine Environment Neutral Drainage
OCWRP	Optimum Coal Water Reclamation Plant
OLC	Open Limestone Channel
PCR	Pulsed Carbonate Reactor
PL	Predictive Learning
RAPS	Reducing and Alkalinity Producing System
RBF	Radial Base Function
REE	Rare Earth Element
RF	Random Forest
RGA	Real-Value Genetic Algorithm
RNN	Recurrent Neural Network
RO	Reverse Osmosis
SAPS	Successive Alkalinity Producing System
SRB	Sulfate Reducing Bacteria
SVM	Support Vector Machine
TDS	Total Dissolved Solids
WNN	Wavelength Neural Network

2.1 Acid Mine Drainage (AMD)

2.1.1 AMD Generation

While mining contributes largely to economic and social development in South Africa, its negative environmental impacts are a challenge not only at a local but also at a regional scale [1]. Environmental impacts include the contamination of streams, groundwater, and agricultural land [2, 3]. Ample studies have been conducted, and there is a large volume of literature available about acid mine drainage (AMD) production and its negative effects on the environment worldwide [4–10]. AMD is characterized by a low pH (<5.6) and high acidity, high concentrations of sulfate (SO_4^{2-}), metals and metalloids [1, 11–16]. Most of these constituents, at high concentrations, qualified as toxic in different environmental media, might be harmful to human life and result in negative effects on the ecosystem [17–24].

AMD is primarily engendered from the exploitation of commodities such as coal, gold, copper, and nickel. Minerals containing these elements usually occur in ore bodies that have acid-forming di-sulfide-bearing minerals which are mainly pyrite and marcasite (FeS_2)

[25]. AMD is generated as a result of the oxidation of di-sulfide minerals in the presence of water and oxygen [3, 26–30]. Research on AMD from other sulfide minerals such as pyrrhotite (FeS), chalcocite (Cu_2S), galena (PbS), sphalerite (ZnS), chalcopyrite ($CuFeS_2$), millerite (NiS), and mackinawite [(Fe,Ni)S] is limited [28]. Many authors illustrate examples of chemical reactions during AMD production using pyrite as a common sulfide mineral [28–36]. The foremost reaction is the oxidation of pyrite into (SO_4^{2-}), ferrous iron (Fe^{2+}), and protons (H^+, Eq. 2.1)

$$FeS_2(s) + 3.5\ O_2(g) + H_2O\ (l) \rightarrow Fe^{2+}(aq) + 2\ SO_4^{2-}(aq) + 2\ H^+\ (aq) \quad (2.1)$$

Fe^{2+} is then oxidized to ferric iron (Fe^{3+}) (Eq. 2.2):

$$4\ Fe^{2+}\ (aq) + O_2\ (g) + 4H^+\ (aq) \rightarrow 4Fe^{3+}\ (aq) + 2\ H_2O\ (l) \quad (2.2)$$

Oxidation of the ferric ion can produce soluble ferrous ion (Fe^{2+}), SO_4^{2-}, and more protons (Eq. 2.3):

$$FeS_2(s) + 14\ Fe^{3+}(aq) + 8\ H_2O\ (l) \rightarrow 15\ Fe^{2+}(aq) + 2\ SO_4^{2-}(aq) + 16\ H^+(aq) \quad (2.3)$$

2.1.2 Factors Controlling AMD Generation

The occurrence and rate of AMD generation are dependent on many factors and are site-specific [27]. The primary factors include the geological and hydrological characteristics of the site, type of sulfide mineral(s) present and their surface area, availability of oxygen, pH and temperature of the interacting water, heat that is being generated as a result of chemical reactions, chemical reactivity of Fe^{3+} in the system, and the availability of bacteria to catalyze the oxidation reaction, among others [27, 38]. For example, easily oxidized sulfide minerals (e.g., framboidal pyrite, marcasite, and pyrrhotite) result in a faster generation of acid. Sulfate compounds also add to the generation of AMD by contributing to the release of constituents such as Fe, Ni, and U in solution [27, 38]. Climate is also an important factor influencing the rate and effects of AMD. The generic process for the generation of AMD incorporates sulfide sources, water sources, and mixing zones where sulfide minerals are exposed to water in an oxidizing environment (Figure 2.1) [23]. South Africa has a prominent east to west climatic slope where annual rainfall ranges between 100 mm and 1,000 mm in the west and east respectively. Annual evapotranspiration potentially increasing from about 1,500 mm in the east to 3,000 mm in the west. Such climatic conditions result in most parts of the country to experience a negative water balance where rainfall is lesser than evapotranspiration. During dry periods, solid efflorescent salts are generated and the solutes that are formed, such as Fe, H^+, and SO_4^{2-}, are released when dissolved during the rainy seasons [39–41].

Younger [43] has described the process of mine flooding and its influence on acid generation, based on experience in British coal mines. During the period of active operation of an underground mine, the open mine workings allow the reaction between oxygen in the

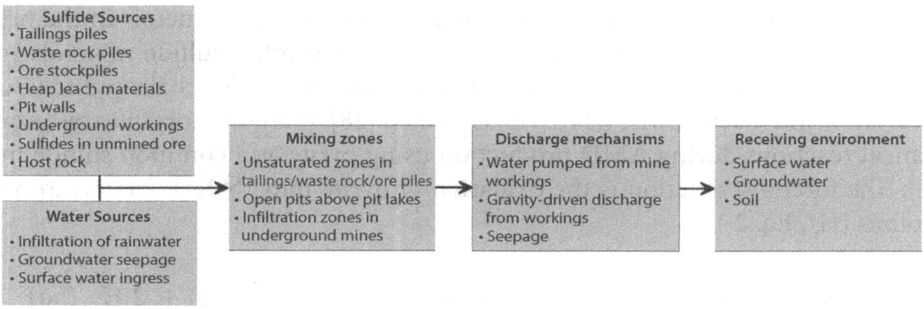

Figure 2.1 Generic process for the generation of AMD incorporating sulfide sources, water sources, and mixing zones where sulfides are exposed to water in an oxidizing environment (modified after the Global Acid Rock Drainage Guide [42]).

air-filled voids, water seeping through the mine and the surrounding rock mass, and sulfide in the mine and rock mass to interact. This oxidizes pyrite, generating AMD as well as solid weathering products in the form of acidic sulfate-bearing minerals. On the cessation of underground operations, as the underground mine workings flood, these minerals are dissolved, resulting in the first flush of acidic water. The flooding of the workings reduces the availability of oxygen, substantially reducing the oxidation of pyrite and the generation of new acidity. Younger [43] describes the two types of acidity generated within a mine void as juvenile acidity—the acid generated by the oxidation of fresh sulfide minerals—and vestigial acidity—the acidity generated by the dissolution of secondary minerals which accumulated during active mining. Mine flooding largely excludes oxygen from the flooded workings, reducing the generation of new juvenile acidity, often leading to substantial improvements in water quality once the first flush, which liberates vestigial acidity from secondary minerals, has dissipated. Where mine workings remain air-filled after mine flooding, juvenile acidity may still be generated for many years following partial mine flooding.

A permanent infiltration zone may develop where the discharge point from a mine void is located at a topographic level below the level of the highest lying shallow mine workings. This will result in a water level in the underground mine workings which is below

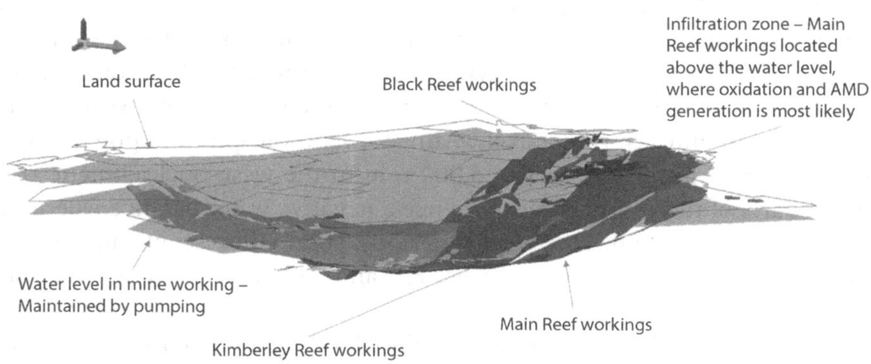

Figure 2.2 Three-dimensional model of South Africa's East Rand Goldfield, showing the presence of unflooded workings above the water level within the voids, as currently maintained by pumping.

the shallow workings in the high-lying areas of the mine or complex of mines which will continue to contribute acidity. The water level may be maintained by pumping, as is currently the case in South Africa's Witwatersrand Goldfields (Figure 2.2), or by gravity-driven discharge via a low-lying shaft or adit.

The Transvaal and Delagoa Bay Colliery in South Africa's Mpumalanga Coalfield is another example of a mine where acid generation continued for many years after flooding [7]. Located on a hillside, the mine workings were allowed to flood. Water entering the mine tended to flow through the workings, leaving a partially air-filled void with the continued generation of acidity. When the effect of this mine was investigated in the early 2000s, low pH values and extremely high sulfate concentrations were still recorded in the discharged water more than half a century after the cessation of mining.

2.2 Remediation of AMD

2.2.1 Introduction

Different AMD control technologies have been investigated, demonstrated, and implemented in different countries such as the United State of America (USA), China, Canada, Australia, Germany, Spain, and South Africa. There are numerous studies and a large volume of literature about AMD treatment techniques available for remediation worldwide. Chemistry and flow rate of the discharge, designated use, space availability, and seasonal fluctuations of the receiving stream are important factors when selecting an appropriate treatment technology [44–47]. There are two broad classes of technologies to treat AMD: passive and active treatment technologies [47–49].

2.2.2 Passive Treatment of AMD

Passive treatment is considered a long-term solution for the management of polluted mine water in many parts of the world [50, 51]. This treatment technique is usually associated with low costs of operation, monitoring, and maintenance [51–53] since it relies on natural ameliorative processes and accessible energy sources to remove contaminants in the water [55]. PIRAMID Consortium [45] defines passive treatment as "the improvement of water quality using only natural available energy sources in gravity-flow treatment systems which are designed to require only infrequent maintenance to operate successfully over their design life".

Different types of passive treatment technologies are available. Examples of passive treatment technologies include anoxic limestone drains (ALDs), aerobic and compost wetlands, reducing and alkalinity producing systems (RAPS, initially called SAPS) and dispersed alkaline systems (DASs). Passive treatment systems range from technologies that were successfully implemented at a full scale, to technologies that are currently being tested at a laboratory scale. The chemistry of AMD discharge (pH, acidity, Fe^{2+}/Fe^{3+}, Al, Mn, SO_4^{2-}), dissolved oxygen (DO), flow rate, and the topography are factors taken into consideration when evaluating and selecting the appropriate passive treatment type [51, 54–57].

Selecting a suitable passive treatment system depends on the chemistry and flow rate of the water to be treated (Figure 2.3) [53–58]. When the pH of water is net alkaline, a settling

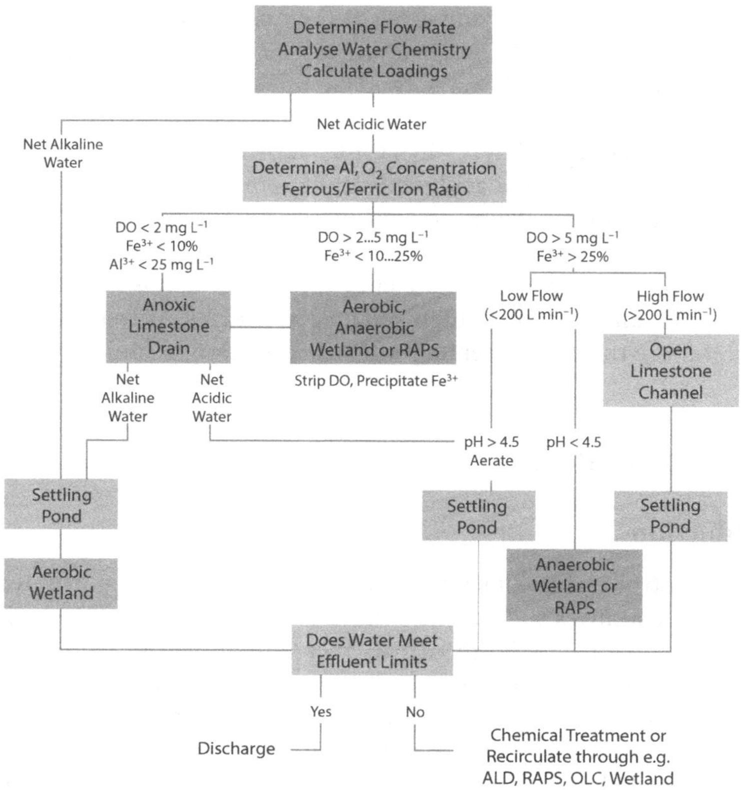

Figure 2.3 Flowchart assisting in the selection of a suitable passive treatment technology for polluted mine water (modified after [53]); image courtesy Christian Wolkersdorfer.

pond may be used to settle most of the suspended solids, followed by an aerobic wetland to oxidise and precipitate metals present in high concentrations (Figure 2.3). In the case of net acidic water, the chemistry of the water should be studied thoroughly to determine the DO concentration, Fe^{2+}/Fe^{3+} ratio, and Al concentrations. If the DO concentration is less than 2 mg/L, the Fe^{3+} concentration is less than 10% of total Fe, and the Al concentration is less than 25 mg/L, ALD will be the best suitable treatment system to use.

For further selection, when AMD has a DO concentration between 2 and 5 mg/L and the Fe^{3+} concentration ranges between 10 and 25% of the total Fe, a SAPS/RAPS can be used. These systems include a combination of an anaerobic or aerobic wetland and ALD. When acidic water (pH less than 4.5 in this case) has a DO concentration of more than 5 mg/L, the Fe^{3+} concentration is more than 25% of the total Fe and a low flow rate (less than 200 mg/L), an anaerobic wetland or SAPS may be used to treat the mine water. However, if the water has a pH of more than 4.5 with the same DO concentration, Fe^{3+} concentration and low flow rate mentioned above, the water can be aerated and transferred to a settling pond for further treatment. Furthermore, if the DO concentration is more than 5 mg/L, the Fe^{3+} concentration is more than 25% of the total Fe and there is a high flow rate (greater than 200 mg/L), open limestone channels (OLC) may be used.

Most of the passive treatment systems mentioned above are followed by a settling pond to settle most of the suspended solids, and when the effluent meets the required water quality standards it can be discharged to the receiving environment. However, when the treated effluent does not meet the water quality standards, retreatment or chemical treatment through RAPS, wetlands, OLC, ALD, or other systems will be needed until the effluents meet the water quality standards. A periodic table for passive treatment, created by Gusek [59], has also been used in the selection of the type of passive treatment (Figure 2.4). From the passive treatment periodic table, it can be seen that high concentrations of Fe, Al, and As can be treated using anaerobic and oxidizing passive systems, whereas Mn can only be treated using oxidizing systems. SO_4^{2-} may also be potentially removed by means of anaerobic systems through microbial reactions. In South Africa, to date, the application of passive treatment for amelioration of contaminated mine water is mainly demonstrated at a pilot and laboratory scale. Therefore, the potential for demonstration of passive systems on a full scale for long-term treatment of AMD is substantial.

2.2.3 Active Treatment of AMD

Active treatment of mine water entails improving the water quality using techniques that requires continuous addition of artificial energy or (bio)chemical reagents, or both [14]. According to the Acid Rock Drainage Prediction Manual, Eger and Wagner, and Watzlaf *et al.* [15, 49, 53], treatment of AMD by conventional treatment technologies is expensive and a long-term commitment. Active treatment uses a range of chemicals such as limestone, soda ash, caustic soda, ammonia, hydrated lime, and pebble quicklime [14, 60]. The selection of the appropriate chemical and suitable active process is influenced by a range of factors (Figure 2.5) including the water chemistry (i.e., total dissolved solids (TDS), total

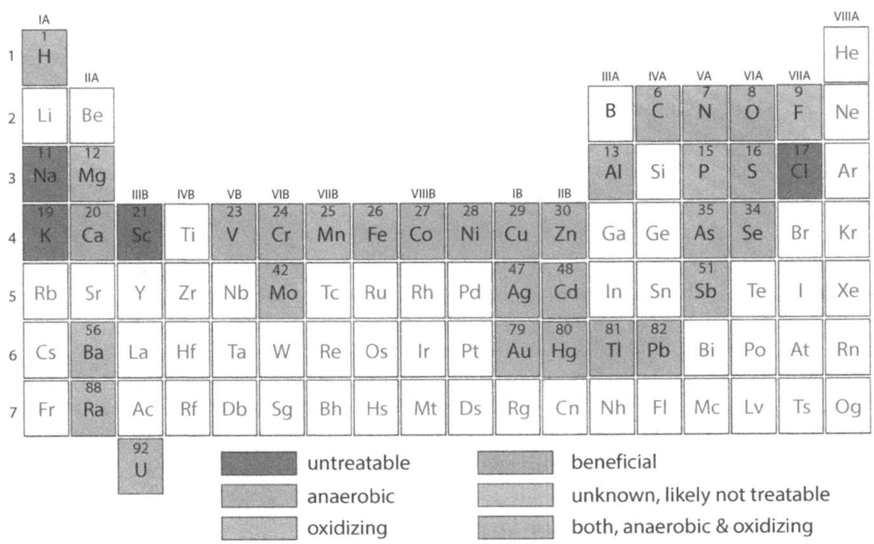

Figure 2.4 Periodic table for passive treatment (modified after Gusek [59]); image courtesy Christian Wolkersdorfer.

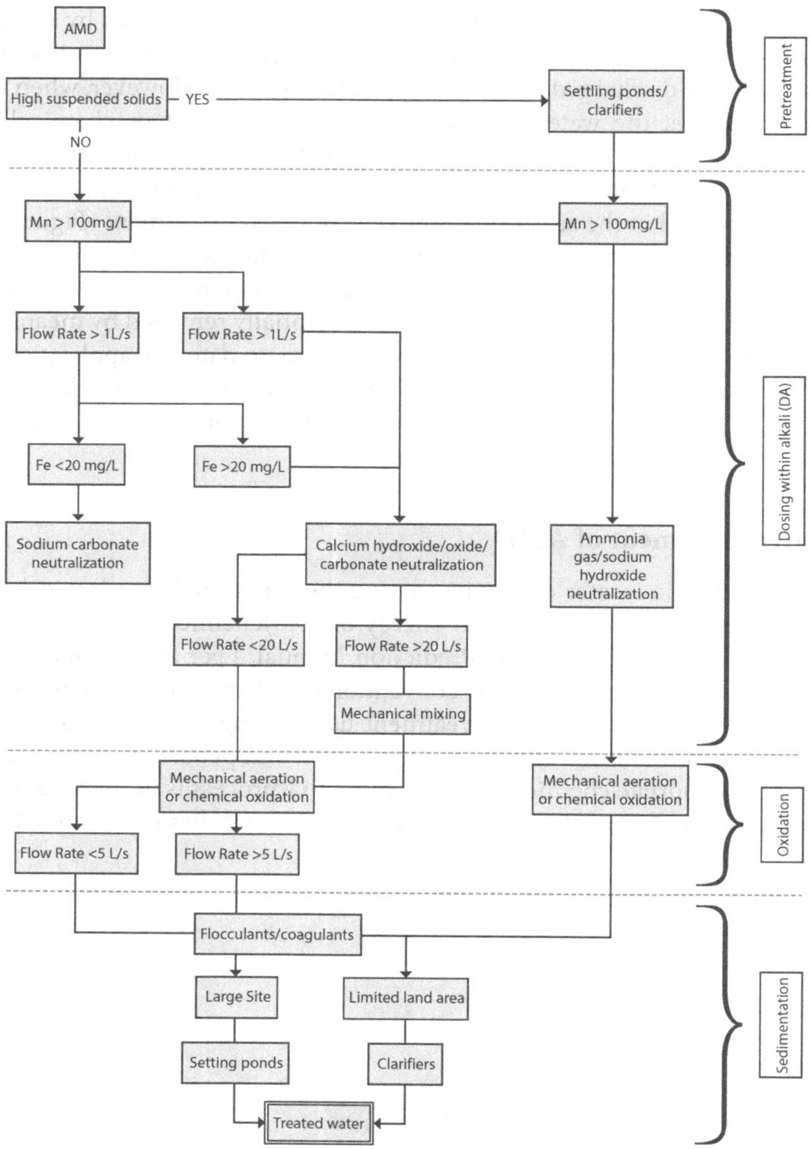

Figure 2.5 Flow chart for the selection of a site-specific active treatment system for AMD (modified after [47]).

suspended solids, and metal concentrations), quantity and flow rate, space, availability, and local climate [61–66]. Although active treatment can ameliorate AMD effectively, they are labor, energy, and maintenance-intensive and therefore not favored for abandoned mines.

There are different active treatment processes that are used for AMD remediation and some of those working at full scale are mentioned as:

a) Low density sludge (LDS): process plants that can accommodate AMD with any acidity loads. According to Taylor *et al.* [11], there are three main treatment stages involved. The first is the reagent mixing and dosing phase where a neutralizing reagent is mixed with AMD to produce a slurry. The second stage includes a reaction phase where mechanical stirrers are used to mix the slurry solution through one or more reactors. Mixing may also take aeration into account to oxidise any reduced metals. The third stage of treatment is the flocculation and clarification stage. This stage includes a thickener or clarifier tank where neutralized water from the second stage enters. For facilitation for sludge settling, a flocculant may be added. The disadvantages of LDS plants for AMD treatment include the large volumes of low-density sludge produced, storage needed for the sludge and the costs associated with sludge handling.
b) High density sludge (HDS): a treatment process that is similar to LDS and involves three stages of treatment, i.e., the reagent mixing stage, reaction stage, and the flocculation and clarification stage. This treatment process is better compared to LDS as the generated sludge contains enough alkalinity and is recirculated back into the treatment process along with the alkaline reagent. Therefore, the final sludge for disposal is concentrated and requires reduced storage space, also maximizing the sludge stability and density [11].
c) Pulsed carbonate reactors (PCRs): a treatment process centered around the principle of increasing the partial pressure of CO_2 in water intensely to improve the solubility of carbonate material such as limestone. Treatment includes CO_2 initially absorbed into AMD at atmospheric pressure and then palpitated over successive reactors and neutralized with calcium carbonate [11].
d) Electrochemical concentration technique: processes that use a combination of chemical, magnetic, electrical, and plasma techniques to recover metals from an AMD solution [11].
e) Biological mediation: a technique that takes AMD neutralization, sulfate reduction and redox control [67] into consideration. Micro bioreactor systems are used for chemical and biological processes [68].
f) Reverse Osmosis (RO): a treatment process that uses a semipermeable barrier (membrane) at a certain pressure to treat mine water. This semipermeable membrane, such as a cellophane-like semipermeable cellulose acetate membrane, impedes the flow of larger molecules (e.g., urea, glucose, and bacteria) and solutions (e.g., Na^+, Ca^{2+}, Cl^-) to pass through [69, 70]. The membrane only allows the passage of water. According to Feng *et al.* [71], membranes during RO processes have been shown to substantially reduce TDS, metals and metalloids, organic pollutants, viruses, bacteria, and other dissolved contaminants.
g) Ion exchange: a process where undesirable or potentially toxic ions in water are exchanged for desirable or less toxic ions [72]. Materials such as zeolites and resins are generally used for ion exchanges processes. According to Johnson and Hallberg [72], only ions with the same electric charge can be exchanged.

Currently, in South Africa, treatment of AMD on a bigger scale is being carried out in six active treatment plants of which three are situated in the East, West, and Central Rand Goldfields of Gauteng Province and three are situated in the coalfields in Mpumalanga Province. The HDS treatment plants in the East, Central, and West Rand Basin plants can treat 50,000 m^3/d, 82,000 m^3/d, and 110,000 m^3/d of mine water respectively, after which the treated water is released into adjacent rivers [73].

Two key-plan high-recovery precipitating reverse osmosis (HiPRO) treatment process plants exist in Mpumalanga Province: the eMalahleni Water Reclamation Plant (EWRP) and the Optimum Coal Water Reclamation Plant (OCWRP). In both cases, the treatment technology involves neutralization, reactors and clarifiers, multistage ultrafiltration, reverse osmosis, and sludge dewatering [74]. Implemented in the eMalahleni area in 2007, the EWRP treats AMD from several mines including Greenside Colliery, Kleinkopje Colliery, South Witbank Colliery, and Navigation Colliery and has been operating successfully since then [75, 76]. Its main aim is to treat AMD, produce high quality potable water for local municipality use and being safe for environmental release, with a current treatment capacity of 50,000 m^3/d [75]. On the other hand, the OCWRP was built in 2010 and has a capacity of 15,000 m^3/d presently. According to the Aveng Group [76], most of the reclaimed water is distributed to the local Hendrina municipality and the remaining is discharged into the Klein Olifants River. Because of the drinking water quality standard of the treated mine water discharge, the plant was bestowed with the prestigious Blue Drop Certification by the Department of Water Affairs (DWA) [77].

In 1997, the Brugspruit Water Pollution Control Works (BWPCW) plant eMalahleni, Mpumalanga Province was constructed by Department of Water Affairs and Forestry (DWAF) to protect Loskop Dam from negative effects of AMD [78]. This plant was treating AMD from the defunct and flooded underground coal mines towards the west and northwest of eMalahleni (Witbank), which started discharging in the mid-1990s [78, 79]. This AMD discharge was one of the contributors to the pollution of the water resources in the upper Olifants River catchment [78, 79]. Water was treated with sodium hydroxide to increase the pH and soda ash to increase the buffering capacity of the final effluent [19, 20]. Before unforeseen circumstances caused the treatment plant to fail, it could treat 10,000 m^3/d and is currently not operational [67].

2.2.4 Challenges With Current AMD Treatment

Passive treatment is capable of improving the water quality of polluted mine water. However, the long-term sustainability of these systems is a common problem and should be ensured by developing appropriate site-specific design criteria (Section 2.1). The depletion rate of the neutralizing agent and organic matter used in different systems is also a key problem [80]. Clogging, due to precipitate build-up and biofilms, is the leading drawback in these treatment techniques affecting the efficiency and longevity of such systems [11, 53]. Clogging results in the passivation of the alkaline substrate slowing down the dissolution rate. This challenge has been addressed by occasional maintenance where periodic flushing of the system is done. Another disadvantage associated with these treatment systems is minimal or no SO_4^{2-} and Mn^{2+} removal. The reasons for this limited or non-performance are lack of sulfate-reducing bacteria (SRB) for SO_4^{2-} reduction and the presence of Fe in the treatment system which competes for oxygen with Mn^{2+}, among others. Moreover, passive treatment systems can only accommodate low-flow volumes, with some exceptions, and cannot cope with high-flow volumes of mine water.

Challenges associated with the existing active treatment systems of AMD include the costs of operation and maintenance, especially in the case of those mines that reached end of life. The major challenge with the active treatment technologies is that they require constant addition of chemicals and artificial energy [81]. High-priced chemicals and operational resources are some of the factors that discourage the continuous active treatment of AMD. Furthermore, some of the neutralizing chemicals such as sodium hydroxide (NaOH) and anhydrous ammonia (NH_3) are hazardous. The extreme use of ammonia may result in challenges like denitrification and nitrification of receiving aquatic environments [82]. Storage and disposal of the slurry produced from the treatment plants are also considered a problem with the active treatment systems.

In the Witwatersrand basin, the mine water is currently being pumped and neutralized to keep the rising water levels under the environmental critical level (ECL). Neutralization is achieved with limestone or lime, or both, to reach a pH value that is suitable for discharge. This process adds alkalinity to the water, increases the pH and precipitates some of the metals as hydroxides. One of the challenges with the existing treatment plants in South Africa are the SO_4^{2-} concentrations in the treated water that are above the South African water quality standards acceptable for discharge [83]. The costs of implementation and operation of the treatment plants are also one of the predicaments of mine water treatment in the country. By the year 2020, R12.3 billion are anticipated to be spent on the implementation of treatment plants in the Eastern, Central, and Western Basins of the Witwatersrand [83, 84]. According to the studies, 67% of this amount will be or have been paid by the operating mining companies in these areas. The envisaged dilemma with these current plants is the continuation of receiving funding for maintenance and running the plants especially once the operating mines that are currently contributing, reach mine closure. Moreover, issues like vandalism are also a major challenge that may lead to failure of mine water treatment plants [67].

2.2.5 Value Recovery From AMD Treatment

Resource recovery and reuse is possible as a holistic sustainable approach in AMD treatment. Nonetheless, not many studies were carried out with regard to AMD-treated water reuse and resource recovery. In South Africa, many researchers have carried out studies on metal precipitation from AMD [3, 68, 86], but only a few have focussed on the recovery of valuable metals [87–89], sulfuric acid [90], and water for reuse [91, 92].

AMD is characterized by high concentrations of metals and metalloids and some of these constituents can be extracted through treatment for economic benefit and to achieve a circular economy. Several authors [65, 93] noted the possibility of recycling Fe-rich precipitates such as jarosite ($KFe_3(SO_4)_2(OH)_6$), schwertmannite ($Fe_{16}O_{16}(OH)_y(SO_4)_z.nH_2O$), goethite (FeOOH), and ferrihydrite ($Fe(OH)_3$) for products such as mineral pigments, depending on the variations in the major and trace metal chemistry. These Fe-hydroxides are acknowledged as sorbents of trace metals and consequently can control the kinetics of the trace metals (e.g., Zn, Co, and Ni) in the environment. Recovery of such trace metal species during remediation presents a case for offsetting treatment costs. Research around the recovery of rare earth elements (REEs) from AMD has also been done as a source for critical materials for batteries, magnets, and other electronic components. Recently, AMD and treated mine water via active and passive treatment were studied as likely economic sources of REE [65, 93–96]. Some of these REEs occur as sulfate species [65] and their recovery is a promising possibility that could potentially offset AMD treatment costs [97].

Partially treated mine water may be suitable for other applications such as in agriculture, sanitation and industries where high-quality water is not often required provided it is treated to acceptable concentrations for various pollutants [65, 73, 97–99]. The advantages of this approach include cost savings from a reduced treatment cost point of view as well as the availability of water as a resource [11, 68, 87].

2.3 Prediction of AMD

In mining environments where AMD may be generated, prediction plays an important role in mine planning, prior to the commencement of mining, environmental management during mining and closure planning during the late stages of mining. In addition, in mining legacy areas where no mine operator is present, predictive tools may be used to estimate the seriousness of current and future effects, as well as the duration of expected ongoing effects.

Three approaches for the prediction of AMD are commonly used:

a) Field measurements, including simple sampling and analysis, as well as controlled field experiments such as the construction and monitoring of lysimeters within the mining environment and field tests such as wall washing [100].
b) Numerical geochemical models, for example, PHREEQC [101], where known chemical parameters of specific minerals are used to predict the reactions between these minerals and solutions. Geochemical models may be coupled with flow models to predict the generation of AMD as well as its transport and fate in the environment.
c) Laboratory experiments, where samples of the material encountered in the mining project are exposed to laboratory-simulated environmental conditions, allowing researchers to monitor process and sample reaction products to develop predictive models of the likely behavior of the materials involved in the mining environment. Since the processes leading to the generation of AMD often operate over periods of decades and even centuries, these laboratory methods generally accelerate the reactions, allowing meaningful conclusions to be drawn in a manageable time-frame, typically days to weeks.

In practice, a meaningful AMD prediction exercise will combine all three of these approaches. Field observations and experiments may be used to develop a reliable conceptual model of the mining environment and to identify physical processes which are likely to influence, accelerate or retard the generation of AMD. A reliable conceptual model is essential in the design and parameterization of geochemical models if the results are to be at all relevant to the real-world situation. Likewise, laboratory tests need to be carefully selected to best represent the real-world conditions which will be experienced during the mining life cycle. A critical consideration for the prediction of mine water quality is that the results of the field investigations, laboratory tests, and numerical models should display sufficient consistency to represent the conditions on the mine and that they should be defined with reference to observed field conditions at all time.

These approaches have been well summarized in the Global Acid Rock Drainage (GARD) Guide [42], while detailed descriptions of laboratory methods have been provided as part of the Canadian Mine Environment Neutral Drainage (MEND) Programs [15]. The basic approach determines [27]:

a) Whether a given sample of material will generate acid drainage, or if it has sufficient neutralization potential to counteract the acidification by sulfide oxidation (e.g., acid-base accounting);
b) What the likely chemistry of the resulting reactions is (various leach tests); and
c) What the behavior of the material is likely to be over time (kinetic tests, including column tests, and humidity cell tests).

Additional laboratory tests such as the Mine Water Leaching Procedure [102] may be employed to determine the interactions between ore and mine residue leachates and other materials.

2.3.1 Limitations of Predictive Tools

The tools used to predict AMD have a number of limitations. In particular, they generally rely on simplified models of the orebody, the mine, and the environment where they exist. In the case of pre-mining predictive modeling, the modeling can also not accommodate changes in the mining programs when the mine operator has to adapt to new information gathered during mining, often rendering the pre-mining predictive models less relevant to the real-world conditions.

Orebodies and mines are complex environments, which are often difficult to reduce to manageable conceptual models. The resulting simplification often fails to provide an adequate prediction of the behavior of the source material, possible pollution dispersal pathways and discharge or fate in the environment. Particularly when these tools are applied in the early phases of a mining project, with the objective of developing mine closure plans and post-closure environmental management programs, the oversimplification may fail to adequately predict real outcomes. Changes in the initial mining program may exacerbate this problem.

While not an inherent quality of predictive modeling, a lack of interdisciplinarity in many environmental impact prediction exercises results in overly narrow predictions being made, with the prediction of water chemistry being done by geochemists and or chemists with insufficient involvement of hydrologists, hydrogeologists, and mining engineers. Constrained budgets in terms of both funding and available time will often exacerbate this problem. The prediction of AMD generation, transport and effect are then often treated as a problem defined by the chemistry of the ore and host rocks and their likely interactions with the environment.

Critical system components are often ignored or underemphasized. These include the topography of a mining area and its influence on the availability of water or oxygen in different parts of an underground or open-pit mine. For example, in the West Rand and Central Rand Goldfields of South Africa, extensive outcrop and near-surface underground operations were developed in the early stages of mining [103]. Within each of these goldfields, the underground workings of multiple mines are interconnected, resulting in water flow from mine to mine. Even if mines are allowed to flood and discharge to surface in the low-lying portions of these goldfields, as happened in the West Rand in 2002 [104], a substantial volume of the mine void will never flood, leading to the ongoing generation of juvenile acidity. It is therefore for the reasons above that people have started looking at alternative techniques to predict mine water quality for better decision-making.

2.4 Application of Artificial Intelligence for AMD Quality Prediction

2.4.1 Introduction

Artificial Intelligence (AI) is the study of making computers to perform functions or activities that are currently deemed "intelligent", such as learning, decision-making, solving problems, among others. The main aim of AI is to produce intelligent machines, computer programs, or embedded systems. Consequently, the field of AI is multidisciplinary in nature, combining various fields such as computer science, psychology, philosophy, and engineering.

The question of whether a machine is able to exhibit human-like intelligence has been a subject of debate for years. Many are against the idea of saying that some behaviors such as creativity, love, and moral values cannot be understood by machines. Others have accepted that machines can indeed exhibit aspects of human-like intelligence. This disagreement remains unresolved. In order to understand AI, it is important to define the concept of "intelligence". Intelligence may be defined in two ways [105]:

a) Intelligence is the ability to interpret and acquire knowledge and skills through a thought process, and
b) Intelligence is the ability to think and comprehend things using instincts.

The first definition does not include machines, whereas the second definition is more flexible to accommodate machines exhibiting intelligence. Thus, for a machine or someone to exhibit intelligence, the ability to think, learn, or understand is required. Thinking may be defined as "the activity of using the brain to consider a problem or to create an idea". Thus, intelligence may be defined as "the ability to think, learn and/or understand to solve problems and to make decisions" [105]. Humans, animals, and machines have this ability. Fundamentally, AI seeks to answer the question as to whether a machine is able to do things which at present are done by humans. The field of AI has been formulated in 1943 and since then has developed steadily until the present day. AI has many branches, all of which are connected and share commonalities and is widely used to build expert systems for natural language processing, robotics, speech, planning, and vision [106].

Uncertainty and subjectivity are inherent in all predictive modeling techniques as these assessments rely on expert opinions. These limitations have to be taken into consideration when formulating predictive intervening, especially if the results are intended to be used for critical applications such as decision-making. By quantifying and incorporating uncertainty and subjectivity, AI systems have been successfully used for prediction in fields such as robotics, science, and engineering [107].

The benefits of using AI for data processing and automation can potentially supplement human mind that has limited ability for processing data and susceptible to subjective bias, AI models can process huge amounts of data relatively faster with greater accuracy. Therefore, predictive models can help reduce computational cost while increasing computational speed, reliability, and consistency in addition to reducing any human-induced uncertainties [108, 109].

2.4.2 Different AI Techniques Used to Predict AMD Quality

The common AI tools used for predictive modeling of water quality can be classified into three board types, namely, the knowledge-driven, the data-driven, and the hybrid type.

According to Swain [110], knowledge-driven systems are computer programs that use knowledge (Expert) base with an inference engine meant to solve problems that typically entail substantial specialized human capability. The knowledge base comprehends an assembly of information while the inference engine infers interactions from the information instituted in the knowledge base. The knowledge-based systems moreover include an interface for users to be able to interrogate the system. Knowledge-driven AI techniques involve articulating computer-based systems that enable extracting and copying intellectual reasoning in making conclusions. Thus, the computer emulates human thought processes [110, 111]. The most common example of knowledge-driven techniques includes fuzzy inference systems, which are often used where the relationship between the input parameters and desired predicted variables is well understood.

Data-driven AI systems are computer programs that solve problems using information derived from data without explicit knowledge of the problem [112]. Modern-day data-driven AI approaches are mainly machine learning in nature. According to the literature, the widely used data-driven tools are regression analysis, artificial neural networks (ANNs), random forest (RF), decision trees, support vector machines (SVMs), radial base functions (RBFs), and genetic algorithms.

Maier and Dandy [112] used ANNs as a viable means of forecasting water quality parameters (salinity) in the Murray River at Murray Bridge (South Australia). The ANN results improved the 14 days forecasting better than the previously used real-time forecasting simulation. Bayatzadeh Fard [113] successfully used ANN and multi-output adaptive neural fuzzy inference systems (ANFISs) to estimate the distribution of (semi-)metals in groundwater of the Lakan lead-zinc mine with a high degree of accuracy.

SVM and back-propagation neural networks (BPNNs) were used to predict the concentration of Ni and Fe generated by the anthropogenic-activated pyrite oxidation in mining areas [114]. In this study, the SVM model provided a better and faster prediction of the Fe and Ni metals. Similar conclusions were also drawn by Tan *et al.* [115] regarding the superiority of the SVM over the ANN in water quality predictions. A hybrid method which is known as the real-value genetic algorithm support vector machine (RGA-SVM) was utilized to predict the quality of aquaculture water [116]. This proved to be an effective approach to predict aquaculture water quality when compared with the traditional SVM and back-propagation neural network models. A machine learning technique, in particular, regression tree induction, was applied to address the difficulty of inferring the chemical parameters from biological parameters of water quality [117]. This proved to be particularly important in enabling selective chemical monitoring of Slovenian River water quality.

Various AI algorithms such as knowledge-based systems, genetic algorithms, ANNs and fuzzy inference systems were used by Chau [118] to simulate and integrate flow and water quality measurements in coastal environments. The use of AI proved to be very useful in incorporating existing heuristic knowledge from various model developers and practitioners and furnished intelligent manipulation of the calibration parameters of numerical modeling systems.

Wavelet-neural network (WNN) and ANN techniques were used to predict water quality parameters (mineralization, temperature, and DO) in Hilo Bay, Pacific Ocean [119]. The outcome of the research shows that the WNN approach is superior to the ANN approach with over 0.98 correlation between model values and the actual values. The research also established that the approach managed to obtain good results even though some of the ocean data was missing due to a lack of regular acquisition or difficulty in data acquisition.

The application of AI for the prediction of mine drainage is not new. Several authors used machine learning techniques such as ANN [120], SVM [121], RBF [121], and K-mean [121] to model and forecast AMD quality using past physico-chemical parameters of a mining area. AI tools such as machine learning offer a rapid and cost-effective solution for mine water quality forecasting.

2.4.3 Limitations of AI Techniques in Prediction of AMD Quality

The AI techniques have their own limitations which include:

a) Time consuming processing and training of the model as optimization is heuristically and repeatedly performed until a good model or learning technique is obtained.
b) Most experts find it difficult to interpret and explain decisions made by AI models, suggesting its black box nature and limiting its successful application.

For any prediction, including mine drainage quality, a thorough investigation and understanding of the fundamental physical processes is compulsory. However, it is often challenging to involve domain expert knowledge directly into the data-driven approaches [122]. Hence, hybrid approaches are developed to mitigate this problem.

Prediction of mine drainage is very complex as the conditions which produce the drainage quality are normally nonlinear, dynamic, and change over time. In addition to this, acquiring good quality mine drainage data is difficult and often contains missing and erroneous values [122]. The approaches mentioned in the previous section often fail to learn the time-variant parameters leading to low accuracy and poor generalization, hence specially designed algorithms that can learn time series data are required. A recurrent neural network (RNN) is used for time series data learning but is mostly not successful when learning the long-term dependences [123], hence the long short-term memory (LSTM) is a specifically designed RNN to overcome this problem. Guazhou Water Source (China) project [124] tested the applicability of using LSTM to predict water quality of the Yangtze River in Yangzhou. The project results show that the trends in the water quality are in agreement with the actual measured values endorsing its potential for utilization as a powerful drinking water prediction tool.

Even with the availability of a well-trained and optimized model, the ability to interpret and analyze the results requires expert domain knowledge, and this can be only achieved by popularizing the predictive learning (PL) models among domain experts [122].

2.4.4 Case Study—Ermelo Coalfield, South Africa

The Ermelo Coalfield study investigates the application of a hybrid system, which combines LSTM and ANN to predict mine water quality with particular emphasis on sulfate concentration from an abandoned underground coal mine in Carolina Town, Mpumalanga Province of South Africa. The full details of this case study are presented in [125].

The methodology involves first the understanding of the likely factors contributing to mine water quality, approximated by the sulfate concentrations in the study environment [126]. According to [127], the most important external factors that contribute to AMD generation at an underground site are recharge (rainfall as a proxy), water table level fluctuations, soil temperature, and water pH. For the study area, the datasets were obtained from various governmental institutions, and hydrogeochemistry data was collected from the discharge test site between 1 November 2014 and 13 June 2018. The forecasting system can be divided into two sub-systems (Figure 2.6): LSTMs and ANN systems. The system architecture is designed to accept historical data (input parameters and output labels) for the training and optimization processes of the ANN system.

To improve the training and avoid under and overfitting, early stopping was used. Early stopping is whereby the training process is stopped when no further decrease in the training root mean square value is observed. LSTMs are used to predict the future values of each input variable (Figure 2.6). The blue dots represent the observation data, the black line is the LSTM fitting model for the observation data and the red line is the output prediction data for the four input parameters (Figure 2.6). Historic data spanning over 1,400 days was used for training and the next 350 days as testing.

The gradient descent optimizer algorithm with a batch size of 100 was used for training the ANN using the historic input parameters (pH, soil temperature, rainfall, and water table). During training, data is fed into the input layer which is connected to one or more "*hidden layers*" where the actual processing is performed by adjusting the weights "*connections*" using the back-propagation algorithm. In a sense, ANNs learn by example, in the same way, as a child learns to walk from observing adults walking. Thus, learning by neural networks may be viewed as a case whereby network weights are updated to produce the desired output, based on a set of

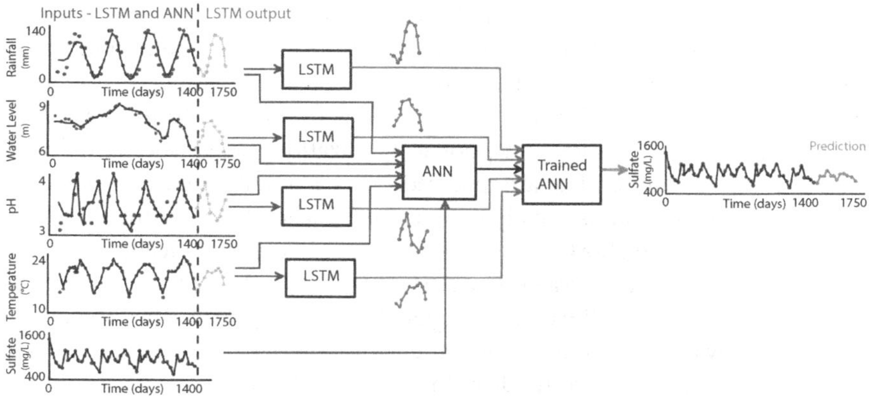

Figure 2.6 Mine water prediction system architecture showing the link between the input parameters, LSTM and ANN systems and the resultant predicted sulfate concentration for the Witkranz site (modified after [125]).

examples such as the beforementioned historical input parameters and the label output being the sulfate in this study. The output values generated by LSTM for each input are fed into the trained ANN model to forecast sulfate concentrations from June 2018 to May 2019 as depicted by the green line ("Prediction" in Figure 2.6). Future work includes testing and calibration of this preliminary AI model using more data collected at the discharge site.

2.5 Conclusions

Poor management of AMD poses a substantial environmental risk at both local and regional scales. Globally, this risk is currently being managed via a variety of interventions including a wide range of treatment technologies. These technologies are broadly categorized into active and passive systems. In the South African context, only active treatment systems are operating at a bigger scale to manage the environmental risk posed by AMD at the Witbank site and the Witwatersrand basins.

Prediction of mine water quality is critical for long-term mine planning and the application of AI techniques in this regard could play a pivotal role in achieving this objective, especially considering the successful application of these tools in other fields such as image processing and financial forecasting. However, in spite of frequently generating effective models, the use of AI for predicting AMD quality is strenuous considering the difficulty with integrating fundamental physical processes into the knowledge base model, the choice of suitable learning techniques and the assessment of the modeling results requires exceptional attention. Other limitations for delivering effective models include availability of limited and good quality data and reluctance to accept PL methods, among others [122].

Taking cognisance of the aforementioned positives and limitations, the case study presented in this chapter, investigated the application of a hybrid system, which combines LSTM and ANN to predict mine water quality with sulfate concentrations as a proxy from an abandoned underground coal mine in Mpumalanga province. A predictive preliminary AI prototype was developed that can be revised as and when new information is available. In conclusion, this chapter provides an overview of this field and, hopefully, will lead to more advances and intense research efforts on issues outlined above.

References

1. Manders, P., Godfrey, L., Hobbs, P., Acid mine drainage in South Africa, Briefing note 2009/02. *Counc. Sci. Innov. Res. (CSIR)*, 2009.
2. Zhao, F., Cong, Z., Sun, H., Ren, D., The geochemistry of rare earth elements (REE) in acid mine drainage from the Sitai coal mine, Shanxi Province, North China. *Int. J. Coal Geol.*, 70, 184–192, 2007. https://doi.org/10.1016/j.coal.2006.01.009.
3. Oelofse, A., Lonvaud-Funel, A., du Toit, M., Molecular identification of *Brettanomyces bruxellensis* strains isolated from red wines and volatile phenol production. *Food Microbiol.*, 26, 377–385, 2009. https://doi.org/10.1016/j.fm.2008.10.011.
4. Johnson, D.B. and Hallberg, K.B., Acid mine drainage remediation options: A review. *Sci. Total Environ.*, 338, 3–14, 2005. https://doi.org/10.1016/j.scitotenv.2004.09.002.
5. Rapson, L.A., *The development of a risk-based computer database to prioritise the environmental rehabilitation of defunct and abandoned collieries in the Witbank area of South Africa, MSc diss*, pp. 1–153, Witwatersrand University, South Africa, 2004.
6. Banks, D., Younger, P.L., Arnesen, R.T., Iversen, E.R., Banks, S.B., Mine-water chemistry: The good, the bad and the ugly. *J. Environ. Geol.*, 32, 157–174, 1997.
7. Inter-ministerial Committee (IMC), *Mine water management in the Witwatersrand Gold Fields with special emphasis on acid mine drainage*, Dept. of Water Affairs Report, Pretoria, RSA, 2010.
8. Mail and Guardian, *Mpumalanga's not-so-clean coal*, Mail and guardian, Johannesburg, 2016.
9. Munnik, V., The social and environmental consequences of coal mining in South Africa—A case study. *Environ. Monit. Group*, 1–24, 2010.
10. Lai, J., *Impact of anthropogenic pollution on selected biota in Loskop Dam. MSc Thesis*, pp. 1–135, University of Pretoria, South Africa, 2013.
11. Taylor, J.S., Pape, S., Murphy, N., A summary of passive and active treatment technologies for acid and metalliferous drainage (AMD). *Proc. 5th Aust. Workshop Acid Drainage*, 29, 1–49, 2005.
12. Durand, J.F., The impact of gold mining on the Witwatersrand on the Rivers and Karst System of Gauteng and North West Province, South Africa. *J. Afr. Earth Sci.*, 68, 24–43, 2012. https://doi.org/10.1016/j.jafrearsci.2012.03.013.
13. Younger, P.L., Banwart, S.A., Hedin, R.S., *Mine water: Hydrology, pollution, and remediation*, p. 464, Springer, Kluwer Academic Publishers, Dordrecht, 2002.
14. Younger, P. and Robins, N., Challenges in the characterization and prediction of the hydrogeology and geochemistry of mined ground, in: *Mine water hydrogeology and geochemistry*, P.L. Younger and N.S. Robins (Eds.), pp. 1–16, Geological Society, London, 2002.
15. Mine Environment Neutral Drainage Program (MEND), *Acid Rock Drainage Prediction Manual: A manual of chemical evaluation procedures for the prediction of acid generation from mine wastes*, pp. 1–77, Dept. of Energy, Mines and Resources, Canada, 2008, Mine Environment Neutral Drainage Program (MEND), Acid Rock Drainage Prediction Manual: A manual of chemical evaluation procedures for the prediction of acid generation from mine wastes. *Department of Energy, Mines and Resources*, 1–77, 2011.
16. Jacobs, J.A., Overview of resources from acid mine drainage and postmining opportunities, in: *Acid mine drainage, rock drainage, and acid sulfate soils: Causes, assessment, prediction, prevention and remediation*, J.A. Jacobs, J.H. Lehr, S.M. Testa (Eds.), pp. 361–375, Wiley, New Jersey, 2014.
17. Mbhele, P.P., *Remediation of soil and water contaminated by heavy metals and hydrocarbons using silica encapsulation. MSc diss*, pp. 1–154, Witwatersrand Unversity, Johannesburg (JHB), RSA, 2007.

18. Environmental Law Alliance Worldwide (ELAW), *Guidebook for evaluating mining projects EIAs*, pp. 8–9, Eugene, Eugene, ELAW, 2010.
19. Bell, F.G., Halbich, T.F.J., Bullock, S.E.T., Environmental impacts associated with an abandoned mine in the Witbank Coalfield, South Africa. *Int. J. Coal Geol.*, 45, 2–3, 195–216, 2001. https://doi.org/10.1016/S0166-5162(00)00033-1.
20. Hobbs, P., Oelofse, S.H.H., Rascher, J., Management of environmental impacts from coal mining in the Upper Olifants River Catchment as a function of age and scale. *Int. J. Water Resour. D.*, 24, 3, 417–431, 2008. https://doi.org/10.1080/07900620802127366.
21. McCarthy, T.S. and Pretorius, K., Coal mining on the Highveld and its implications for future water quality in the Vaal River system, in: *Water Institute of Southern Africa, International Mine Water Association (Ed.), Proceedings of the International Mine Water Association, Pretoria, South Africa, October 19–23, 2009*, pp. 56–65, 2009.
22. Oberholster, P.J., Genthe, B., Hobbs, P., Cheng, P.H., de Klerk, A.R., Botha, A.M., An ecotoxicological screening tool to prioritize acid mine drainage impacted streams for future restoration. *Environ. Pollut.*, 176, 244–253, 2013. https://doi.org/10.1016/j.envpol.2013.01.010.
23. McCarthy, T.S., The impact of acid mine drainage in South Africa. *S. Afr. J. of Sci.*, 107, 5/6, 1–7, 2011. https://doi.org/10.4102/sajs.v107i5/6.712.
24. Novhe, N.O., Evaluation of applicability of the passive treatment for the management of polluted mine water in Witwatersrand Goldfields, South Africa, in: *International Mine Water Association Symposium, Bunbury, Australia, September 30-October 4, 2012*, McCullough, C.D., Lund, M.A., Wyse, L. (Eds.), pp. 417–424, 2012.
25. Hancox, P.J., The coalfields of South-Central Africa: a current perspective. *Int. J. Geosci.*, 39, 2, 407–428, 2016.
26. Norton, P.J., The control of acid mine drainage with wetlands. *Mine Water Environ.*, 11, 3, 27–34, 1992.
27. United States Environmental Protection Agency (USEPA), *Technical Document: Acid mine drainage prediction, EPA 530-R-94-036*, USEPA, Washington DC, pp. 1–48, 1994.
28. Wolkersdorfer, C., *Water management at abandoned flooded underground mines: Fundamentals, tracer tests, modelling, water treatment*, p. 465, Springer, Verlag Berlin Heidelberg, 2008.
29. Rawlings, D.E., Characteristics and adaptability of iron-and sulphur-oxidizing microorganisms used for the recovery of metals from minerals and their concentrates. *Microb. Cell Fact.*, 4, 1–15, 2005. https://doi.org/10.1186/1475-2859-4-13.
30. Feris, L. and Kotze, L.J., The regulation of acid mine drainage in South Africa: Law and governance perspectives. *Afri. J. Online*, 17, 5, 2105–2160, 2014.
31. Kleinmann, R.L.P., Hedin, R.S., Nairn, R.W., Treatment of mine drainage by anoxic limestone drains and constructed wetlands, in: *Acid mining lakes: Acid mine drainage, limnology and reclamation*, A. Geller, H. Klapper, W. Solomons (Eds.), pp. 303–319, Springer, Berlin, 1998.
32. Rose, A.W. and Cravotta, C.A., III, Geochemistry of coal mine drainage, in: *Coal mine drainage prediction and pollution prevention in Pennsylvania*, K.B.C. Brady, M.W. Smith, J. Schueck (Eds.), p. 371, Harriburg, 5600-BK-DEP2256, Department of Environmental Protection, Harrisburg, PA, Pennsylvania, 1998.
33. Singer, P.C. and Stumm, M.W., Kinetics of the oxidation of ferrous iron, in: *Symposium on coal mine drainage*, Mellon Institute (Ed.), Pittsburgh, Pennsylvania, 14–16 May 1968.
34. Singer, P.C. and Stumm, W., Acidic Mine Drainage – Rate-Determining Step. *Science*, 167, 1121–1123, 1970.
35. Sobek, A.A., Schuller, W.A., Freeman, J.R., Smith, R.M., Field and laboratory methods applicable to overburdens and minesoils. *U.S. Environ. Prot. Agency*, 1–2, 1978. EPA-600/2-78-054.
36. Skousen, J., Overview of passive systems for treating acid mine drainage. *Green Lands*, 27, 4, 34–43, 1997.

37. Parker, G.K. and Robertson, A., Acid drainage, in: *AMEEF Occasional Paper no. 11*, pp. 101–117, Australian Minerals and Energy Environment Foundation, Melbourne, 1999.
38. Lefebvre, R., Hockley, D., Smolensky, J., Gelinas, P., Multiphase transfer processes in waste rock piles producing acid mine drainage1: Conceptual model and system characterization, U.S.A. *J. Contam. Hydrol.*, 52, 1-4, 137–164, 2001.
39. Olyphant, G.A., Bayless, E.R., Harper, D., Seasonal and weather-related controls on solute concentrations and acid drainage from a pyritic coal-refuse deposit in southwestern Indiana, U.S.A. *J. Contam. Hydrol.*, 7, 219–236, 1991.
40. Hornberger, R.J., Smith, M.W., Friedrich, A.E., Lovell, H.L., Acid mine drainage from active and abandoned coal mines in Pennsylvania, in: *Water Resources in Pennsylvania-Availability, Quality, and Management*, S.K. Majumdar, E.W. Miller, R.R. Parizek (Eds.), pp. 432–451, Penn. Ac. Sci., Easton, Pa, 1990.
41. Cravotta, C.A., III, Dugas, D.L., Brady, K.B.C., Kovalchuk, T.E., Effects of selective handling of pyritic, acid-forming materials on the chemistry of pore gas and groundwater at a reclaimed surface coal mine, Clarion County. *U.S. Bureau Mines Spe. Publ.*, SP 06A, 365–374, 1994.
42. International Network of Acid Prevention (INAP), *The Global Acid Rock Drainage Guide*, International Network for Acid Prevention (INAP), Skelleftea, Sweden, 2013.
43. Younger, P.L., The longevity of minewater pollution: A basis for decision-making. *Sci. Total Environ.*, 194–195, 457–466, 1997.
44. Tuttle, L.H., Dugan, P.R., Randles, C.I., Microbial sulfate reduction and its potential utility as an acid mine water pollution abatement procedure. *Appl. Microbiol.*, 17, 2, 297–302, 1969.
45. PIRAMID Consortium, *Engineering Guidelines for the Passive Remediation of Acidic and/or Metalliferous Mine Drainage and similar Wastewaters – "PIRAMID Guidelines"*, pp. 1–166, University of Newcastle Upon Tyne, Newcastle Upon Tyne, UK, 2003.
46. Costello, C., Acid mine drainage: Innovative treatment technologies. United States Environmental Protection Agency (USEPA). *Technol. Innov. Off.*, 52, 1–47, 2003. http://www.clu-in.org/download/studentpapers/costello_amd.pdf.2014.08.14.
47. Trumm, D., Selection of active treatment systems for acid mine drainage—Flow charts for New Zealand conditions. *N. Z. J. Geol. Geophys.*, 53, 2–3, 195–210, 2010.
48. Sheoran, A.S. and Sheoran, V., Heavy metal removal mechanism of acid mine drainage in wetlands: A critical review. *Min. Eng.*, 19, 2, 105–116, 2006. https://doi.org/10.1016/j.mineng.2005.08.006.
49. Eger, P. and Wagner, J., Wetland treatment systems—How long will they really work? *Minnesota Dept. Nat. Resour.*, 1–14, 1997.
50. Younger, P.L., The adoption and adaptation of passive treatment technologies for mine waters in the United Kingdom. *Mine Water Environ.*, 19, 2, 84–97, 2000.
51. Skousen, J. and Ziemkiewicz, P., Performance of 116 passive treatment systems for acid mine drainage, in: *ASMR, Proceedings of American Society of Mining and Reclamation, Breckenridge, CO*, pp. 1100–1133, 2005.
52. Hedin, R.S., Watzlaf, G.R., Nairn, R.W., Passive treatment of acid mine drainage with limestone. *J. Environ. Qual.*, 23, 6, 338–345, 1994.
53. Hedin, R.S. and Nairn, R.W., Designing and sizing passive mine drainage treatment systems. In: *Proceedings of the 13th Anniversary West Virginia Surface Mine Drainage Task Force Symposium*. pp. 1–4, Morgantown, WV, 1992. https://wvmdtaskforce.files.wordpress.com/2015/12/92-hedin.pdf.
54. Watzlaf, G.R., Schroeder, K.T., Kleinmann, R.L.P., Kairies, C.L., Nairn, R.W., The passive treatment of coal mine drainage. *Report*, 71, 2, 136–143, 2004. U.S. DOE/NETL-2004/1202.
55. Kuyucak, N., Chabot, F., Martschuk, J., Successful implementation and operation of a passive treatment system in an extremely cold climate, Northern Quebec, Canada, in: *Proceedings of*

the 7th International Conference on Acid Rock Drainage (ICARD), R.I. Barnhisel (Ed.), pp. 980–992, American Society of Mining and Reclamation (ASMR), Lexington, 2006.

56. Neculita, C.M., Zagury, G.J., Kulnieks, V., Short-term and long-term bioreactors for acid mine drainage treatment. *Environ. Health Sci.*, 12, 1–10, 2007. https://scholarworks.umass.edu/soilsproceedings/vol12/iss1/2.2020.02.007. 22nd Annual International Conference on Contaminated Soils, Sediments and Water.
57. Neculita, C., Zagury, G.J., Bussiere, B., Passive treatment of acid mine drainage in bioreactors: Short review, applications, and research needs. *J. Environ. Qual.*, 36, 1, 1439–1446, 2007.
58. Ford, K.L., Passive treatment systems for acid mine drainage. *Nat. Sci. Tech. Centre*, 1–20, 2003.
59. Gusek, J.J., A periodic table of passive treatment for mining influenced water, in: *National Meeting of the American Society of Mining and Reclamation*, Barnhisel, R. (Ed.), pp. 550–562, 2009.
60. Skousen, J.G., Sexstone, A., Ziemkiewicz, P.F., Acid mine drainage control and treatment, in: *Reclamation of Drastically Disturbed Lands, Agronomy*, vol. 41, Barnhisel, R. (Ed.), pp. 131–168, 2000.
61. The Office of Surface Mining AMD, *Bureau of the U.S. Dept. of the Interior*, Federal register, Washington DC, 2004.
62. Kuyucak, N., Acid Mine Drainage—Treatment options for mining effluents. *J. Min. Environ. Manage.*, 14–17, 2001.
63. Groudev, S.N., Komnitsas, K., Spasova, I.I., Paspaliaris, I., Clean-up of mine waters from a uranium deposit by means of a constructed wetland, in: *Proceedings of the International Biohydrometallurgy Symposium, IBS 2003, Athens, Hellas, September 14–19, 2003*, Tsezos, M., Hatzikioseyian, A., Remoudaki, E. (Eds.), pp. 533–539, 2004.
64. Murck, H.B., Progress toward pollution prevention and waste minimization in the North American gold mining industry. *J. Clean. Prod.*, 9, 405–415, 2001.
65. Grewar, T., Reuse of treated mine-impacted water as a potential resource for accelerated carbon sequestration, in: *Proceedings of the International Mine Water Association*, Wolkersdorfer, C., Sartz, L., Weber, A., Burgess, J., Tremblay, G. (Eds.), pp. 1056–1057, 2018.
66. De Vegt, A.L. and Buisman, C.J.N., Sulfur compounds and heavy metal removal using bioprocess technology, in: *EPD Congress- The Minerals, Metals, and Material Society*, Warren, G.W. (Ed.), pp. 995–1004, 1996.
67. Colvin, C., Burns, A., Schachtschneider, K., Mahery, A., Charmier, J., de Wit, M., Coal and Water Futures in South Africa: The case for protecting headwaters in the Enkangala grasslands. *Report, World Wide Fund (WWF)*, 1–82, 2011.
68. Kalin, M., Fyson, A., Wheeler, W.N., The chemistry of conventional and alternative treatment systems for the neutralization of acid mine drainage. *Sci. Total Environ.*, 366, 2, 395–408, 2006.
69. Vikram, S., Yadav, H.L., Srivastav, S.K., Jamal, A., Overview of active and passive systems for treating acid mine drainage. *Int. J. Adv. Sci. Eng. Tech.*, 4, 5, 2017. https://doi.org/10.17148/IARJSET.2017.4525.
70. Cartwright, P., Membranes separations technology for industrial effluent treatment—A review. *Desalination*, 56, 17–35, 1985. https://doi.org/10.1016/0011-9164(85)85012-8.
71. Feng, D., Aldrich, C., Tan, H., Treatment of acid mine water by use of heavy metal precipitation and ion exchange. *Miner. Eng.*, 13, 6, 623–642, 2000. https://doi.org/10.1016/S0892-6875(00)00045-5.
72. Johnson, D.B. and Hallberg, K.B., Acid mine drainage remediation options: A review. *Sci. Total Environ.*, 338, 3–14, 2005.
73. Grewar, T., South Africa's options for mine-impacted water reuse: A review. *J. S. Afr. Inst. Min. Metall.*, 119, 3, 321–331, 2019. https://doi.org/10.17159/2411-9717/2019/v119n3a12.

74. Hutton, B., Kahan, I., Naidu, T., Gunther, P., Operating and maintenance experience at the Emalahleni Water Reclamation Plant, in: *Water Institute of Southern Africa & International Mine Water Association, Proceedings of International Mine Water Conference*, pp. 415–430, 2009.
75. Gunther, P. and Mey, W., Selection of mine water treatment technologies for the Emalahleni (Witbank) water reclamation project. *Water Inst. SA*, 1–14, 2006. http://www.waterinformation.co.za/literature/files/122%20Gunther.pdf.2015.06.015.
76. Cogho, V.E. and Niekerk, A.M., Optimum Coal Mine Water Reclamation Project, in: *Water Institude of Southern Africa & International Mine Water Association, Proceedings of International Mine Water Conference*, pp. 130–140, 2009.
77. The Aveng Group (TAG), Recycling water with new technology. *Annu. Rep.*, 1, 1–258, 2010.
78. Department of Water Affairs and Forestry (DWAF), *South African Water Quality Guidelines: Industrial Water Use*, vol. 3, pp. 1–160, The Government Printer, Pretoria, 1995.
79. Oelofse, S.H.H., Roux, S., De Lange, W.J., Mahumane, B., Le Rouw, W.J., Du Preez, M., Greben, M.A., Steyn, M., A comparison of the cost associated with pollution prevention measures to that required to treat polluted water resources. *Counc. Sci. Innov. Res. (CSIR)*, 1–101, 2012. http://hdl.handle.net/10204/7297.2019.05.025.
80. Gibert, O., de Pablo, J., Cortina, J.L., Ayora, C., Evaluación de materia orgánica como material para su utilización en barreras permeables reactivas, in: *Proceedings of the Conference "Las caras del agua subterránea", Barcelona, Spain*, Medina, A., Carrera, J., Vives, L. (Eds.), pp. 43–48, 2001.
81. Skousen, J. and Ziemkiewicz, P., *Acid mine drainage control and treatment, National Research Center for Coal and Energy*, 2nd Ed., National mine land reclamation center, West Virginia University, Morgan Town, p. 362, 1996.
82. Hilton, T., *Handbook-Short course for taking a responsible environmental approach towards treating acid mine drainage with anhydrous ammonia*, Mining Recl. Asso, West Virginia, 1990.
83. Thorius, T., *The effect of Grootvlei mine water on the Blesbokspruit. MSc Diss*, Rand Afri. University of Johannesburg, 2004.
84. *Mines to carry 67% of acid mining clean-up costs - minister. News24*, News 24, Johannesburg, 2018.
85. Mining.com, *South African miners to pay 67% of acid drainage clean-up costs*, 2016.
86. RoyChowdhury, A., Sarkar, D., Datta, R., Remediation of Acid Mine Drainage-Impacted Water. *Curr. Pollution Rep.*, 1, 131–141, 2015. https://doi.org/10.1007/s40726-015-0011-3.
87. Kefeni, K.K., Msagati, T.A.M., Mamba, B.B., Acid mine drainage: Prevention, treatment options, and resource recovery. *J. Clean. Prod.*, 151, 475–493, 2017. https://doi.org/10.1016/j.jclepro.2017.03.082.
88. Batterham, R., *In Situ* Leaching—How Far in the Future?, in: *Biohydrometallurgy: Biotech Key to Unlock Mineral Resources Value*, G. Qui, T. Jiang, W. Qin, X. Liu, Y. Yang, H. Wang (Eds.), pp. 21–22, Central South University Press, Changsha, China, 2011.
89. Tangeh, A., Geochemistry of Rare Earth Elements in a neutral mine drainage. *Int. J. Coal Geol.*, 183, 120–135, 2017.
90. Nleya, Y., Simate, G.S., Ndlovu, S., Sustainability assessment of the recovery and utilisation of acid from acid mine drainage. *J. Clean. Prod.*, 113, 17–27, 2016.
91. Masindi, V., Gitari, W.M., Tutu, H., De Beer, M., Nekhwevha, N., Neutralization and attenuation of metal species in acid mine drainage and mine leachates using magnesite: A batch experimental approach, in: *An Interdisciplinary Response to Mine Water Challenges*, Sui, Sun, Wang (Eds.), pp. 640–644, China University of Mining and Technology, Xuzhou, 2014.
92. Pulles, W., Howie, D., Otto, D., Easton, J., A manual on mine water treatment and management practices in South Africa. 2, 527, 2006. Water Research Commission Rep No. 700/1/01.

93. Hedin, R.S., Nairn, R.W., Kleinmann, R.L.P., Passive treatment of coal mine drainage. *Bur. Mines Inf. Circ.*, 9389, 1–44, 1994.
94. Cravotta, C.A. III, Dissolved metals and associated constituents in abandoned coal-mine discharges, Pennsylvania, USA—Part 2: Geochemical controls on constituent concentrations. *Appl. Geochem.*, 23, 203–226, 2008.
95. Ayora, C., Macius, F., Torres, E., Nieto, J.M., Rare earth elements in acid mine drainage. *J. Environ. Sci. Technol.*, 3, 1, 3–22, 2015.
96. Ziemkiewicz, P.F., Skousen, J.G., Lovett, R.J., Open limestone channels for treating AMD: A new look at an old idea. *Green Lands*, 24, 4, 36–41, 1994.
97. Ayora, C., Caraballo, M.A., Macías, F., Rötting, T.S., Carrera, J., Nieto, J.M., Acid mine drainage in the Iberian Pyrite Belt: 2. Lessons learned from recent passive remediation experiences. *Environ. Sci. Pollut. Res.*, 20, 7837–7853, 2013.
98. Stoakley, A., Alternative water management in Pretoria, South Africa: An investigation into public perceptions of water recycling, in: *Proceedings of People and the Planet 2013 Conference: Transforming the Future*, P. James, C. Hudson, S. Carroll-Bell, A. Taing (Eds.), Global Cities Research Institute, Melbourne, Australia, 2013.
99. Annandale, J., Beletse, Y., De Jager, P., Jovanovic, N., Steyn, J., Benade, N., Lorentz, S., Hodgson, F., Usher, B., Vermeulen, D., Aken, M., Predicting the environmental impact and sustainability of irrigation with coal mine water. Water Research Commission (WRC) Report No. 1149/01/07, p. 1–178, 2007.
100. Pope, J., Christenson, H., Newman, N., Gordon, K., Trumm, D., Wall wash samples to predict AMD longevity at coal mines in New Zealand, in: *Proceedings of 11th ICARD | IMWA | WISA MWD 2018 Conference - Risk to Opportunity, Pretoria, South Africa*, Wolkersdorfer, C., Sartz, L., Weber, A., Burgess, J., Tremblay, G. (Eds.), pp. 1–6, 2018.
101. Parkhurst, D.L. and Appelo, C.A.J., User's guide to PHREEQC - A computer program for speciation, batch-reaction, one-dimensional transport, and inverse geochemical calculations, Water-Resources Investigations Report. *U.S. Geol. Surv. Tech. Methods*, Version 2, 1–312, 1999. https://doi.org/10.3133/wri994259.
102. Ziemkewicz, P.F., Simmons, J.S., Knox, A.S., The mine water leaching procedure: Evaluating the environmental risk of backfilling mines with coal ash, in: *Chemistry of Trace Elements in Fly Ash*, K.S. Sajwan, A.K. Alva, R.F. Keefer (Eds.), pp. 75–90, Springer, Boston, MA, 2003.
103. McCarthy, T.S., The decanting of acid mine water in the Gauteng city-region: Analysis, prognosis, and solutions. Gauteng City Region Observatory, 1–28, 2010. https://www.gcro.ac.za/m/documents/gcro_terence_mccarthy_amd__final_version.pdf.2019.09.023.
104. Coetzee, H., Croukamp, L., Venter, J., de Wet, L., *Contamination of the Tweelopiespruit and environs by water from the Western Basin decant point on Harmony Gold's property*, Council For Geoscience, Pretoria, RSA, Council of Geoscience Report No. 2005-0148, pp. 1–28, 2005.
105. Logica, N., Artificial Intelligence in Law: The state of play 2016 part 1. White Paper, World Economic Forum 1–6, 2015. https://www.neotalogic.com/wp-content/uploads/2016/04/Artificial-Intelligence-in-Law-The-State-of-Play-2016.pdf.2019.08.028.
106. PWC, *Artificial Intelligence and Robotics: Leveraging artificial intelligence and robotics for sustainable growth*, Pricewaterhouse Coopers Private Limited (PWCPL), India, 2017.
107. Fijani, E., Nadiri, A.A., Asghari Moghaddam, A., Tsai, F.T.C., Dixon, B., Optimization of drastic method by supervised committee machine artificial intelligence to assess groundwater vulnerability for Maragheh-Bonab Plain aquifer. *J. Hydrol.*, 503, 89–100, 2013. https://doi.org/10.1016/j.jhydrol.2013.08.038.
108. Porwal, A. and Carranza, E.J.M., Introduction to the Special Issue: GIS-based mineral potential modelling and geological data analyses for mineral exploration. *Ore Geol. Rev.*, 71, 477–483, 2015. https://doi.org/10.1016/j.oregeorev.2015.04.017.

109. Swain, M., Knowledge-based system, in: *Encyclopedia of Systems Biology*, O. Wolkenhauer (Ed.), pp. 65–141, Springer, New York, 2013.
110. Porwal, A., Carranza, E.J., Hale, M., Knowledge-driven and data-driven fuzzy models for predictive mineral potential mapping. *Nat. Resour. Res.*, 12, 1, 1–25, 2003. https://doi.org/10.1023/A:1022693220894.
111. Solomatine, D., See, L.M., Abrahart, R.J., Data-driven modelling: Concepts, approaches and experiences, in: *Practical Hydroinformatics. Water Science and Technology Library*, vol. 68, R.J. Abrahart, L.M. See, D.P. Solomatine (Eds.), pp. 17–31, Springer, Heidelberg, 2008.
112. Maier, H.R. and Dandy, G.C., The use of artificial neural networks for the prediction of water quality parameters. *Water Resour. Res.*, 32, 4, 1013–1022, 1996. https://doi.org/10.1029/96WR03529.
113. Bayatzadeh Fard, Z., Ghadimi, F., Fattahi, H., Use of artificial intelligence techniques to predict distribution of heavy metals in groundwater of Lakan lead-zinc mine in Iran. *J. Min. Environ.*, 8, 1, 35–48, 2017.
114. Gholami, R., Ziaii, M., Ardejani, F.D., Maleki, S., Specification and prediction of nickel mobilization using artificial intelligence methods. *J. Geosci.*, 3, 4, 375–384, 2011. https://doi.org/10.2478/s13533-011-0039-x.
115. Tan, G., Yan, J., Gao, C., Yang, S., Prediction of water quality time series data based on least squares support vector machine. *Proc. Eng.*, 31, 1194–1199, 2012.
116. Liu, S., Tai, H., Ding, Q., Li, D., Xu, L., Wei, Y., A hybrid approach of support vector regression with genetic algorithm optimization for aquaculture water quality prediction. *Math. Comp. Model.*, 58, 3–4, 458–465, 2013. https://doi.org/10.1016/j.mcm.2011.11.021.
117. Džeroski, S., Demšar, D., Grbović, J., Predicting Chemical Parameters of River Water Quality from Bioindicator Data. *Int. J. Appl. Intel.*, 13, 1, 7–17, 2000. https://doi.org/10.1023/A:1008323212047.
118. Chau, K., A review on integration of artificial intelligence into water quality modelling. *Mar. Pollut.*, 52, 7, 726–733, 2006. https://doi.org/10.1016/j.marpolbul.2006.04.003.
119. Alizadeh, M.J. and Kavianpour, M.R., Development of wavelet-ANN models to predict water quality parameters in Hilo Bay, Pacific Ocean. *Mar. Pollut.*, 98, 1–2, 171–178, 2015. https://doi.org/10.1016/j.marpolbul.2015.06.052.
120. Rooki, R., Doulati Ardejani, F., Aryafar, A., Bani Asadi, A., Prediction of heavy metals in acid mine drainage using artificial neural network from the Shur River of the Sarcheshmeh porphyry copper mine, Southeast Iran. *Environ. Earth Sci.*, 64, 5, 1303–1316, 2011. https://doi.org/10.1007/s12665-011-0948-5.
121. Betrie, G.D., Tesfamariam, S., Morin, K.A., Sadiq, R., Predicting copper concentrations in acid mine drainage: A comparative analysis of five machine learning techniques. *Environ. Monit. Assess.*, 185, 5, 4171–4182, 2013. https://doi.org/10.1007/s10661-012-2859-7.
122. Cherkassky, V., Krasnopolsky, V., Solomatine, D.P., Valdes, J., Computational intelligence in earth sciences and environmental applications: Issues and challenges. *Neural Networks*, 19, 2, 113–121, 2006. https://doi.org/10.1016/j.neunet.2006.01.001.
123. Liu, P., Wang, J., Sangaiah, A.K., Xie, Y., Yin, X., Analysis and prediction of water quality using LSTM deep neural networks in IoT environment. *Sustainability*, 7, 1–14, 2019.
124. Li, L., Jiang, P., Xu, H., Lin, G., Guo, D., Wu, H., Water quality prediction based on recurrent neural network and improved evidence theory: A case study of Qiantang River, China. *Environ. Sci. Pollut. Res.*, 26, 19, 19879–19896, 2019. https://doi.org/10.1007/s11356-019-05116-y.
125. Sakala, E., Novhe, O., Vadapalli, V.R.K., Application of Artificial Intelligence (AI) to predict mine water quality, a case study in South Africa, in: *Proceedings of the Mine Water Association Conference: Technological and Ecological Challenges, International Mine Water Association Annual Conference*, Wolkersdorfer, Ch., Khayrulina, E., Polyakova, S., Bogush, A. (Eds.), pp. 140–145, 2019.

126. Sakala, E., Fourie, F., Gomo, M., Coetzee, H., GIS-based groundwater vulnerability modelling: A case study of the Witbank, Ermelo and Highveld Coalfields in South Africa. *J. Afr. Earth Sci.*, 137, 46–60, 2018. https://doi.org/10.1016/j.jafrearsci.2017.09.012.
127. Devasahayam, S., Dowling, K., Mahapatra, M.K., *Sustainability in the Mineral and Energy Sectors*, vol. 1, p. 730, CRC Press, Boca Raton, 2019.

3

The Prediction of Acid Mine Drainage Potential Using Mineralogy

Deshenthree Chetty*, Olga Bazhko, Veruska Govender and Samuel Ramatsoma

Mintek, Randburg, South Africa

Abstract

The generation of acid mine drainage (AMD) occurs through oxidation of sulfide minerals exposed to air and water, a process exacerbated by mining and extraction activities. Although the assessment of AMD potential is often performed through static geochemical tests, mineralogical information produces complementary data to more thoroughly assess the risk of AMD formation. This chapter focuses on mineralogy and geochemistry as tools to generate input for the formulation of AMD prediction models, specifically addressing mineral acid formation and consumption potential, mineral abundance, mineral reactivity, and mineral liberation in process wastes. Mineral liberation is defined here as the amount of mineral surface exposed to make the mineral amenable to reaction with air, water, or acidic solutions. Adding a mineral liberation parameter builds on previous models for the prediction of AMD potential of wastes. These newly developed models were applied to gold and copper ore process residues and results compared with traditional static geochemical test results. Both, the effects of modal mineralogy and mineral liberation were further evaluated for two base metal sulfide (BMS) tailings compositions. To conclude, this predictive approach lends itself to improved process design and better-informed waste management strategies to mitigate detrimental effects of AMD, as part of a geometallurgical program, with possibilities to extend into prediction of AMD potential over time. In this respect, when applied to waste rocks, the predictive approach could be a desirable alternative to time-consuming and costly kinetic tests.

Keywords: Acid mine drainage, prediction, mineralogy

3.1 Introduction

Sustainable use of natural resources is a long-held ideal for the mining and ore processing industries. When adequate sustainability measures are not implemented, environmental and social issues arise that require urgent attention [1]. Acid mine drainage (AMD) is one such issue, for which several treatment options have been developed, but which remains a formidable challenge worldwide. A geometallurgical approach to the mining and processing of ores reduces technical, economic, environmental, and social risk for operations and

**Corresponding author*: deshc@mintek.co.za [ORCID: 000-0002-1913-7598]
[ORCID: Bazhko, 000-0002-1598-0123; Govender, 000-0001-6848-6527; Ramatsoma, 000-0002-8079-8424]

Elvis Fosso-Kankeu, Christian Wolkersdorfer and Jo Burgess (eds.) Recovery of Byproducts from Acid Mine Drainage Treatment, (49–72)

therefore must incorporate sustainability as a key aspect. The approach relies on a high confidence in predictive capability, which is underpinned by an understanding of the orebody with respect to how it will be mined and processed to reduce such risks. As part of a predictive approach, not only the process efficiencies must be considered, but also the management of waste that will be generated from mining and processing of the ore, to reduce environmental and, by extension, social and economic risk.

Mineralogy is a key factor in considering value extraction from an orebody. In this respect, the relationship between ore and gangue minerals is important for determining appropriate process flowsheet options. Such options must provide optimized grades and recoveries to concentrates that will be further refined through e.g., smelting or hydrometallurgical process routes. While concentrates must meet certain specifications, the same is not necessarily true for the wastes generated at each step in the concentration process.

Acid mine drainage is the detrimental consequence of inadequate planning for the release of acid (defined as H_2SO_4 in this study) and metals from the oxidation of (di-)sulfide minerals present in the ore, a natural process that is exacerbated by mining and processing [2–5]. This results from each step of the mining and processing chain of activities, in which rock is broken up into progressively smaller particles, leading to increased exposure of (di-) sulfide minerals to air and water. Consequently, this leads to increased mineral surface area that is amenable to oxidation. Whereas value-bearing sulfides are largely extracted through processing, non-value-bearing sulfides remain in waste products and will be subject to oxidation, which is accelerated by bacteria [4]. The mitigation of such effects is aided by the presence of associated, potentially neutralizing minerals in the ore [4]. It is therefore evident that mineralogy must play an integral role in sustainable use of the resource.

Presently, much of the effort in addressing AMD generation in South Africa is targeted at remedial processes to reverse the damage already done [6–8], but lesser attention has been given to prediction and prevention of AMD generation and associated metal contamination in the environment. Yet, predictive capabilities would be of benefit not only to new operations, but also brownfield expansion programs, and indeed, waste already generated from ongoing and spent operations.

Over the last two decades, many strides have been made in complementing traditional geochemical techniques like static acid-base-accounting (ABA) methods, as well as kinetic geochemical tests, with mineralogical characterization of waste rock materials [5, 9–15]. The mineralogical characterization has included modal mineralogy and increasingly quantitative assessment of mineral association in defining lithotypes [9, 16], assessing accuracy of static geochemical tests [11], and understanding reactions in kinetic tests [10, 17, 18] in predicting AMD potential.

In many of these instances, liberation of minerals, whether sulfides or other minerals, has not been given much attention. Mineral liberation describes the degree of surface exposure of the mineral, i.e., the amount of mineral grain surface that is amenable to interaction with air, water, or other reactants. Only a few instances have recently been shown where mineral liberation is reported from automated scanning electron microscopy (SEM) [17, 18] in the context of AMD studies. However, these data have not been expressly incorporated into a descriptor for AMD potential or risk. Mineral liberation is a function of the particle size of rock, and in macro-scale observations [9] of a few centimeters to meter-sized rocks, sulfide mineral liberation will not play an overtly large role for acid production, as the particle size is much larger than the constituent mineral grain size, resulting in partial exposure of the mineral grains for reaction. However, where particles occur at the micrometer-scale of observation (i.e., the typical particle size after processing to extract valuable sulfide), liberation of the same constituent mineral grains will increase, and therefore, mineral liberation becomes more important for the mineral reactivity. In the base metal sulfide (BMS) and gold mining industries, the waste generated from such processes as physical separation, flotation, and leaching generally contains notable concentrations of sulfide minerals that may oxidize and release metal contaminants.

This chapter therefore describes the use of modal mineralogy and mineral liberation data in the calculation of acid production and neutralization potential (NP), specifically with regard to process wastes, for the prediction of AMD risk. The chapter further briefly discusses the incorporation of such efforts in geometallurgical programs, including mineralogical prediction as a potential alternative to kinetic testing. In the latter case, the application of mineralogical studies to ores and waste rock can be coupled with mine planning and spatial distribution of potential acid forming rock types to inform not only mine production efficiency but also the potential for wastes to form AMD over time. Temporal aspects include when each rock type will be mined and, from there on, when AMD can be expected to be released, as well as the rate at which this release can take place.

3.2 Mineralogical Approach for Prediction of AMD Potential

In applying a mineralogical approach for prediction of AMD potential, various factors must be considered. The first is maximum potential of minerals to form or consume acid, the second is the proportion of such minerals in the sample, and the third is the mineral reactivity. An additional parameter, liberation, is considered in this contribution. These aspects are individually discussed in the sections below.

3.2.1 AMD Chemistry for Maximum Acid Generation or Consumption Potential

The AMD potential of waste is the balance between the acid potential (AP) of the waste, determined by the acid-producing components, and the NP, determined by the acid-consuming components. Generally, the evaluation of AMD potential involves static geochemical tests to ascertain the potentially acidic and alkaline nature of samples [19–22]. Classification of AMD potential is based on the ratio of AP and NP, derived from such tests. Commonly used static procedures for characterization of mining wastes to predict AMD are the ABA test procedure of Sobek *et al.* [19] and its modified version [20, 21]. In the method of Sobek *et al.* [19] the AP of a sample is determined by calculation of the theoretical amount of acid that can be produced if the total amount of sulfur in the sample is oxidized to form sulfuric acid. The NP of the sample is determined experimentally by digestion in excess hydrochloric acid at boiling, and back titration of remaining acid to pH 7. The disadvantage of the Sobek method is the overestimation of AP and NP values [22, 23]. The modified Sobek procedure of Lawrence [20] determines NP at lower temperature, with the end point of back titration being pH 8.3, which prevents overestimation of NP. This modified procedure uses sulfide, as opposed to total sulfur, for the AP calculation [23]. Lapakko [22] proposed measuring NP by the amount of acid reacted with a sample to achieve a stable pH 6. This conservatively determined the required NP to meet water quality standards, which typically require pH in excess of 6, and showed better accuracy than the Sobek *et al.* [19] procedure.

The chemistry of AMD generation can be complex, as various environmental factors can influence chemical reactions, which will result in different products. For example, metal ions such as Fe^{3+}, released during (di-)sulfide oxidation, along with Al^{3+}, may hydrolyze, releasing additional acid. Therefore, pyrite oxidation may be described by the following equations [24]:

$$FeS_2 + 7/2\ O_2 + H_2O \rightarrow Fe^{2-} + 2\ SO_4 + 2\ H^+ \tag{3.1}$$

$$Fe^{2+} + 1/4\ O_2 + H_+ \rightarrow Fe_{3+} + 1/2\ H_2O \tag{3.2}$$

$$Fe^{3+} + 3\ H_2O \rightarrow Fe(OH)_3 + 3\ H^+ \tag{3.3}$$

Other sulfides show different acid-producing potential. Iron-bearing sulfides readily generate acidity, whereas sulfides that do not contain Fe in their crystal lattices (e.g., galena, PbS) are not capable of generating substantial acid [25–27]. Sulfides like sphalerite [(Zn,Fe)S], which can show solid solution substitution of Zn^{2+} with Fe^{2+}, will either not produce acid upon oxidation or can produce considerable acid with increased Fe^{2+} content [28].

The precipitation of Fe^{3+} and Al^{3+} hydroxides is a substantial source of acidity in mine waste [11]. Secondary Fe^{3+}-hydroxy-sulfate minerals (e.g., schwertmannite, $Fe_{16}O_{16}(OH)_{12}(SO_4)_2$; jarosite, $KFe_3(SO_4)_2(OH)_6$) commonly occur in pyrite or pyrrhotite ($Fe_{1-x}S$) mine waste [5], and dissolution or precipitation of these phases is potentially acid-producing (Eq. 3.4).

$$KFe_3(SO_4)_2\ (OH)_6 + 3\ H_2O \rightarrow 3\ Fe(OH)_3 + K^+ + 3\ H^+ + 2\ SO_4^{2-} \tag{3.4}$$

Acid produced by (di-)sulfide oxidation can be consumed during reaction with gangue minerals. Neutralization is primarily achieved through dissolution of carbonate minerals; calcite ($CaCO_3$) is the most effective at this [12]. Some neutralization can be achieved from silicate mineral dissolution, particularly the dissolution of olivine [$(Mg,Fe)_2SiO_4$], serpentine [$(Mg,Fe)_3Si_2O_5(OH)_4$], and wollastonite ($CaSiO_3$), as well as clay minerals, mica (Eq. 3.5) and calcic plagioclase (Eq. 3.6). However, practical NP values can be much lower than expected, as the contributions are minimized from those acid neutralizing minerals that would dissolve only under acidic conditions, i.e., such minerals are relatively stable under neutral conditions.

$$KAl_2[AlSi_3O_{10}](OH)_2 + H^+ + 3/2\ H_2O \rightarrow K^+ + 3/2\ Al_2Si_2O_5(OH)_4 \tag{3.5}$$

$$CaAl_2Si_2O_8 + 2\ H^+ + H_2O \rightarrow Ca^{2+} + Al_2Si_2O_5(OH)_4 \tag{3.6}$$

Identified minerals are broadly grouped as acid forming, acid consuming or buffering, and inert [4, 29–34]. Acid formers include most sulfides, which react with oxygen and water to produce metal cations, sulfate and H^+. The process is, however, a complex, multistage one, as indicated above, and can include metal cation oxidation and hydrolysis, during which precipitates may be formed.

In building on mineralogical-based protocols for the prediction of AMD potential, acid formation (and acid consumption) of different minerals are considered, as per previous workers [11, 15, 27]. Acid formation (or consumption) is calculated according to Eq. 3.7:

$$AP = \text{acid formation (consumption)} \times 98.097 \times \text{molecular mass of mineral} \times 1{,}000 \quad (3.7)$$

Where

AP = acid production potential of the mineral, based on its oxidation (or acid consumption) reaction, in kg H_2SO_4/t;
acid formation (consumption) is in mol acid per mol mineral reacted;
98.097 is the molecular mass of H_2SO_4, in g mol^{-1};
molecular mass of the mineral is in g mol^{-1};
1,000 is the factor to convert to kg H_2SO_4/t.

This parameter is therefore solely a descriptor of the amount of acid that is formed per mol of mineral that reacts; alternatively, it is the amount of acid that is consumed per mol of mineral that reacts, as in the case of carbonates. Positive values indicate acid production, whereas negative values indicate acid consumption, based on the reactions involved. To simplify geochemical reactions and calculate acid formation potential, complete oxidation and formation of products and a neutral pH have been assumed in the present calculations. The chemical reactions and acid formation or consumption for assessed minerals show that Fe^{2+}, Sb^{3+}, and As^{3+} containing sulfides generate notable acid, compared with BMS minerals lacking these elements [35]. Examples are provided with arsenopyrite (Eq. 3.8) and galena (Eq. 3.9), which have acid production potentials of 601 kg/t H_2SO_4 and 0, respectively, as per Eq. 3.7.

$$2\ FeAsS + 7\ O_2 + 2\ H_2O \rightarrow 2\ FeAsO_4 + 2\ H_2SO_4 \quad (3.8)$$

$$PbS + 2\ O_2 \rightarrow PbSO_4 \quad (3.9)$$

Tailings from metallurgical processes often contain metal hydroxyl sulfates (particularly Fe^{3+} and Al^{3+}) that may also release acid. On the other hand, carbonate, phosphate, and some hydroxide and oxide minerals will consume acid. Silicate alteration reactions will all be acid consuming, as metals are replaced by H^+ with cation release in the forms of the metal cation (e.g., K^+, Ca^{2+}, Na^+), oxidation (of Fe^{2+}) and hydrolysis (Fe^{3+} and Al^{3+}), with potential formation of more stable minerals like clays [36].

For the evaluation of AMD potential, complete hydrolysis reactions with formation of SiO_2 and metal salts or hydroxides are assumed. It should be highlighted that the reactions make available the maximum NP of the mineral. However, the NP offered by these reactions is often not attained because mineral dissolution is commonly accompanied by neoformed phases [37]. The nature of the neoformed phases is dependent on the abundance, composition, and chemical properties of the solution. The latter include Eh, pH, and solute concentration.

3.2.2 Mineral Modal Abundance

In order to calculate the amount of acid that may be produced (or consumed) by the mineral as part of its contribution to the total behavior of the sample, consideration is given to the proportion of minerals present in the sample [15]. The mass proportions of key acid producers (i.e., sulfides) and consumers (i.e., carbonates) may then be determined by such mineralogical techniques as quantitative X-ray diffraction, using Rietveld refinement. Other gangue minerals like silicates are also quantified by this method. However, the technique, while rapid, suffers from high detection limits, and so automated SEM techniques like QEMSCAN (Quantitative Evaluation of Minerals using Scanning Electron Microscopy) [38], MLA (Mineral Liberation Analyzer) [39], and Mineralogic [17], among others, may be used [11, 13, 17, 18]. These methods provide typical detection limits of 0.1 mass% of mineral present.

3.2.3 Mineral Reactivity

Mineral reactivity is the third aspect to consider in determining propensity for the sample to produce (or consume) acid. Therefore, not only the total amount of each mineral consuming or producing acid, but also the rate at which acid generation or neutralization will occur, must be determined. Different minerals have different reactivities. Carbonate minerals such as calcite ($CaCO_3$) and dolomite [$CaMg(CO_3)_2$] are the most effective neutralizing minerals because they have relatively high reactivities and neutralize acidity over a wide pH range. For low reactivity minerals, their presence in a waste will not necessarily provide protection against AMD if the rate of neutralization is lower than the rate of acid generation. Silicate minerals can be important neutralizers because of their abundance, but their relative reactivity ranges from intermediate to extremely slow relative to that of carbonates.

Sverdrup [40] suggested that minerals can be divided into different groups (carbonate, silicates, and others) in order of relative reactivity in acidic solution (Table 3.1), with the reaction rates at pH 5.0 relative to a calcite reactivity of 1.00 as calculated by Kwong [41]. Sulfides are not listed in this scheme.

The abrasion pH value has been previously used for comparing the reactivity of acid generating and neutralizing minerals. Noble and Lottermoser [42] measured the abrasion pH value for a wide variety of minerals. To convert the abrasion pH to reactivity, Parbhakar-Fox *et al.* [11] used the deviation of the pH value from a $CaCl_2$ solution with pH 6.4. According to this normalization of the abrasion pH values, BMS minerals should have low reactivity, which is not reflected by reality. These minerals are oxidized but may not necessarily produce acid. The oxidation, however, results in sulfate and metal release, which contributes to toxicity and is important to consider for AMD generation.

Alternative approaches were adopted by Bouzahzah *et al.* [43] and Chopard *et al.* [27, 44] for determining sulfide oxidation rates as a measure of sulfide reactivity. The approach proposed by Chopard *et al.* [44] is preferred for determining sulfide reactivity, as it is based on measurements of oxidation rates of various sulfide minerals, which consider mass and surface area of individual sulfides of different compositions. Table 3.2 shows the results for sulfide mineral oxidation from their work, along with relative reactivity calculated from their data. The relative reactivity can be calculated by assigning a reactivity of 1 for pure pyrite.

Table 3.1 Reactivity of selected minerals [40, 41].

Mineral group	Typical minerals	Relative reactivity (at pH 5)
Dissolving	Calcite, aragonite, dolomite, magnesite, brucite	1.00
Fast weathering	Anorthite, nepheline, forsterite, olivine, garnet, jadeite, leucite, spodumene, diopside, wollastonite	0.6
Intermediate weathering	Sorosilicates (epidote, zoisite), pyroxenes (enstatite, hypersthene, augite, hedenbergite), amphiboles (hornblende, glaucophane, tremolite, actinolite, anthophyllite), phyllosilicates (serpentine, chrysotile, talc, chlorite, biotite)	0.4
Slow weathering	Plagioclase feldspars (albite, oligoclase, labradorite), clays (vermiculite, montmorillonite)	0.02
Very slow weathering	K-feldspars, muscovite	0.01
Inert	Quartz, rutile, zircon	0.004

It can be further noted that the same mineral may show different oxidation rates as a result of differing minor and trace element contents, e.g., pyrrhotite and pyrite [27]. Other sulfides, e.g., sphalerite, are prone to solid solution substitution of Zn^{2+} by Fe^{2+} and Mn^{2+}, which increase the oxidation rate. In this respect, it is worth investigating sulfide mineral compositions to assign appropriate reactivity, particularly where large concentrations of such minerals are evident. Thus, mineral chemistry, which is useful for understanding recovery and grade from processing of ores, could also prove useful for understanding oxidation of minerals in a tailings storage facility. Different compositions will be governed in part by the type of ore deposit in which the given sulfide occurs. The small database of Chopard *et al.* [44] is a useful start in this regard. In the current study, the reactivities of sulfides were calculated as per equation 1.10, based on the experimental results of these authors.

Table 3.2 Oxidation rate for selected (di-)sulfides and normalization to pyrite [44].

Mineral	$g/m^{-2}/day$	Normalized to pyrite = 1
pyrrhotite	0.638×10^{-3}	8.43
Fe sphalerite	1.51×10^{-4}	1.99
pyrite	7.57×10^{-5}	1.00
galena	6.52×10^{-5}	0.86
chalcopyrite	3.3×10^{-5}	0.44

$$\text{Reactivity of sulfide} = \text{sulfide oxidation rate/pyrite oxidation rate} \quad (3.10)$$

The relative reactivity for (di-)sulfides and non-sulfides applied in the calculation of AMD risk was therefore based on the data provided by Chopard *et al.* [44] and Kwong [41]. Where no data are available in the literature, assumptions were required, which will influence the accuracy of the data. It is noted that the reactivity of sulfides and carbonates or silicates are based on different mechanisms for reaction. In the case of sulfides, reactivity is based on the oxidation potential, whereas for carbonates and silicates, reactivity is the ability to react with acid. Where pyrite produces acid, calcite will immediately consume it, whereas silicates will take longer to do so. Thus, the scaling of relative reactivity is done with calcite and pyrite both at 1.

3.2.4 Mineral Liberation

A fourth aspect to be considered in AMD studies is mineral liberation, which has been used to interpret kinetic test results of waste rocks [17, 18]. Mineral liberation is expressed in two ways. The first is the mass of the mineral of interest (e.g., sulfide) as a proportion of the total mass of the particle in which it occurs. This is referred to as liberation by grade. Intuitively, liberation by grade will increase as the particle size decreases. Thus, liberation by grade may be defined as the degree to which particle size (also mass) and mineral grain size (also mass) assume unity. A completely liberated particle will be the same composition as the mineral grain, whereas a small mineral grain locked in a particle containing other, larger mineral grains, will assume a low grade approaching 0 (Figure 3.1).

The second way is liberation expressed as free surface or exposure, i.e., the proportion of the surface of the mineral that is exposed to the atmosphere. As per the example above, the small mineral grain locked in the particle will not have any exposed surface and will therefore not be amenable to oxidation (except through possible galvanic interactions where two sulfides are in direct contact) or acid-base reactions interaction, whereas the completely liberated mineral will have its entire surface available for reaction (Figure 3.1).

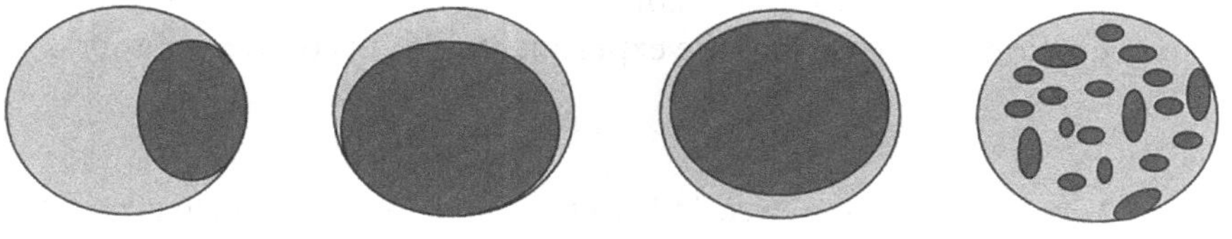

Figure 3.1 Liberation by grade and exposure concepts, using identical particle sizes (particles composed of dark and light minerals), but different mineral-of-interest grain size (dark) and location in the particle. The free or exposed surface of the dark mineral is the same in the first two images (≈45% of the dark total surface, expressed as perimeter in 2D), even though their liberation by grade is different (higher in the second image than in the first). The second and third images show identical liberation by grade (the areas of the dark mineral are the same) but very different liberation by exposure (≈45% versus 0). The last image shows an identical liberation by grade to the first image (same area covered by one grain versus several smaller grains of the dark mineral) but rather different liberation by exposure (<5% for the several smaller grains versus 45% for the single large grain).

Note that the two concepts are rather different but related. Both are very much dependent on the original texture of the ore versus the particle size to which it has broken. For example, an ore particle containing a finely disseminated sulfide mineral in a gangue matrix might have a higher liberation by grade value than a liberation by exposure value. This is because the mass of total sulfide may be more concentrated in the interior of the particle than at the surface, leading to lower exposure at the particle surface. Similarly, in the case of process wastes, passivation effects that prevent complete oxidation of minerals will likely result in high liberation by grade values but will give zero liberation by exposure (Figure 3.1). Generally, however, higher liberation by grade may be expected to yield higher liberation by exposure as particles are broken down.

Liberation by grade and by exposure are well-established descriptors derived from automated SEM. In this contribution, liberation by exposure is used, and this descriptor can be generated from a mineral association list, in which association with free surface (e.g., MLA) [39] or background (i.e., mounting medium of resin, QEMSCAN) [45] can be used.

3.2.5 Calculation of the AMD Potential

Oxidation of minerals for acid formation or acid consumption, as may be the case, relies on available mineral surface for reaction. It is therefore proposed here that liberation by exposure be included with the acid formation or consumption potential and mineral reactivity, to assess AMD risks. Using the information gleaned, an AMD value and an AMD ratio are calculated to assess the AMD potential (Eqs. 3.11 and 3.12):

$$\text{AMD value} = \text{AP} \times m \times \text{R} \times \text{L} \tag{3.11}$$

Where

the AMD value is the amount of H_2SO_4 produced, in kg/t;
m is the mass fraction of the mineral, i.e., mass%/100;
AP is the acid potential, calculated based on maximum acid generation/consumption (mol acid produced or consumed per mol mineral reacting), expressed as H_2SO_4 in kg/t;
R is the relative reactivity of the mineral;
L is the liberation of the mineral, i.e., %free surface/100.

$$\text{AMD ratio} = -\Sigma\text{AMD value (ACM)}/\Sigma\text{AMD value (AGM)} \tag{3.12}$$

Where

ACM are the acid consuming minerals;
AGM are the acid generating minerals.

The AMD value will be negative if the mineral consumes acid but will be positive if the mineral generates acid. If no acid is generated or consumed, the AMD value is 0. Similarly, if a mineral is not liberated, the AMD value is 0. The sum of all mineral AMD values will give the AMD value for the sample. Again, a positive AMD value would be obtained for a sample with acid generating potential whereas a negative value indicates acid-neutralizing potential of the sample. A higher value (positive or negative) indicates a larger acid generation or neutralization capacity.

The AMD ratio assists in determining the risk of acid formation. It is ranked based on the correlation between the AMD ratios calculated for real samples with outcomes from environmental test work. Where the AMD ratio is <1, AMD risk is evident, where >1, no AMD risk is evident. This ratio differs from the CARD risk ratio of Parbhakar-Fox *et al.* [11] as it incorporates mineral liberation in the AMD value, in addition to mineral acid producing or consuming potential and mineral reactivity. In this way, a texture-based ABA protocol is better realized.

The steps for producing a liberation-based AMD ratio were therefore carried out for different process wastes as follows:

- identification and quantification of the modal proportions of acid-generating and neutralizing minerals in a sample using mineralogical techniques (QEMSCAN analysis);
- identification of maximum acid generation or consumption potential of these minerals based on chemical reactions involving the release or consumption of H_2SO_4;
- evaluation of mineral reactivity based on information available in published literature or obtained during gold leaching test work carried out at Mintek;
- assessment of mineral liberation to further evaluate availability for acid generation or neutralization;
- calculation of sample AMD values and ratios.

Comparison with ABA and Net Acid Generating (NAG) static geochemical results was additionally conducted on four residues.

An AMD predictor software package was further developed to take as inputs, the ABA and NAG geochemical data and the mineralogical data (mineral proportions, reactivity, and liberation). Generated outputs were the geochemical and mineralogical results, providing a quick comparison of traditional static geochemical tests and mineralogical-based AMD prediction.

3.3 Application of the AMD Predictive Protocol

High concern for AMD generation in South Africa emanates from its gold processing tailings, for which AMD treatment options are continuously tested. In this regard, QEMSCAN mineral modal abundance and liberation data, carried out on three residues from different gold processing projects (Au res 1, Au res 2, Au res 3), were assessed using the AMD predictive protocol. The results were compared with static geochemical tests conducted on those samples. Additionally, a copper-bearing residue (Cu res 1) was assessed for comparison against static geochemical tests. Two further compositions represent typical BMS process tailings that may be generated, a Zn tailing (BMS Tails 1) and a Ni tailing (BMS Tails 2). Together, the data are presented to illustrate the influence of liberation, modal mineral abundance, and mineral reactivity on the AMD value and ratio of different waste types.

3.3.1 Experimental Procedures

To facilitate modal analysis, the samples were screened into appropriate size fractions, with polished sections made from each fraction and subjected to particle mineral analysis (PMA) [38] using a QEMSCAN automated scanning electron microscope. At the particle size analyzed, more than 10,000 particles are typically scanned in each fraction from which polished sections are made. With at least four fractions derived, at least 40,000 particles are assessed over the full sample, ensuring good representivity. Mineral modal abundance and liberation data were extracted for the minerals identified, as a function of the total sample (i.e., combined size fractions), which is standard practice with samples of this nature. The AMD value and the AMD ratio were calculated for each sample, using the extracted mineralogical data.

Static geochemical tests were conducted on the gold residues and one copper residue. These included ABA tests according to the modified method of Lawrence and Wang [21]. Fizz tests were additionally conducted to determine the quantity of acid required for the ABA procedure. From this, the neutralization potential (NP, expressed in kg $CaCO_3$/t) and acid generating potential (AP) were calculated (Eqs. 3.13 and 3.14):

$$NP = [(N_{HCl} \times V_{HCl}) - (N_{NaOH} \times V_{NaOH}) \times 50]/\text{mass of sample (g)} \tag{3.13}$$

Where

N is the normality (Eq/L) of HCl and NaOH, and

V is the volume (mL) HCl and NaOH

$$AP = \text{Sulfide S} \times 31.25 \tag{3.14}$$

Where

Sulfide S is determined from wet chemical analysis (LECO method), and 31.25 is a given constant to account for unit conversion

Other parameters calculated were:

Net NP (in kg $CaCO_3$/t):

$$NNP = NP - AP \tag{3.15}$$

Net acid production potential (in kg H_2SO_4/t):

$$NAPP = -NNP \times 0.98 \tag{3.16}$$

NP ratio:

$$NPR = NP/AP \tag{3.17}$$

Net acid generation (NAG) tests were additionally conducted on the samples, according to the method of Miller *et al.* [46], and the NAG was calculated as follows:

$$NAG = 49 \times V \times M/m \tag{3.18}$$

Where
NAG = net acid generation (in kg H_2SO_4/t)
V = volume of base NaOH titrated (mL)
M = molarity of base NaOH (mol/L)
m = mass of sample reacted (g)

3.3.2 Results and Discussion

Particle size distributions (PSDs) among the gold residues are, respectively, 56%, 64%, and 77% passing 75 μm for Au res 1, Au res 2, and Au res 3. The BMS 1 Tails and BMS 2 tails, representing very different mineral assemblages, are 64% and 50% passing 75 μm, respectively. The Cu res 1 tails shows a PSD around 80% passing 75 μm.

Various minerals were identified in the samples, along with calculated acid generation potentials and assigned relative reactivities (Table 3.3). AP is based on reactions of minerals with O_2 and water, releasing H_2SO_4 as the acid and therefore resulting in a positive potential. On the other hand, minerals like carbonates can consume the released acid, leading to a negative AP. As relative reactivity is assigned, it is noted that obtained acid production or consumption values are not absolute, and only comparative trends can be discussed.

Modal mineralogy and liberation data are presented for each of the six samples examined (Table 3.4). Modal mineral proportions, together with individual mineral chemical compositions, can be used to calculate a sample chemical composition. This calculated sample element composition can be compared against measured bulk assays in assessing accuracy of the modal mineral abundance. For the samples in this study, accuracy on Si, at concentrations of 25–45 mass%, was typically within 3–10% of the measured assay. For the BMS tails compositions, Mg, at concentrations of 5–15 mass%, was also within 3–10% of the measured assay. Mineral maps of particles show the associations and liberation of minerals, with liberation increasing as the particle size decreases (Figure 3.2).

The summarized results of the ABA and NAG tests, as well as the calculated AMD values and AMD ratios for each sample submitted to the geochemical tests, are also presented (Table 3.5). In this respect, it is noted that the AMD value is similar in meaning to the NAPP, as it considers the summed potential acid consuming minerals relative to summed potential acid producing minerals. Similarly, the AMD ratio may be considered equivalent to the NPR, as it considers the ratio of potential acid consuming to potential acid producing minerals. The AMD value was plotted against the NAPP, and AMD ratio against NPR, in assessing deviations between static geochemical tests and mineralogical-based predictions for each sample (Figure 3.3). The AMD values and ratios were also considered for liberation as observed, versus a calculation with liberation omitted, i.e., minerals considered all entirely liberated, in which case the liberation factor would be 1 (Table 3.6).

Table 3.3 Identified minerals in process wastes, calculated acid potential, and assigned relative reactivity.

Mineral	Formula	AP (H_2SO_4, kg/t)	Relative reactivity
Amphibole	$NaCa_2Mg_2Fe_3Si_7AlO_{22}(OH)_2$	−383	0.40
Apatite	$Ca_5(PO_4)_3(OH,F,Cl)$	0	0.004*
Calcite	$CaCO_3$	−980	1.00
Carrollite	$CuCo_2S_4$	317	1.00*
Chalcocite	Cu_2S	−616	0.07#
Chalcopyrite	$CuFeS_2$	534	0.44
Chlorite	$(Mg_3Fe_2)Al(AlSi_3O_{10})(OH)_8$	−455	0.40
Clay	$Al_2Si_2O_5(OH)_4$	0	0.02
Dolomite	$CaMg(CO_3)_2$	−1065	1.00
Fe oxide	$FeO(OH)$	<0	1.00*
Feldspar	$NaAlSi_3O_8$	−187	0.02
Galena	PbS	0	0.86
Muscovite	$KAl_3Si_3O_{10}(OH)_2$	−123	0.01
Pentlandite	$Ni_{4.5}Fe_{4.5}S_8$	443	1.00*
Pyrite	FeS_2	1633	1.00
Pyrophyllite	$Al_2Si_4O_{10}(OH)_2$	0	1.00*
Pyroxene	$Mg_2Si_2O_6$	−980	0.40
Pyrrhotite	Fe_7S_8	1210	8.43
Quartz	SiO_2	0	0.00
Serpentine	$Mg_3Si_2O_5(OH)_4$	−1061	0.40
Sphalerite Fe	$Zn_{0.9}Fe_{0.1}S$	145	0.96
Talc	$Mg_3Si_4O_{10}(OH)_2$	−778	0.40

*No data available; reactivity assigned based on assumed similar reactivity to pyrite, calcite, or quartz; #assigned based on observations of lower reactivity to pyrite; mineral formulae are taken from [47] and are also based on general formulae for mineral groups from [48].

Table 3.4 Modal mineral abundance (modal, mass%) and liberation (lib., % free surface) for the investigated process wastes.

Mineral	Au res 1		Au res 2		Au res 3		Cu res 1		BMS Tails 1		BMS Tails 2	
	modal	**lib.**	**modal**	**lib.**	**modal**	**lib.**	**modal**	**lib.**	**modal**	**lib.**	**modal**	**lib.**
Amphibole	0.1	92.8			0.3	97.2	0.4	35.2			34.5	61.1
Ankerite												
Apatite									0.3	50.3		
Calcite	0.2	99.3							0.8	56.8	1.0	67.2
Carrollite							0.3	77.8				
Chalcocite							0.6	30.5				
Chalcopyrite									0.3	73.4	0.4	72.7
Chlorite	2.5	99.4	0.3	63.4	2.6	99.7	4.4	32.2			15.7	44.0
Covellite							0.2	18.7				
Clay	0.3	n.d.					8.7	n.d.	0.7	n.d.	2.8	n.d.
Dolomite									43.8	76.1	0.7	50.2
Fe Oxides	1.1	n.d.	0.2	n.d.	0.5	n.d.	4.6		0.1	n.d.	5.6	n.d.
Feldspar	3.2	98.9	0.5	74.7	2.7	99.2	3.0		8.3	52.9	7.6	43.5
Galena									0.9	77.7		

(Continued)

Table 3.4 Modal mineral abundance (modal, mass%) and liberation (lib., % free surface) for the investigated process wastes. (*Continued*)

Mineral	Au res 1		Au res 2		Au res 3		Cu res 1		BMS Tails 1		BMS Tails 2	
	modal	**lib.**	**modal**	**lib.**	**modal**	**lib.**	**modal**	**lib.**	**modal**	**lib.**	**modal**	**lib.**
Mica	2.6	99.4	0.3	73.8	2.6	99.7	21.6	50.7	4.1	41.3	2.3	44.1
Pentlandite											0.8	67.0
Pyrite	1.9	99.9	0.5	80.7	0.2	99.8	11.2	75.9	12.8	71.8	2.1	58.4
Pyrophyllite	5.6	n.d.	2.5	n.d.	1.7	n.d.						
Pyroxene							0.1	11.2	0.6	23.1	4.3	25.4
Pyrrhotite	0.4	98.1	0.1	80.2	0.1	99.2			0.3	9.9	5.8	74.2
Quartz	82.2	n.d.	95.6	n.d.	89.4	n.d.	45.0	n.d.	25.8	n.d.	2.7	n.d.
Serpentine									0.1	51.6	6.1	28.4
Sphalerite									1.2	77.3		
Talc											7.7	34.8

n.d.: not determined. For inert minerals, liberation values were not extracted from the automated SEM data, as these values are inconsequential to the AMD value. The latter will be zero, with or without a liberation factor, as it is determined by the AP, which is zero (Table 3.3).

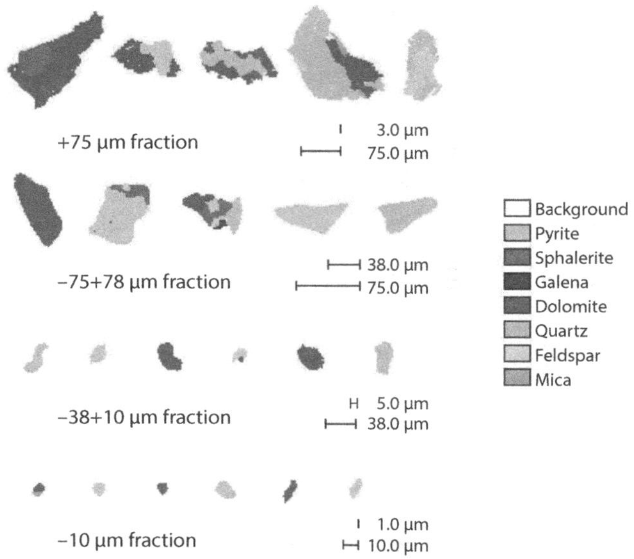

Figure 3.2 Mineral maps of selected particles from different size fractions of BMS Tails 1 analyzed by QEMSCAN, showing improved mineral liberation from coarser to finer particles.

Table 3.5 Calculated AMD parameters from static geochemical tests and mineralogical predictions on residue samples.

Sample	AMD value, kg H_2SO_4/t	AMD ratio	NNP, kg $CaCO_3$/t	NAPP, kg H_2SO_4/t	NPR	NAG pH
Au res 1	60.5	0.06	−21.3	20.83	0.15	2.23
Au res 2	15.4	0.02	3.4	−3.37	3.2	2.52
Au res 3	7.6	0.41	−2.6	2.57	0.62	3.07
Cu res 1	136.6	0.02	−145.5	142.63	0.02	2.36

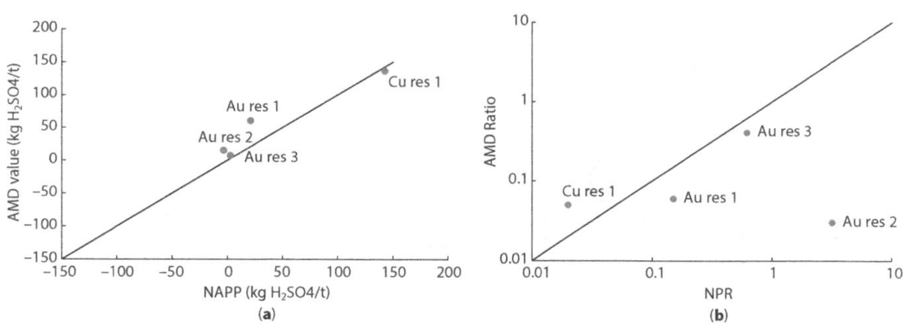

Figure 3.3 NAPP versus AMD value (a) and NPR versus AMD ratio (b) for the four samples subjected to ABA and NAG tests.

Table 3.6 Liberation (lib) effects on AMD values and ratios.

	AMD value	AMD value with lib = 1	Difference	AMD ratio	AMD ratio with lib = 1
Au res 1	60.5	61.2	0.7	0.06	0.06
Au res 2	15.4	19	3.6	0.02	0.03
Au res 3	7.6	7.7	0.1	0.41	0.41
Cu res 1	136.6	174.1	37.6	0.02	0.05
BMS Tails 1	−204.2	−233.8	29.6	2.31	1.97
BMS Tails 2	386.6	464.1	77.54	0.16	0.26

For the gold residue samples, overall, very good liberation of pyrite and pyrrhotite, as key acid producers, is observed. Silicates in these residues play virtually no role in NP, owing to the dominance of inert quartz. Differences in AMD values are due primarily to modal proportions of pyrite and pyrrhotite, as well as the pyrite:pyrrhotite ratio in influencing reactivity, given the very high reactivity assigned to pyrrhotite. Using a ranking scheme to compare the AMD values from these residues (Table 3.7), the Au res 1 residue reports to the range of values classified as extremely acid forming. Based on the ABA/NAG test classification scheme [49], this would be classified as potentially acid forming, but the relatively high pyrrhotite and pyrite content and liberation, coupled with the high reactivity of the pyrrhotite, would render the residue extremely acid forming instead.

The Au res 2 residue is acid forming, whether complete or actual liberation of pyrite and pyrrhotite is considered. Using the ABA/NAG scheme [49], this would be classified as uncertain. The Au res 3 residue is marginally lower in AMD value and is classified as potentially acid forming (Table 3.7), in agreement with ABA/NAG classification.

The Cu res 1 residue is classed as extremely acid forming (Table 3.7), owing to the high abundance of pyrite, as well as its relatively high liberation, at ≈76%. If liberation were not considered (liberation = 1, i.e., 100%), the AMD value increases substantially. The absence of carbonate, along with the presence of poorly neutralizing silicates, contributes to this high AMD value. Based on the ABA/NAG scheme, this residue is classified as potentially acid forming, but from the ABA/NAG scheme [49], it is high risk acid generating.

For the compositions of BMS tailings, composition BMS Tails 1 is potentially alkaline generating, owing to the dominance of reactive dolomite, compared with pyrite as the dominant sulfide. As the same reactivity is assigned, and both display similar liberation, the modal proportions become the key factor in determining the AMD value. Where complete liberation is assumed, the AMD value would decrease substantially; this appears to be more prevalent where liberation is typically <70% (Tables 3.4 and 3.6). Composition BMS Tails 2 is extremely acid producing, with a substantial increase in AMD value with complete liberation assumed, again because of liberation typically <70% for most minerals. Relatively high content of highly reactive pyrrhotite, along with good liberation, accounts for the classification. Reactive silicates, despite their abundance, cannot mitigate the effects of the reactive pyrrhotite in this instance.

Table 3.7 Classification based on the AMD value (Eq. 3.11).

AMD value	Classification
<−100	Potentially alkaline generating
−100 to −50	Acid neutralizing
−50 to −10	Inert
−10 to +10	Potentially acid forming
+10 to +50	Acid forming
>+50	Extremely acid forming

Table 3.8 Classification based on the AMD ratio (Eq. 3.12).

AMD ratio	Classification
<0.1	Extremely high risk of AMD
0.1 to 0.5	High risk of AMD
0.5 to 1.0	Risk of AMD
1.0 to 5	Low risk of AMD
5 to 10	Very low risk of AMD
>10	No risk of AMD

In terms of the AMD ratio, the smaller the effect of the acid consumers on the AMD value, the smaller the AMD ratio. Given the absence of carbonate in three of the residues, along with very small contribution from neutralizing silicates, these report to the extremely high risk of AMD formation class (Table 3.8). Note that this is not an indication of how much AMD could form but the risk of it forming. One gold residue and one of the BMS tailings composition report to the high risk of AMD formation class, and the tailings composition high in dolomite reports to the low risk of AMD formation class. The AMD ratios do not change dramatically with a full liberation assumed, compared with measured liberation.

The examples illustrate a complex interplay of mineral type (acid producer, neutralizer, inert), mineral proportion, reactivity (oxidation rate versus acid consumption reactivity), and liberation in contributing to the amount of acid that may be produced by a waste sample, as well as the risk of that acid being produced. The addition of reactivity and liberation to mineral proportions brings a kinetic aspect to the determination of AMD potential.

Based on the general particle size representing process wastes in this study, it is evident that mineral liberation plays a vital role in quantifying exposed surface available for mineral reactivity. As particle size increases relative to the grain size of the minerals in the particles of a given ore, liberation will approach zero, because less surface is typically exposed (Figures 3.1 and 3.2). In such a case (e.g., particle size at the mm scale, containing mineral grains at the micron scale), liberation decreases, and thus, the potential and risk to form AMD is decreased. The difference in AMD value using measured liberation (i.e., liberation <1) as opposed to full liberation (i.e., liberation = 1) is particularly pronounced in the BMS process wastes in this study (Table 3.6). This will be a function of the mineral grain size distribution at the PSD studied. Mineral grain size distribution is a function of the original texture of the ore. Larger grains in the original ore liberate at coarser particle sizes than smaller grains. This therefore gives rise to potentially different mineral liberation properties for a given PSD of two different ores or waste compositions containing the same mineral. Reaction kinetics will be consequently affected.

As part of a geometallurgical approach, feed material to a process will have a given PSD and particle tracking models have been developed to help simulate the behavior of such particles in the process [50]. Which particles will report to the tails is a key consideration in optimizing grade and recovery to the concentrate and forms a cornerstone of many process mineralogical investigations. Using this approach from the angle pursued in this study, the same said particles reporting to the final tails after optimization for recovery of value should be assessed for the potential to produce acid over time. Indeed, current efforts are aimed at desulfurization processes, in which bulk sulfide flotation of tailings streams is considered to "clean" the tailings of residual sulfides and the resultant material can be used as cover over tailings storage facilities [51, 52]. This is costly and should be a process design consideration upfront rather than a mitigation option after processing.

Analysis of a judiciously selected set of samples can yield information pertinent to not only process efficiency (grade and recovery optimization versus market conditions and mining plan progression) but also tailings production (AMD potential, considering environmental conditions for the tailings storage facility and temporal changes). Indeed, design considerations would aim at integrated strategies for both process efficiency optimization and AMD mitigation. In this respect, detailed mineralogical information would not come at extra cost, as the ore deposit definition will be serving multiple purposes and will provide more value to the operations from a prediction perspective.

3.4 Conclusions and Further Work

To complement traditional static geochemical tests, an approach using mineral proportions, mineral acid production or consumption potential, mineral reactivity, and mineral liberation was applied for the prediction of AMD potential of gold and copper process residues, as well as BMS tailings compositions, representing process wastes. The mineral reactivity, together with relative abundance, plays a major role in determining AMD values of samples. Liberation appears to be an important factor when minerals occur with exposure less than 70%. Abundance of higher reactivity silicates like chlorite, amphibole, and talc may be more influential over time but are unlikely to mitigate more immediate sulfide reaction.

Mineral proportions and liberation data were obtained using automated scanning electron microscopy techniques, with a database developed for mineral acid production or consumption potential and mineral reactivity. It should be noted that certain assumptions are made where published information on the latter two parameters is sparse. Nevertheless, comparative analysis is possible and provides an indication of AMD potential. As empirical data are obtained from test work and complemented by mineral chemical compositional information, the database can be strengthened through addition of observed values for use with quantitative mineralogy.

The approach has application in the prediction of AMD potential of process wastes which typically show particle sizes in the micron range. At these particle sizes, liberation effects of acid producing sulfides are more pronounced than at the centimeter or meter scales of waste rock and therefore will influence the acid production potential of the waste.

Although mineralogical observations have also aided in the understanding of changes during kinetic or long-term testing [17, 18], these observations presently do not predict how much acid will be formed, and when, if at all. Building on the concepts around liberation and mineral reactivity presented here, work is underway on iterative modeling and simulation of changing particle properties over time. As an example, a strong acid producer, where liberated, should produce acid faster than the same strong acid producer locked entirely within a poorly reactive silicate, based on the surface area available for oxidation. This concept would be suited to prediction of AMD production over time for process wastes, and finer particles of waste rocks (e.g., –3 mm) and would be a desirable alternative to time-consuming and costly kinetic tests that are presently performed. Consideration of the sulfide suite of minerals, together with liberation and association information, should yield clues on the extent to which galvanic interaction may be expected, and their contribution to possible passivation effects. Together with the modeling of environmental effects like temperature and fluid flow dynamics the outcomes present an attractive alternative to costly and time-consuming kinetic tests to predict how much AMD will form, and when.

Acid mine drainage prediction studies should form part of an integrated strategy that optimizes process efficiency while mitigating environmental problems so that sustainable operations may be pursued in line with, and as part of, a geometallurgical program. The consideration of mineralogy, mineral modal abundance, mineral reactivity, and mineral liberation, as added in this work, should improve these predictive studies.

References

1. Sutton, P.A., Perspective on environmental sustainability? A paper for the Victorian Commissioner for Environmental Sustainability, 2004. http://www.green-innovations.asn.au/A-Perspective-on-Environmental-Sustainability.pdf.
2. Younger, P.L., Banwart, S.A., Hedin, R.S., *Mine Water – Hydrology, Pollution, Remediation*, p. 464, Kluwer, Dordrecht, 2002.
3. Wolkersdorfer, C., *Water Management at Abandoned Flooded Underground Mines – Fundamentals, Tracer Tests, Modelling, Water Treatment*, p. 465, Springer, Heidelberg, 2008.
4. Jamieson, H.E., Walker, S.R., Parsons, M.B., Mineralogical characterization of mine waste. *Appl. Geochem.*, 57, 85–105, 2015.
5. Dold, B., Acid rock drainage prediction: A critical review. *J. Geochem. Expl.*, 172, 120–132, 2017.
6. McCarthy, T.S., The impact of acid mine drainage in South Africa. *S. Afr. J. Sci.*, 107, 5/6, 7, 2011. Art. #712, https://doi.org/10.4102/sajs.v107i5/6.712, 2011.
7. van Rooyen, M., Acid Mine Drainage Water Treatment using the SAVMIN Process, in: *SAIMM Hydrometallurgy 2016*, pp. 52–60, South African Institute of Mining and Metallurgy, Cape Town, South Africa, 2016.

8. Masindi, V., Madzivire, G., Tekere, M., Reclamation of water and the synthesis of gypsum and limestone from acid mine drainage treatment process using a combination of pre-treated magnesite nanosheets, lime and CO_2 bubbling. *Water Resour. Ind.*, 20, 1–14, 2018.
9. Parbhakar-Fox, A.K., Edraki, M., Walters, S., Bradshaw, D., Development of a textural index for the prediction of acid rock drainage. *Miner. Eng.*, 24, 1277–1287, 2011.
10. Parbhakar-Fox, A., Lottermoser, B.G., Bradshaw, D., Evaluating waste rock mineralogy and microtexture during kinetic testing for improved acid rock drainage prediction. *Miner. Eng*, 52, 111–124, 2013.
11. Parbhakar-Fox, A., Lottermoser, B., Hartner, R., Berry, R.F., Noble, T.L., Prediction of Acid Rock Drainage from Automated Mineralogy, in: *Environmental Indicators in Metal Mining*, B. Lottermoser (Ed.), pp. 139–156, Springer, Cham, 2017, https://doi.org/10.1007/978-3-319-42731-7_8.
12. Parbhakar-Fox, A. and Lottermoser, B.G., A critical review of acid rock drainage prediction methods and practices. *Miner. Eng.*, 82, 107–124, 2015.
13. Becker, M., Dyantyi, N., Broadhurst, J.L., Harrison, S.T.L., Franzidis, J.P., A mineralogical approach to evaluating laboratory scale acid rock drainage characterization tests. *Miner. Eng.*, 80, 33–36, 2015.
14. Jambor, J.L., Dutrizac, J.E., Groat, L.A., Raudsepp, M., Static tests of neutralization potentials of silicate and aluminosilicate minerals. *Environ. Geol.*, 43, 1–17, 2002.
15. Paktunc, A.D., Mineralogical constraints on the determination of neutralization potential and prediction of acid mine drainage. *Environ. Geol.*, 39, 103–112, 1999.
16. Abrosimova, N., Gaskova, O., Loshkareva, A., Edelev, A., Bortnikova, S., Assessment of the acid mine drainage potential of waste rocks at the Ak-Sug porphyry Cu–Mo deposit. *J. Geochem. Expl.*, 157, 1–14, 2015.
17. Brough, C., Strongman, J., Bowell, R., Warrender, R., Prestia, A., Barnes, A., Fletcher, J., Automated environmental mineralogy; the use of liberation analysis in humidity cell testwork. *Miner. Eng.*, 107, 112–122, 2017.
18. Brough, C., Parbhakar-Fox, A., Fletcher, J., Barnes, A., Griffiths, R., Strongman, J., Bowell, R., Garner, C., Becker, M., Automated Environmental Analysis: Multiple size fraction analysis of kinetic tests and implications for geochemical interpretations, in: *Proceedings: Process Mineralogy 18 conference*, MEI, Cape Town, 2018.
19. Sobek, A.A., Schuller, W.A., Freeman, J.R., Smith, R.M., *Field and Laboratory Methods Applicable to Overburdens and Minesoils*, p. 203, U.S. Environmental Protection Agency, EPA, Washington, D.C., 600-2-78–054, 1978.
20. Lawrence, R.W., Prediction of the behaviour of mining and processing wastes in the environment, in: *Proceedings, Western Regional Symposium on Mining and Mineral Processing Wastes*. F. Doyle (Ed.), pp. 115–121, Society for Mining, Metallurgy, and Exploration, Littleton, CO, 1990.
21. Lawrence, R.W. and Wang, Y., Determination of Neutralization Potential in the Prediction of Acid Rock Drainage, in: *Proceedings of the 4th International Conference on Acid Rock Drainage*, pp. 449–464, Vancouver, BC.1997.
22. Lapakko, K., Evaluation of neutralization potential determinations for metal mine waste and a proposed alternative, in: *Proc. International Land Reclamation and Mine Drainage Conference*, pp. 129–137, 1994, https://doi.org/10.21000/JASMR94010129.
23. Maest, A.S., Kuipers, J.R., Travers, C.L., Atkins, D.A., *Predicting Water Quality at Hardrock Mines: Methods and Models, Uncertainties, and State-of-the-Art*, Kuipers & Associates and Buka Environmental, Washington, D.C., 2005.
24. Perkins, E.H., Nesbitt, H.W., Gunter, W.D., St-Arnaud, L.C., Mycroft, J.R., *Critical review of geochemical processes and geochemical models adaptable for prediction of acid drainage from waste rock*, p. 283, MEND report, Ontario, 1995.

25. Parbhakar-Fox, A., *Establishing the benefit of an integrated geochemistry-mineralogy-texture approach for acid rock drainage prediction*, PhD thesis, University of Tasmania, 2012.
26. Nordstrom, D.K., Blowes, D.W., Ptacek, C.J., Hydrogeochemistry and microbiology of mine drainage: An update. *Appl. Geochem.*, 57, 3–16, 2015.
27. Chopard, A., Benzaazoua, M., Plante, B., Bauzahzan, H., Marion, P., Kinetic Tests to Evaluate the Relative Oxidation Rates of Various Sulfides and Sulfosalts, in: *Proceedings, 10th International Conference on Acid Rock Drainage and IMWA annual conference*, 2015.
28. Stanton, M.R., Taylor, C.D., Gemery-Hill, P.A., Shanks, W.C. III, Laboratory studies of sphalerite decomposition: Applications to the weathering of mine wastes and potential effects on water quality, in: *The 7th International Conference on Acid Rock Drainage (ICARD)*, R.I. Barnhisel (Ed.), The American Society of Mining and Reclamation (ASMR), Lexington, KY, 2006.
29. Plumlee, G.S., The Environmental Geology of Mineral Deposits, in: *The Environmental Geochemistry of Mineral Deposits, Part A: Processes, Techniques and Health Issues*, vol. 6A, G.S. Plumlee and M.J. Logsdon (Eds.), pp. 71–116, Society of Economic Geologists, Littleton, Colorado, Reviews in Economic Geology, 1999.
30. The International Network for Acid Prevention (INAP), Global Acid Rock Drainage Guide (GARD Guide), 2009. http://www.gardguide.com.
31. Dold, B., Evolution of Acid Mine Drainage Formation in Sulphidic Mine Tailings. *Minerals*, 4, 621–641, 2014.
32. Edraki, M., Baumgartl, T., Manlapig, E., Bradshaw, D., Designing mine tailings for better environmental, social and economic outcomes: A review of alternative approaches. *J. Clean. Prod.*, 84, 411–420, 2014.
33. Khorasanipour, M., Environmental mineralogy of Cu-porphyry mine tailings, a case study of semi-arid climate conditions, Sarcheshmeh mine, SE Iran. *J. Geochem. Explor.*, 153, 40–52, 2015.
34. Moncur, M.C., Ptacek, C.J., Lindsay, M.B.J., Blowes, D.W., Jambor, J.L., Long-term mineralogical and geochemical evolution of sulfide mine tailings under a shallow water cover. *Appl. Geochem.*, 57, 178–193, 2015.
35. Jambor, J.L., Ptacek, C.J., Blowes, D.W., Moncur, M.C., Acid drainage from the oxidation of iron sulfides and sphalerite in mine wastes, in: *Proceedings from: Lead & Zinc '05*, vol. 1, T. Fujisawa (Ed.), The Mining and Materials Processing Institute of Japan, Tokyo, 2005.
36. Aspandiar, M.F. and Eggleton, R.A., Weathering of Chlorite: I. Reactions and Products in Microsystems Controlled by the Primary Mineral. *Clay Clay Miner.*, 50, 685–698, 2002.
37. Dos Santos, E.G., de Mendonça Silva, J.C., Duarte, H.A., Pyrite Oxidation Mechanism by Oxygen in Aqueous Medium. *J. Phys. Chem.*, 120, 2760–2768, 2016.
38. Gottlieb, P., The revolutionary impact of automated mineralogy on mining and mineral processing, in: *24th International Mineral Processing Congress*, pp. 165–174, Science Press, Beijing, 2008.
39. Fandrich, R., Gu, Y., Burrows, D., Moeller, K., Modern SEM-based mineral liberation analysis. *Int. J. Miner. Process.*, 84, 310–320, 2007.
40. Sverdrup, H.U., *The kinetics of base cation release due to chemical weathering*, Lund University Press, Lund, 246 pp, 1990.
41. Kwong, Y.T.J., *Prediction and prevention of acid rock drainage. MEND project*, p. 47, Canadian Centre for Mineral and Energy, Ottawa, 1993.
42. Noble, T.L. and Lottermoser, B.G., Modified Abrasion pH and NAG pH Testing of Minerals. *Environ. Indic. Met. Mining*, 19 October 2016, 211–220, 2016.
43. Bouzahzah, H., Benzaazoua, M., Bussiere, B., Plante, B., Prediction of Acid Mine Drainage: Importance of Mineralogy and the Test Protocols for Static and Kinetic Tests. *Mine Water Environ.*, 33, 54–65, 2014.

44. Chopard, A., Benzaazoua, M., Bouzahzah, H., Plante, B., Marion, P., A contribution to improve the calculation of the acid generating potential of mining wastes. *Chemosphere*, 175, 97–107, 2017.
45. Goodall, W.R., Scales, P.J., Butcher, A.R., The use of QEMSCAN and diagnostic leaching in the characterisation of visible gold in complex ores. *Miner. Eng.*, 18, 877–886, 2005.
46. Miller, S., Robertson, A., Donahue, T., Advances in Acid Drainage Prediction Using the Net Acid Generation (NAG) Test, in: *Proceedings of the 4th International Conference on Acid Rock Drainage*, pp. 533–549, 1997.
47. Mineralogy Database: http://www.webmineral.com
48. Deer, W.A., Howie, R.A., Zussman, J., *An Introduction to the Rock-Forming Minerals*, 2nd edition, p. 696, Longman Scientific and Technical, New York, 1992.
49. De Wet, L., Acid base accounting of mining ore and waste. *Test Meas. Conf.* 2012. September, http://nla.org.za/webfiles/conferences/2012/Presentations
50. Lamberg, P. and Vianna, S.M.S., A technique for tracking multiphase mineral particles in flotation circuits, in: *VII meeting of the southern hemisphere on mineral technology*, pp. 195–203, Ouro Preto, 2007.
51. Skandrani, A., Demers, I., Kongolo, M., Desulfurization of aged gold-bearing mine tailings. *Miner. Eng.*, 138, 195–203, 2019.
52. Benzaazoua, M., Bouzahzah, H., Taha, Y., Kormos, L., Kabombo, D., Lessard, F., Bussi, B., Demers, I., Kongolo, M., Integrated environmental management of pyrrhotite tailings at Raglan Mine: Part 1 challenges of desulphurization process and reactivity prediction. *J. Clean. Prod.*, 162, 86–95, 2017.

4

Oxidation Processes and Formation of Acid Mine Drainage from Gold Mine Tailings: A South African Perspective

Bisrat Yibas

Department of Geology, University of the Free State, Bloemfontein, South Africa

Abstract

Detailed multidisciplinary investigation of selected tailings storage facilities (TSFs) from the Witwatersrand gold mines generated datasets of mineralogy, chemistry, hydraulic properties, oxygen content with depth and leachate pore water chemistry. Three oxidation zones, namely mature oxidation (MOZ), transitional oxidation (TZ) and unoxidized (UZ) are recognized and mapped.

Pyrite, the major primary sulfide mineral, generally increases with depth, whereas jarosite appears at shallow depth at the expense of pyrite and decreases with depth, consistent with the progression of the oxidation front. Slight modifications of grain size distribution due to oxidation is evident. Hydraulic conductivity and moisture content increases and residual water content decreases with depth. Oxygen diffusion varies from 2 to 5 m with an average of about 4 m depth, correlating well with the measured oxidation zone thicknesses.

Sulfide minerals paragenesis, relative depletion and enrichment of certain major and trace elements in the MOZ and the TZ, respectively, and the pore water pH in the OZ, all indicate oxidation process reached "Late Stage" in the TSFs.

This study indicates that integrated investigation provides insights in the kinetics of oxidation processes and establishes a relationship between the volume of oxidized part of a TSF and the volume and quality of generated AMD.

Keywords: Tailings storage facilities, oxidation process, oxidation zone, mine drainage, oxygen diffusion, pore water chemistry

4.1 Introduction

Mine residue deposits are substantial contributors of pollution that adversely affects water resources, soil, and the surrounding ecosystem. As a result, studies to investigate the potential of mine residue deposits to generate pollution and its qualities and quantities are frequently conducted. Tailings storage facilities (TSF) generate voluminous, mining-affected

**Corresponding author*: yibasbabsob@ufs.ac.za [ORCID: 0001-7985-1681]

Elvis Fosso-Kankeu, Christian Wolkersdorfer and Jo Burgess (eds.) Recovery of Byproducts from Acid Mine Drainage Treatment, (73–96)

water of varying quality and metal concentrations. The quality and chemical composition of the drainage from a TSF is directly related to the volume of the tailings portion subjected to oxidation processes.

Although the oxidation zone thickness in a TSF is site specific, geochemical prediction of drainage from a TSF is in most cases based on extrapolations from other sites. The oxidation processes and the resultant oxidation zone thickness is governed by the interplay of various factors such as particle size distribution, mineralogy and chemistry, water and oxygen availability, temperature and infiltration, and diffusion rates, as well as the age of the TSF after decommissioning.

South Africa has been mining gold for over 130 years and was the largest gold mining country for over a century. This gold mine operation created of over 270 mine residue deposits such as TSFs and waste rock deposits over an area of 400 km^2 (Figures 4.1 and 4.2). It is estimated that billion tons of mine residues are contained in these mine residue deposits [1].

This chapter presents and discusses data such as oxidation zone thickness, physical, mineralogical, geochemical, and geohydrological characteristics of tailings, as well as oxygen diffusion with depth based on the findings of two research projects to investigate kinetic oxidation processes and water balance methodologies on mine residue deposits [2, 3]. The interplay of the factors critical to oxidation processes of tailings material and advances in understanding the quality and quantity of the resultant effluent from the Witwatersrand mine TSFs is discussed here.

4.2 Weathering and Oxidation of the Witwatersrand Gold Tailings

The oxidation processes and oxidation profiles of selected TSFs from the Witwatersrand gold mines were studied [2]. These TSFs were selected using criteria such as operational status (operational, closed, decommissioned, reclamation), age after decommissioning, geographic distribution, and geological variation. Available historical information such as design and construction, physical dimension and volume, seepage volume and quality, phreatic surface level data for the selected TSFs were collected [2]. Detailed site investigation of several TSFs and cross section mapping of exposed tailings due to reclamation were undertaken to investigate the lateral and vertical extent of oxidized and unoxidized zones [2, 4, 5]. Geochemical,

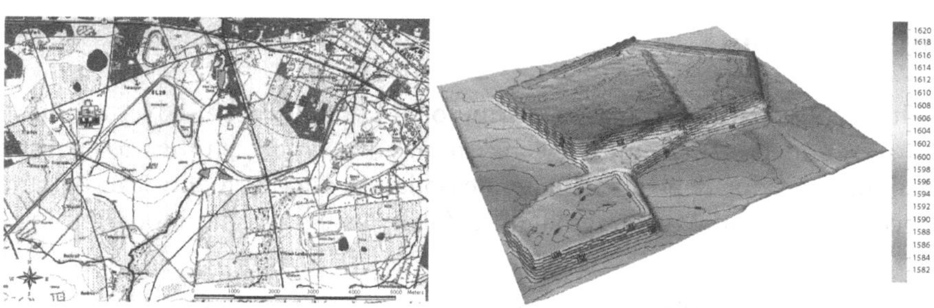

Figure 4.1 (a) Local topography map for TSF1; (b) 3D visualisation of TSF1, representative of the investigated TSFs (from [2], courtesy Anglo Gold Environmental).

Figure 4.2 Exposed mature oxidation zone (MOZ) with underlying Transitional Zone (TZ) (TSF5) with a vegetation cover on top.

mineralogical, and hydraulic properties such as grain size distribution, porosity, permeability, and water content and water retention characteristics of the tailings' samples were conducted on representative samples collected from the exposed cross sections and from auger holes [3]. The occurrence and effect of secondary permeability inducing phenomenon such as preferential flow pathways (e.g., cracks, fractures, mega-pores, bedding parallel flows) on oxygen and water infiltration and therefore on oxidation processes of tailings materials were investigated. A multilevel oxygen flux measurement with depth in selected TSFs was conducted following the methodology of [6] as discussed in detail in [2, 5, 7].

Most of the Witwatersrand gold mine TSFs are deposited in bench configuration of variable areal extent and height. The height of the investigated TSFs ranges from 1.1 to 34 m (Figures 4.1 and 4.2) with benches ranging from 1 to 4 (Figures 4.3 and 4.4). Cross-sections of profiles exposed during reclamation together with logging of several auger holes up to 10 m depth each were used to characterize variations of oxidation and identify three depth dependent oxidation zones (Figure 4.4). The lateral extent, thicknesses, and relative position of these zones are governed by the number and configuration of depositional benches (Figure 4.4). These three oxidation zones are: the mature oxidation zone (MOZ), transitional or active oxidation zone (TZ), and primary or unoxidized zone (UZ). The MOZ is amorphous with its most part and is characterized by leached buff gray top layer that grades into a light brown bottom layer. Itis interlayered with dark gray organic thin layers. The TZ is characterized by alternation of oxidized and partially oxidized or entirely unoxidized layers which are in most cases parallel to the depositional layers. However, irregularly shaped oxidation layers grading into partially oxidized and unoxidized parts of the zone are not uncommon. The MOZ thickness ranges from 1.5 to 3.5 m and the TZ thickness ranges from 3.0 to 4.5 m. The combined depth of the MOZ and TZ represents the total oxidation zone thickness and ranges from 5.5 to 9 m. The UZ is the thickest zone in all the TSFs forming

Figure 4.3 Detailed views of exposed tailing sections: (a) oxidation zone profile (≈8-m high) with oxidation zone (OZ) thickness of about 4.5 m; (b) exposed section of TSF5 (≈9-m thick) with about 3.5-m thick MOZ and over 5-m thick TZ with alternating oxidized and unoxidized layers reflecting compositional and textural variations, (c) exposed section showing the gradational contact between the MOZ and the TZ.

up to 70 % of the thickness of the TSFs. It is gray to dark gray and moist with little or no visible oxidation.

4.3 Water Infiltration and Oxygen Diffusion *vs* Oxidation Processes

4.3.1 Hydrogeology of Tailings Storage Facilities

4.3.1.1 Introduction

Tailings storage facilities are variably unsaturated porous media consisting of an unsaturated upper part and saturated lower part separated by a phreatic surface. The proportion of the saturated zone decreases and may even disappear with time in non-operational/decommissioned TSFs.

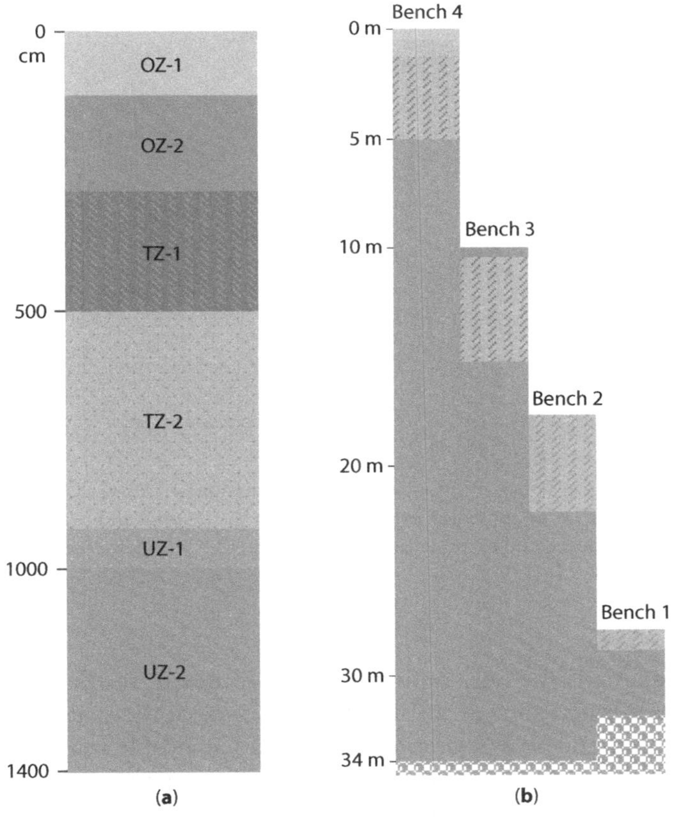

Figure 4.4 Oxidation profiles (a) OZ (MOZ), TZ, and UZ and subdivisions within the oxidized zones due to internal variation in oxidation intensity in TSF1, (b) oxidation zone development mimicking the depositional bench configuration wherein the vertical positions of the MOZ, TZ, and UZ are different in the different benches.

The saturated zone water balance components of a TSF are relatively well understood and can be estimated or calculated with relative ease. However, the estimation or calculation of the surface flux components above the phreatic surface is not well understood. Measurement of these surface flux components is expensive and time consuming and complicated further by the lateral thickness variation of the unsaturated zone from the edge to the center of a TSF due to the bell-shaped curve of the phreatic surface (Figure 4.5). This results in the spatial variation of the water balance components from the edge of the pool to the center of the TSF [8].

In operational TSFs, infiltration is maximum at the edge of the pool and minimum at the center of the pool, while evaporation is maximum at the center of the pool and minimum at its edge. Decommissioned TSFs have spatially varying unsaturated zones with the position of the phreatic surface and rate of drop of this surface difficult to estimate accurately which makes simple physical water balance calculations inappropriate (Figure 4.5). The main hydraulic properties that govern water flow paths in a porous unsaturated medium are particle size distribution, moisture content, bulk density, permeability/hydraulic conductivity, and water retention characteristics.

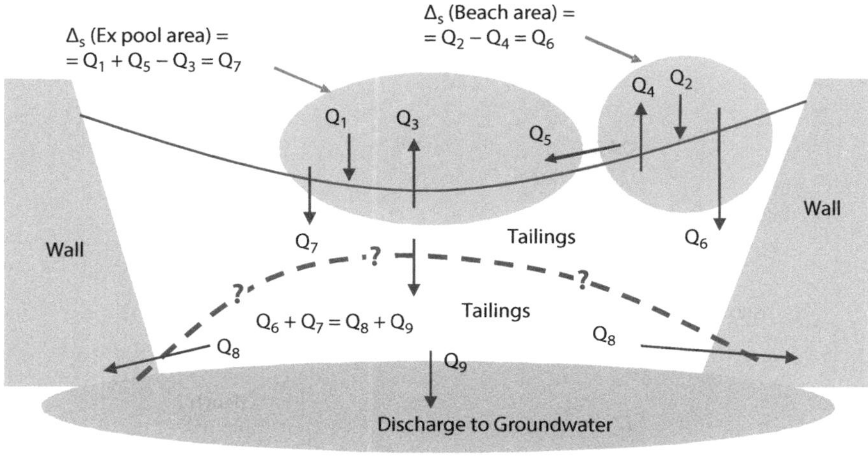

Figure 4.5 Schematic of water balance components for decommissioned TSF: Q_1, Q_2: precipitation on ex-pool and beach areas, respectively; Q_3, Q_4: potential evaporation from the ex-pool and beach area, respectively; Q_6, Q_7: infiltration from the beach and ex-pool areas, respectively; Q_8, Q_9: sideway seepage and discharge into groundwater, respectively (after [2]).

Numerical modeling is the only option available to determine the water balance components of the unsaturated zone (e.g., [8]). The existence of the spatial distribution of the surface fluxes (infiltration and evaporation) along the TSF profile could be used to establish a methodology to calculate these fluxes, which paves the way to formulate a more rigorous multidimensional unsaturated/saturated flow water balance modeling [8, 9].

4.3.1.2 Primary Hydraulic Characteristics

The depositional particle size distribution of the Witwatersrand gold mine tailings was relatively uniformly graded and dominantly sandy silt and rarely silty sand. In operational TSFs, particle size distribution tends to be finer from the edge to the center (pond or pool area) of the TSF during deposition forming alternate layers of relatively finer and coarser tailings vertically as the TSF grows in thickness and areal extent. Such alternate layers create capillary break, which complicates flow of water in tailings materials [3].

The particle size distribution of the tailings material in the different oxidation zones does not show substantial differences. However, detailed observation of the hydrometer analyses shows that the TZ materials are slightly coarser than that of the MOZ and both are coarser than the underlying UZ, the latter may reflect the original depositional particle size distribution (Table 4.1 and Figure 4.6). The slight grain size increase at TZ compared to the MOZ indicates the reworking of grain size distribution due to the weathering and oxidation processes.

Commonly, the water content of the original slimes and slurries of gold mines during deposition are approximately 1:1 [1]. Moisture content of tailings material then becomes variable in time and space. Measured moisture content data range from 7 to 20 wt. % with an overall increase of moisture content with depth. The lowest water content recorded

Table 4.1 Summary of the particle size distribution and water content data of selected Witwatersrand gold mine TSFs.

		TSF1		
Zone	**OZ**	**TZ1**	**TZ2**	**UZ**
Depth (m)	0–2	2–5	5–9	9–10
% Clay	0	0	0	0
% Silt	68	63	68	61
% Sand	32	37	32	39
% Gravel	0	0	0	0
Material	Sandy silt	Sandy silt	Sandy silt	Sandy silt
Moisture (%)	15	16.7	14.1	19.5
		TSF2		
Zone	**OZ**	**TZ**	**UZ-A**	**UZ-B**
Depth (m)	0–1.5	1.5–4.5	4.5–8	8–10
% Clay	0	0	0	2
% Silt	45	45	47	55
% Sand	55	55	53	43
% Gravel	0	0	0	0
Material	Silty sand	Silty sand	Silty sand	Silty sand
Moisture (%)	13.9	18.9	16.4	19.1
		TSF3		
Zone	**OZ1**	**OZ2**	**TZ**	**UZ**
Depth (m)	0–2	2–4	4–5	5–10
% Clay	0	0	0	0
% Silt	59	60	43	64
% Sand	41	40	57	36
% Gravel	0	0	0	0
Material	Sandy silt	Sandy silt	Silty sand	Sandy silt
Moisture (%)	7.3	8.2	13.2	10.8

(Continued)

Table 4.1 Summary of the particle size distribution and water content data of selected Witwatersrand gold mine TSFs. (*Continued*)

TSF5				
Zone	**OZ1**	**OZ2**	**TZ**	**UZ**
Depth (m)	0–2	2–4	4–5	8.5–10
% Clay	0	0	0	0
% Silt	59	56	56	56
% Sand	41	43	44	44
% Gravel	0	1	0	0
Material	Sandy silt	Sandy silt	Sandy silt	Sandy silt
Moisture (%)	10.6	7.6	13.5	15.3
TSF4				
Zone	**OZ2**	**OZ1**	**TZ**	**UZ**
Depth (m)	1.5–2.5	0–1.5	2.5–5	5–8
% Clay	0	2	2	2
% Silt	76	66	75	74
% Sand	24	31	23	24
% Gravel	0	1	0	0
Material	Sandy silt	Sandy silt	Sandy silt	Sandy silt
Moisture (%)	16.1	19.1	20.3	27.9

is from the TSF3 tailings and ranges from 7 to 13 wt. %, whereas TSF4 shows the highest moisture content which ranges from 19–28 wt. % (Table 4.1). Water retention capacities and residual water content decreases with depth consistent with the general particle size distribution pattern (Figure 4.7). Residual water content ranges widely from 5.5 to 20 wt. % consistent with increasing bulk densities and decreasing porosities with depth. The hydraulic conductivities of samples from the TSFs ranges from 7.1×10^{-5} to 1.2×10^{-2} cm/s consistent with uniformly graded and well sorted nature of the tailings and the formation of hardpans leading to the formation of relatively flat beaches and hence relatively uniform permeability.

4.3.1.3 Geological Structures as Preferential Flow Paths

Discontinuous oxidized layers and irregular bodies were observed within UZ far below the boundaries of the UZ and the TZ (Figure 4.8a). The oxidation process that formed

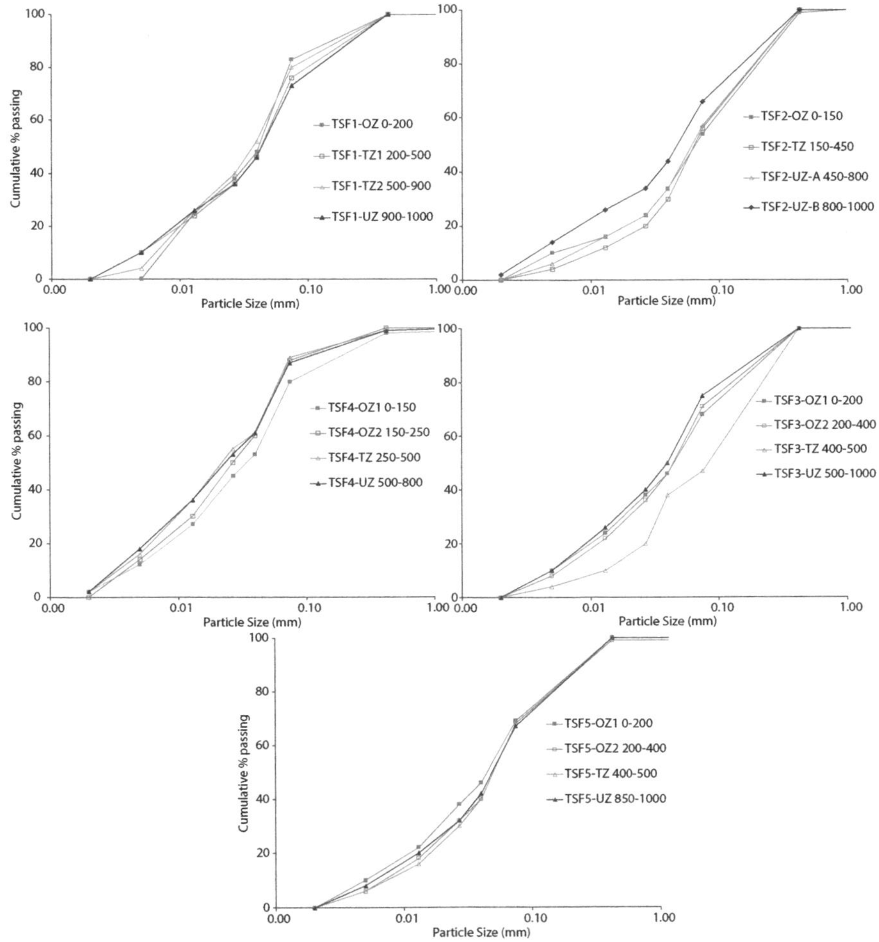

Figure 4.6 Particle size distribution of the different TSF oxidation zones.

these bodies is attributed to the existence of geological structures such as fractures, desiccation cracks, and bedding planes which acted as preferential flow paths to water and oxygen below the normal oxidation fronts (Figure 4.8b).

Although, generally, gold tailings are not the best growth medium due to their mineralogical composition which dominated by quartz and due to their low pH after weathering and oxidation, the effect of vegetation in the formation of preferential flow paths was evident in rare cases in TSFs where proper rehabilitation using soil cover and grass growth was effective for the growth of dense grass cover. The development of up to 20 cm of organic rich dark gray soil with grass roots penetrating up to 15 cm into the TSF were observed in these TSFs. In rare cases, where trees are planted (e.g., TSF3), root induced cracks are evidenced which could be used as preferential flow paths for water and oxygen infiltration in these TSFs.

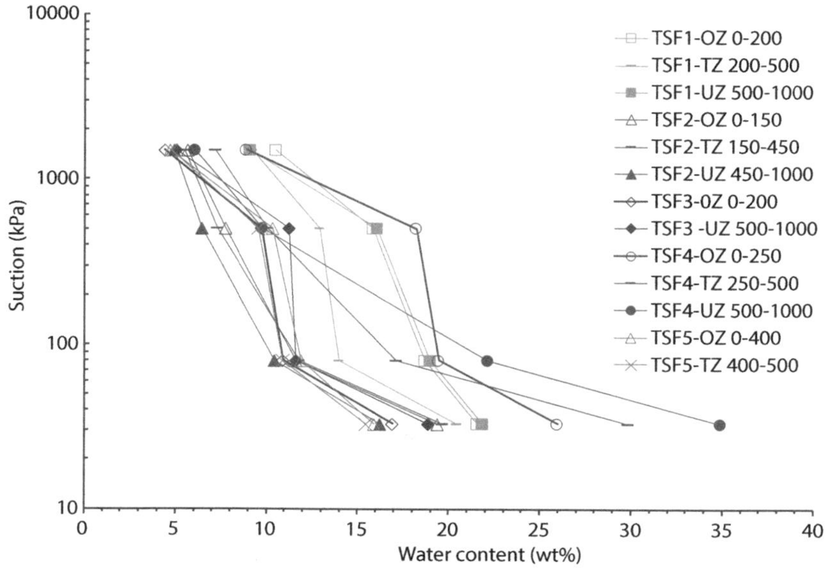

Figure 4.7 Water retention characteristics of the TSFs with depth as measured on four pressures (after [3]).

Figure 4.8 (a) Heterogeneity in oxidation intensity in a transitional oxidation zone (TZ); (b) Desiccation crack formation, which might lead to preferential flow paths for water and oxygen when interlayered and interconnected with other planar features.

4.3.2 Oxygen Diffusion

Oxidation in TSFs can occur in an aqueous or sub aqueous system where water content and oxygen flux will govern the rate of oxidation of minerals such as pyrite. As the phreatic surface above the saturated zone migrates downwards, the oxidation system will change from subaqueous to sub aerial, whereby the partial pressure of oxygen and vapor pressure are the main factors controlling oxidation processes.

Transportation of oxygen into the tailings deposit could occur in one of three mechanisms: (1) advective transport with water that contains oxygen, (2) free or forced air convection through the top layer of the tailings deposit, and (3) diffusion in the gaseous and aqueous phases through pores in the material [10]. Diffusion is the dominant mechanism for transporting atmospheric oxygen from the surface of the tailings to the depth where oxidation takes place [11, 12, 15, 22]. Mechanisms (1) and (2) could be applicable in operational TSFs and at the top thin layers of all TSFs, whereas the atmospheric oxygen transport in the subaerial and aerial system of most TSFs is dominantly by diffusion in the gaseous and aqueous phases.

Inside a TSF, the degree of water saturation, the rate of water infiltration, the rate of oxygen consumption and temperature govern variations in oxygen concentration. The shape of an oxygen profile is controlled by the rate and the way oxygen diffuses into the TSF and the rate at which the oxygen is consumed within the TSF. The evolution of oxygen profiles as oxidation process progresses is the function of the consumption of oxygen [14].

Field studies conducted on uncovered unsaturated tailings deposits in Canada, has shown that the pore gas oxygen concentration decreased from atmospheric concentrations (21 vol. %) to less than 5 vol % within the upper 60 cm of the TSFs [15]. This showed that oxygen gradients exist in unsaturated upper part of TSFs. At water-saturated conditions, i.e., below the ranges widely from 5.5 to 20 vol. % (phreatic surface), the oxygen concentration is limited to the solubility of oxygen in water (i.e., ≈ 21 vol. %). In general, however, the dissolved oxygen concentrations are very low in the saturated part of TSFs. Measurement of oxygen concentration in a TSF and establishment of oxygen profiles have become part of mine residue characterization studies. Earlier studies of TSFs have demonstrated that the consumption of pore-gas O_2 through *in situ* reactions results in decreasing O_2 concentrations with depth where transport is controlled by vertical diffusion of O_2 to the atmosphere [16].

In situ pore gas oxygen concentration with depth was measured at 1 m interval up to 10 m depth in the selected TSFs and the results are discussed in detail elsewhere [2, 4, 9]. The TSFs exhibit variable diffusion rates with an averaging oxygen concentration of 10.62 vol. % at 1 m depth and decreases substantially to an average concentration of 0.49 vol. % at 6 m depth and drops to 0.41 vol. % at 9 m depth. The oxygen diffusion depth ranges from about 2 to 5 m with an average of about 4 m (Figure 4.9), in agreement with the observed depth of oxidation front [2, 4, 9].

O_2 concentrations in all the TSFs show a negative correlation with porosity, permeability, and percentage of sand size grains and moisture content and a positive correlation with increasing silt percentages. High positive correlations exist between O_2 concentration and vegetation density. The apparent negative correlation with porosity, permeability, and moisture content may indicate that the dominant diffusion mechanism in the upper 3 to 4 m of these TSFs is not diffusion through pores in the gaseous and aqueous phases but rather by free or forced air convection through the top layer of the deposit following oxidation progression.

A decrease of O_2 concentration within the pore spaces of the tailings with increasing depth is consistent with findings of similar studies elsewhere [17–19]. The most rapid sulfide oxidation occurs shortly after tailings deposition ends, whereupon O_2 diffuses into

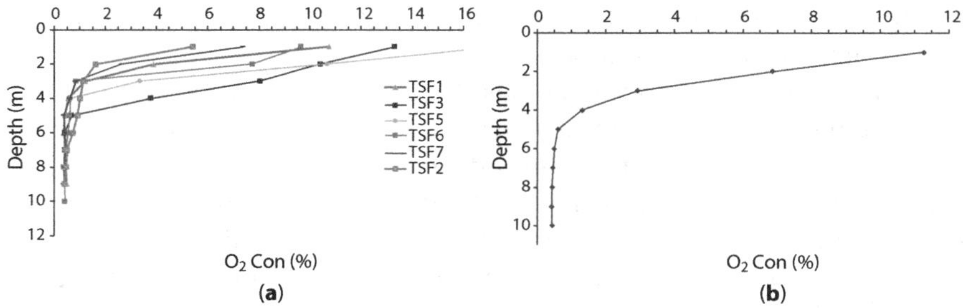

Figure 4.9 O_2 gas concentration profile in the Witwatersrand gold mine TSFs: (a) oxygen concentration with depth for all the TSFs; (b) average oxygen concentration with depth; Note the depth where all the curves converge which marks the active oxidation front.

the tailings and the bacterial population within the tailings becomes established [20]. As oxidation proceeds, more (di-)sulfide will be consumed, and oxygen migrates deeper into the tailings [4]. Blowes and Jambor [18] have shown that there is a positive correlation between pore-gas O_2 concentrations and oxidation of sulfides in the Waite Amulet (Quebec, Canada) tailings. In the shallow tailings, where sulfide minerals have been extensively depleted, gas-phase O_2 concentrations are high and decline sharply deeper in the tailings, as unaltered (di-)sulfide minerals become available. This is reflected in the Witwatersrand TSFs showing a positive strong correlation between depth of MOZ and O_2 gas concentration with depth (Figure 4.9). This comparison highlights the potential of O_2 concentration measurement with depth to decipher the depth of active oxidation in TSFs [2].

4.4 Geochemical and Mineralogical Evolution

4.4.1 Tailings Geochemistry and Mineralogy

Chemistry and mineralogy are the two main principal factors together with availability of oxygen and water that govern the rate and intensity of oxidation. Generally, the UZ shows high SiO_2 contents for all the TSFs, suggestive of the high quartz content of the mined ore. Enrichment of Al, Fe, Mg, Ca, in the TZ and depletion in the MOZ is evident. Ni, Co, Zn, Cu, and Cr show enrichment in the TZ at the expense of depletion in the MOZ. The concentration of trace metals such as Cr, Ni, Co, Cu, Zn, and Pb depends mainly on the relative concentration of sulfide minerals.

The characteristics principal minerals of the Witwatersrand gold mine tailings are quartz, mica, and chlorite/chloritoid with pyrophyllite and traces of K-feldspar present in some. In addition, 2 wt. % to <0.5 wt. % of pyrite and jarosite are common. The effect of oxidation in the mineralogy of the tailings is shown by the appearance of jarosite as a secondary mineral and increases whereas pyrite decreases and then disappears as oxidation process intensified. Pyrite then reappears in the UZ (Table 4.2).

The oxidation evolution of the Witwatersrand gold TSFs shows that: 1) oxidation is intense and reached "Late stage" in all the MOZ as evidenced by absence of sulfide minerals

Table 4.2 Mineralogical profile across the three oxidation zones of the investigated TSFs.

Zone	Pyrite	Jarosite	Gypsum	Chloritoid	K-Feldspar	Quartz	Mica	Chlorite	Pyrophyllite
TFS1-OZ	0.00	2	1		0	80	8	8	
TSF1-TZ	1.00	1	0		1	82	9	7	
TSF1-UZ	1.00	0	0		1	80	11	7	
TSF2-OZ	0.00	1		22	0	65	4	3	5
TSF2-TZ	0.25	0		16	0	74	4	4	3
TSF2-UZ	0.00	0		19	1	58	8	6	8
TSF3-OZ	0.00	1		13	1	72	6	6	–
TSF3-TZ	1.00	0		10	Trace	75	6	6	–
TSF3-UZ	0.00	1		22	Trace	61	7	8	–
TSF4-OZ	–	2		7	1	74	10	6	–
TSF4-TZ	1.00	1		10	1	67	10	10	–
TSF4-UZ	1.00	–		9	–	71	10	9	–
TSF5-OZ	–	2		2	–	64	8	3	22
TSF5-TZ	1.00	2		3	–	66	5	3	20
TSF5-UZ-A	2.00	1		2	–	76	3	3	14
TSF5-UZ-B	–	1		2	–	78	6	3	10

Table 4.3 Schematic representation of the progressive oxidation of a mine residue containing sulfides pyrite (py) and pyrrhotite (po) (after [24]).

Early (1)	Early (2)	Maturing	Late
Oxidation of Fe sulfides	Acceleration	Slowing	Consumed
Native sulfur	–	–	–
Marcasite	–	–	–
Fe oxyhydroxides	Fe oxyhydroxides	Geothite	Goethite
Fe sulfates	Fe^{2+} sulfates to Fe^{2+}, Fe^{3+}	Fe^{3+} sulfates	–
–	Jarosite	Jarosite	Jarosite
Po + Py	Py > Po	Py	–

and concomitant presence of jarosite, 2) the presence of both pyrite and jarosite in the TZ in all the TSFs show "Early-2 stage" to "Mature Stage" of oxidation in TZ, and 3) most of the UZ show a varying degree of oxidation dominantly "Early-1" and "Early-2" stages and rarely the "Maturing Stage" (Table 4.4) following the classification of Jambor [21] based on the paragenetic relationships in the oxidized mine residue deposits (Table 4.3).

4.4.2 Pore Water Geochemistry

The pH and paste pH of the pore water increase with depth. SO_4 and total metals concentrations, EC, and redox values are consistently high in the TZ, intermediate in the MOZ, and low in the UZ (Table 4.5 and Figure 4.10). This trend indicates pore water migration with soluble secondary minerals from MOZ into the TZ. The quality of the pore water depicts the mature stage of the oxidation process attained in the TSFs when considered in terms of the total oxidation (MOZ + TZ) thickness. This trend is distinct in terms of pH, sulfate, and total metals concentrations which are high in the oxidized zone (MOZ + TZ) and consistently low in UZ (Figure 4.11).

The low pH values (2.5–4.1) of the pore water are indicative of the final stage of oxidation whereby weathering of both sulfide and neutralizing minerals progressed beyond the carbonates and oxyhydroxides, having reached the aluminosilicates buffer range with Al^{3+} and Si^{4+} in the pore water. Zn, Ni, and Co become mobile when the pH decreases to below 6, while Al, Cr, V, and Pb remain immobile until the pH decreases to 4.0, consistent with results of laboratory investigation [13]. This sequence of mobility is governed by the pH plateaus and can be tied to the times when the specific acid-neutralization reactions predominate.

The pore water chemistry therefore mirrors the acid mine drainage that would have been discharged from these TSFs subjected to extensive "Late Stage" oxidation processes and provides useful information on the progress, depth, and rate of oxidation consistent with the mineral paragenesis discussed earlier.

Table 4.4 Progressive oxidation mapping based on the mineralogical classification in the three oxidation zones for the investigated TSFs.

Zone	Pyrite	Jarosite	Gypsum	Chloritoid	K-Feldspar	Quartz	Mica	Chlorite	Pyrophyllite	Oxidation stage
TFS1-OZ	0.00	2	1		0	80	8	8		Late
TSF1-TZ	1.00	1	0		1	82	9	7		Early (2)
TSF1-UZ	1.00	0	0		1	80	11	7		Early (1)
TSF2-OZ	0.00	1		22	0	65	4	3	5	Late
TSF2-TZ	0.25	0		16	0	74	4	4	3	Early (1)
TSF2-UZ	0.00	0		19	1	58	8	6	8	Early (1)
TSF3-OZ	0.00	1		13	1	72	6	6	–	Late
TSF3-TZ	1.00	0		10	Trace	75	6	6	–	Early (1)
TSF3-UZ	0.00	1		22	Trace	61	7	8	–	Early (2)–Maturing
TSF4-OZ	–	2		7	1	74	10	6	–	Late
TSF4-TZ	1.00	1		10	1	67	10	10	–	Early (2)–Maturing
TSF4-UZ	1.00	–		9	–	71	10	9	–	Early (1)
TSF5-OZ	–	2		2	–	64	8	3	22	Late
TSF5-TZ	1.00	2		3	–	66	5	3	20	Maturing
TSF5-UZ-A	2.00	1		2	–	76	3	3	14	Maturing
TSF5-UZ-B	–	1		2	–	78	6	3	10	Early (2)–Maturing

Table 4.5 Pore water chemistry of the three oxidation zones for the investigated TSFs, BDL: below detection limit.

Sample ID	pH	Paste pH	EC, mS/m	Redox, mV	SO_4	Al	Ca	Fe	Mg	Mn	ΣBase metals*
TSF1-OZ	3.28	3.6	314	218	2,934	233	265	157	101	4	22
TSF1-TZ	3.92	4.1	204	235	1,215	46	237	49	60	6	19
TSF1-UZ	4.11	6.4	115	217	584	2	193	30	24	5	11
TSF2-OZ	3.06	3.2	345	456	1,791	25	750	9	23	3	3
TSF2-TZ	2.56	2.8	483	606	2,347	50	643	226	26	30	24
TSF2-UZ-A	2.52	2.8	469	618	2,280	43	571	346	36	24	29
TSF2-UZ-B	6.16	7.0	190	324	1,084	6	400	BDL	23	3	1
TSF3-OZ	3.32	3.5	241	505	1,234	73	239	3	47	2	9
TSF3-TZ	2.52	2.9	269	522	945	50	15	119	3	2	16
TSF3-UZ	3.50	4.0	204	452	1,138	12	340	2	33	6	19
TSF4-OZ	3.31	3.5	323	475	1,770	125	188	3	112	6	20
TSF4-TZ	3.28	3.6	344	404	2,149	78	242	45	138	9	22
TSF4-UZ	3.25	3.8	204	546	887	18	174	4	43	5	19
TSF5-OZ	3.41	3.6	315	454	1,576	81	401	4	85	4	11
TSF5-TZ	3.22	4.1	405	531	2,114	101	564	4	71	7	45
TSF5-UZ	3.27	4.0	237	529	1,213	19	364	2	52	8	18

*ΣBase metals = Zn + Pb + Ni + Cu + Co.

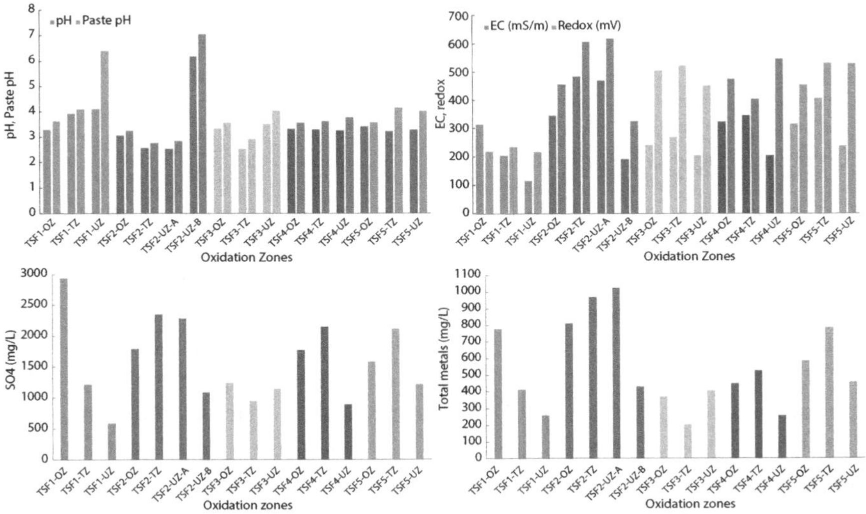

Figure 4.10 Pore water chemistry in the three oxidation zones.

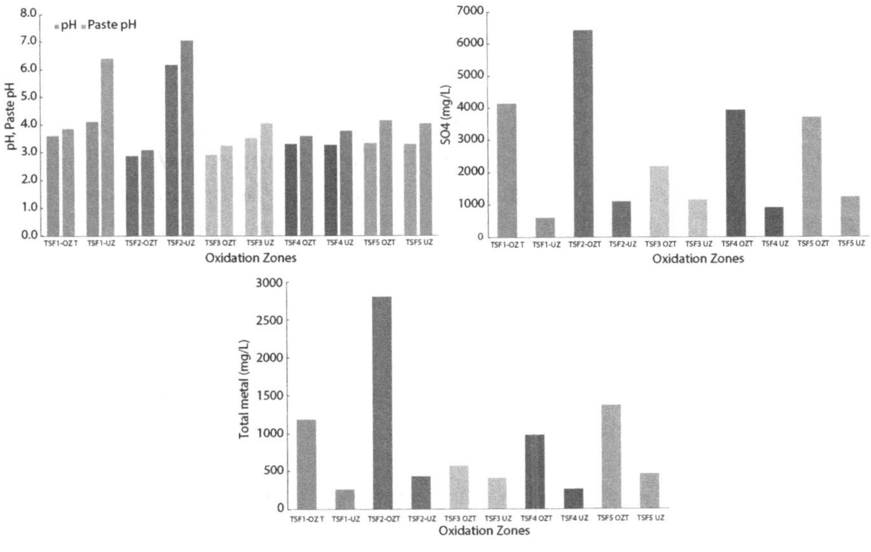

Figure 4.11 Comparison of pore water quality in the total oxidation zone (MOZ and TZ) and UZ.

4.5 Discussion, Conclusion, and Recommendations

4.5.1 Discussion

4.5.1.1 Mapping of the Oxidation Zones in Tailings Dams

Detailed site investigation and mapping of tailings profiles exposed by reclamation operations, logging of several auger-holes each up to 10-m deep in selected Witwatersrand TSFs provided classification of the TSFs into: MOZ, TZ, and UZ. The OZ and TZ constitute the total oxidizing zone and are separated from the UZ by the active oxidation front. The volume of oxidized mine residue deposits depends on the combined thickness of the MOZ and active oxidizing zone (transition zone), which ranges from 4.5 to 9 m.

4.5.1.2 Hydrogeological Situation

Tailings storage facilities are a variably unsaturated porous medium with an unsaturated upper part and a saturated lower part with phreatic surface marking the boundary between the two. The proportion of the saturated zone shrinks with time in non-operational and decommissioned TSFs. Therefore, unsaturated flow processes dominate in such TSFs where pore water flow is mainly vertical and governed by the interplay of hydraulic properties of the tailings. Hydraulic conductivity with depth in unsaturated upper part of a TSF is dependent on the volumetric water content of the tailings which varies with depth. Other factors that complicate unsaturated flow processes in TSFs are grain size, porosity, bulk densities, and mineralogy.

Laboratory and *in situ* investigations of the hydraulic properties provided insight on the relationship between hydraulic properties and oxidation processes. Although the tailings were dominantly uniformly graded sandy silt when deposited, detailed field observation and hydrometer analyses showed alternations of fine and coarse layers. This was attributed to the lateral grading of the tailings from coarse grained to fine grained from the edges to the center of the TSFs during deposition. Such depositional variations create capillary breaks, which complicates the flow of water in the tailings materials [3]. Tailings samples from the different oxidation zones show slight change in grain size distribution in that the TZ materials are coarser than those of the MOZ. The UZ materials are all sandy silt and finer than the overlying TZ and MOZ materials and reflects the original depositional particle size distribution. The consistent coarser grain size at TZ compared to the MOZ indicates reworking of the vertical grain size distribution due to oxidation and weathering.

The slimes and slurries of gold mines during deposition commonly have moisture contents approximately 1:1 [23], which after deposition naturally decrease in time and space due to various factors such as climatic variation (rainfall, temperature, and evaporation rate) as well as mineralogy, geochemistry, and time after closure and decommissioning. Measured moisture content ranges from 7 to 28 wt. % with an overall increase with depth. Water retention capacities and residual water content decreases with depth consistent with the general particle size distribution pattern, increasing bulk densities, and decreasing porosities with depth. An order of magnitude of variation in hydraulic conductivity values with depth is observed for most tailings, with higher K values recorded for the deeper parts. This is consistent with initial moisture content data, which also increases with depth.

The formation of fractures, cracks, desiccation cracks, and bedding planes provide preferential flow for water and oxygen movement for oxidation to occur in the deeper part of the UZ beyond the active oxidation front [2]. However, this oxidation process is probably not relevant compared to the normal oxidation processes.

4.5.1.3 Oxygen Diffusion With Depth

The TSFs exhibit variable oxygen concentrations with depth ranging from 10.62 vol. % at 1-m depth and decreasing substantially to an average concentration of 0.49 vol. % at 6-m depth and drops to 0.41 vol. % at 9-m depth. The optimum oxygen diffusion depth in the TSFs for oxidation to proceed varies from 2 to 5 m with an average depth of about 4 m, correlating well with the observed depth of active oxidation front.

4.5.1.4 Mineralogical and Geochemical Evolution of Tailings

Quartz, mica, and chlorite or chloritoid are the major minerals with rare pyrophyllite and K-feldspar presence as trace minerals. Pyrite is the major primary disulfide mineral with jarosite as the prevalent secondary sulfide mineral ranging from 2 wt. % to <0.5 wt. % in abundance. Pyrite generally increases and jarosite decreases with depth, consistent with oxidation progression. The oxidation evolution in the TSFs was established based on the mineral paragenetic relationships as follows:

- the MOZ in all TSFs shows complete consumption of sulfide minerals, and presence of jarosite indicating that the zone has reached "Late Stage" oxidation;
- the TZ shows the concomitant presence of pyrite and jarosite in all the TSFs supporting "Early advanced (Early-2)" to "Mature oxidation" stages;
- most of the UZ zones show varying degrees of oxidation ranging from "Early-1" to "Maturing Stage" of oxidation and none of the UZ show "Late Stage" oxidation wherein complete consumption of sulfides is expected.

There are evidences of geochemical reworking of the tailings due to the oxidation and leaching processes. Although the tailings are high in SiO_2 due to the high quartz content of the mined ore which is also reflected in the high SiO_2 content of the UZ, there is a slight increase in the proportion of SiO_2 in the MOZ compared to the TZ. Al, Fe, Mg, and Ca are enriched in the TZ and depleted in the MOZ. Ni, Co, Zn or Cu, and Cr are enriched in the TZ and are depleted in the MOZ.

4.5.1.5 Evolution of Pore Water Chemistry

The pore water chemistry in general shows a pH-increase with depth. SO_4, electrical conductivity and total metal concentrations are all consistently high in the TZ and low in UZ with intermediate values recorded in the MOZ. Due to the migration of secondary minerals in pore water from the MOZ into the TZ, there is a general increase of total metal concentrations with depth. Because the total oxidation zone is the summation of the MOZ and the TZ, the total sulfate and metal concentrations in the pore water indicate the potential drainage quality if discharged from these TSFs.

4.5.1.6 Oxidation Processes and Drainage Formation

The mineralogical and geochemical evolution of the tailings presented in the preceding sections show that oxidation has reached "Late-stage" and is active at "Maturing" and "Advanced (Early-2)" stages of oxidation. The presence and absence and the amount of sulfides and acid neutralizing minerals could be estimated indirectly by the pH of tailings material.

The pore water pH which is confined between 3 and 4 in the total oxidized zone (MOZ + TZ) supports the mineralogical evidence that the oxidation processes reached "Late-stage" and attained the aluminosilicates buffering stage as shown by the Al^{3+} and Si^{4+} concentrations in the pore water. When the pore water pH is below 6, the mobility of Zn, Ni, and Co increases, whereas Al, Cr, V, and Pb become mobile only after the pH decreases to 4. Such mobility patterns are associated with pH plateaus and related to periods of predominance of certain buffer reactions.

The leachate generated from a TSF which is represented herein by the pore water quality reflects the stage of oxidation that governs the acid generating and acid neutralizing reactions. The oxidation of sulfide minerals releases protons, SO_4^{2-}, Fe^{2+}, and metals. The amount of metals released is a function of the tailings mineralogy, sulfide mineral oxidation rates, and formation of secondary minerals by the removal of metals from the pore water.

In regard to neutralization reactions in the Witwatersrand tailings the contribution of carbonate minerals is very limited due to the minor concentration of carbonates in the original mineralogy of the tailings. No gibbsite or other Al bearing secondary phase were evident. This may be due to the difficulty in identifying gibbsite or other Al bearing precipitates [4]. The precipitation of Al bearing phases is, however, supported by other indications. The breakdown of Al bearing primary minerals in the tailings of the Witwatersrand basin is evidenced by the presence of Al^{3+} in the tailings pore water and by the relatively high Al_2O_3 in the bulk samples of the unoxidized tailings as compared to the oxidized zone tailings. The oxidation of Fe^{2+} to Fe^{3+} causes in most tailings the precipitation of ferric hydroxides, the most common of which is goethite [$Fe(OH)_3$]. In the absence of carbonates, the near constant pore water pH should only be attained due to the primary buffering capacity of Al and ferric hydroxides by releasing Al and Si into the pore water system. Blowes *et al.* [20] and Jambor and Blowes [24] argued that the dissolution kinetics of aluminosilicates varies but is generally slow compared to the groundwater flow rate.

4.5.2 Conclusions

All factors governing oxidation processes and drainage formation from the Witwatersrand TSFs have been investigated. The data collected from the TSFs have improved our understanding of mine drainage formation processes and the accuracy of drainage predictions for TSFs.

Oxidation processes and its variation with depth as well as the total oxidation thickness and the position of the active oxidation front are established. Modification of the grain size distribution takes place during oxidation and leaching processes. Although the principal mechanism of oxygen diffusion and moisture inflow into the tailings is progression of oxidation processes, the interplay of primary and secondary hydraulic properties (geological structures as preferred potential follow paths) improves our understanding of oxygen and water infiltration mechanisms and oxidation processes at the active oxidation front and deeper below the front. The measured rate of oxygen diffusion has shown to vary in the different TSFs from about 2 to 5 m in depth with an average of 4 m, consistent with the observed and measured oxidation fronts.

The mineralogical and geochemical evolution of the tailings' materials and the pore water chemistry has been established and is well understood. The effect of oxidation in the mineralogy of the tailings is evidenced by the appearance of jarosite as a secondary mineral and its increase as oxidation process is intensified whereas pyrite decreases and disappears in the AOZ but reappears in the UZ. Oxidation process reached "Late-stage" in in all the TSFs as evidenced by the mineral paragenesis (absence of sulfide and presence of jarosite) while still active at "Early advanced (Early-2)" to "Maturing" oxidation stages as evidenced by the concomitant presence of pyrite and jarosite.

The low pH values the pore water (2.5–4.1) and the total metal concentrations of the pore water mark the final stage of acid generation (aluminosilicate buffering). The various pH buffer plateaus (6, 4, and 2.5–3) of the pore water indicate the times when specific acid-neutralization reactions predominate which in turn determine the leaching sequence of metals. Zn, Ni, and Co leach into the pore water when the pH of the pore water decreases to below 6, whereas the decrease of pH to 4 starts leaching of Al, Cr, V, and Pb.

The extent of the acidity and the metal concentrations reflected in the pore water chemistry are by far lower than what would be expected if all the products of the oxidation and leaching processes remained in the TSFs as pore water.

4.5.3 Recommendations

Key findings that have practical application are as follows:

1) The depth of active oxidation zone varies between 1.5 and 5 m in depth with an average value of around 4 m. This suggests that, in the absence of more site-specific data, geochemical prediction assessments for tailings dams should consider the active oxidizing layer to be around 4-m thick. A further approximately 4 m of depth marks the active oxidation front;
2) Oxygen diffusion depth correlates very well with the actual observed active oxidation front and it is recommended that where a higher degree of confidence in the results of the prediction are required, oxygen diffusion measurement be undertaken to determine the depth of oxidation front;
3) It is expected that in 10–20 years, most TSF would have established an active oxidation zone of around 4m thickness. Variations in this depth are more likely to be linked to physical properties such as particle size distributions and chemical composition (e.g., available sulfides) than to TSF age after decommissioning.

Acknowledgements

The author thanks the Water Research Commission (WRC) of South Africa for funding the projects K5/1554 and K5/1460 and for permission to publish the research findings. He also thanks all the members of the research teams in these projects: William Pulles, Colbert Nengovhela, Simon Lorentz, and Bethania Maiyana. Permission to reuse parts of an earlier publication of the author [9] has been provided by the copyright holders (WRC and IMWA). The reviewers of this book are acknowledged for their review which improved the content of this chapter. Christian Wolkersdorfer is specially acknowledged for assisting in improving the quality of the figures over and above his editorial responsibilities.

References

1. Rosner, T., Boer, R., Reyneke, R., Aucamp, P., Vermaak, J., A preliminary assessment of pollution contained in the unsaturated and saturated zone beneath reclaimed gold mine residue deposits. *WRC Rep.*, 2001. *No* 797/1/01, p314.
2. Yibas, B., Pulles, W., Nengovhela, C., Kinetic Development of Oxidation Zones In Tailings Dams With Specific Reference To The Witwatersrand Gold Mine Tailings Dams. *Water Res. Commission Rep.*, 2010. No. 1554/1/10, p105.
3. Yibas, B., Pulles, W., Lorentz, S., Maiyana, B., Development of water balances for operational and post-closure situations for gold mine residue deposits to be used as input to pollution prediction studies for such facilities. *Water Res. Commission Rep.*, 2011. No. 1460/1/11, p104.
4. Nengovhela, A.C., Yibas, B., Ogola, J.S., An investigation into the availability of oxygen in gold tailings dams of the Witwatersrand basin with reference to their acid mine drainage potential. *Water SA*, 33, 2, 271–274, 2007.
5. Nengovhela, A.C., Yibas, B., Ogola, J.S., Characterisation of gold tailings dams of the Witwatersrand Basin with reference to their acid mine drainage potential, Johannesburg, South Africa. *Water SA*, 32, 4, 409–506, 2006.
6. Sibanda, L.K. and Broadhurst, J.L., Exploring an alternative approach to mine waste management in the South African gold sector, in: *11th ICARD/IMWA conference proceedings*, Wolkersdorfer, Ch., Sartz, L., Weber, A., Burgess, J., Tremblay, G. (Eds.), pp. 1130–1135, 2018.
7. Nicholson, R.V., Gillham, R.W., John, A.C., Reardon, E.J., Reduction of acid generation in mine tailings through the use of moisture-retaining cover layers as oxygen. *Can. Geotech. J.*, 26, 1, 1–8, 1989. https://doi.org/10.1139/t89-001.
8. Wilson, G.W., Williams, D.J., Rykaart, E.M., The Integrity of Cover Systems – An Update, in: *Proceedings of the 6th International Conference for Acid Rock Drainage,* T. Farrell and G Taylor (Eds), pp. 445–451, Australian Institute of Mining & Metallurgy, Queensland, Australia, 2003.
9. Yibas, B., Pulles, W., Lorentz, S., Maiyana, B., Nengovhela, C., Oxidation process and hydrology of tailings dams: Implication for acid mine drainage from TSFs management – The Witwatersrand experience, South Africa, in: *Proc. Intern. Mine Water Assoc. Annual Conf.*, McCullough, C.D., Lund, M.A., Wyse, L. (Eds.), pp. 245–255, 2012.
10. Werner, K., Soil-cover remediation of mill tailings deposits: Effects on oxygen transport and hydrological conditions, in: *Licentiate Thesis*, p. 111, Royal Institute of Technology, Stockholm, 2000.
11. Jaynes, D.B., Pionke, H.B., Rogowski, A.S., Acid mine drainage from reclaimed coal strip mines, 2. Simulation results of model. *Water Resour. Res.*, 20, 243–250, 1984.
12. Pantelis, G. and Ritchie, A.I.M., Macroscopic transport mechanisms as a rate-limiting factor in dump leaching of pyretic ores. *Appl. Mathem. Model.*, 15, 3, 136–143, 1991.
13. Yanful, E.K., Oxygen diffusion through soil covers on sulphidic mine tailings. *J. Geotech. Eng.*, 119, 1207–1228, 1993.
14. Ritchie, A.I.K., Sulfide oxidation mechanism-Control and rates of oxygen transport, in: *The Environmental Geochemistry of sulfide Mine-Wastes*, vol. 22, J.L. Jambor and D.W. Blowes (Eds.), pp. 210–245, Mineral. Assoc. Canada, Short Course Ser, Quebec, Qc, Canada, 1994.
15. Elberling, B., Nicholson, R.V., David, D.J., Field evaluation of sulphide oxidation rates. *Nordic Hydrol.*, 24, 323–338, 1993.
16. Shaw, S.A., Determining the long-term persistence of mercury releases to the environment from cyanide-rich gold mine tailings, in: *Unpublished MSc thesis*, The University of New Brunswick, Canada, 2004.

17. Smyth, D.J.A., Hydrogeological and geochemical studies above the water table in an inactive uranium tailings impoundment near Elliot Lake, Ontario, in: *MSc thesis*, University of Waterloo, Waterloo, 1981.
18. Blowes, D.W. and Jambor, J.L., The pore water geochemistry and mineralogy of the vadose zone of sulphide tailings, Waite Amulet, Quebec, Canada. *Appl. Geochem*, 5, 327–346, 1990.
19. Blowes, D.W., Reardon, E.J., Jambor, J.L., Cherry, J.A., The formation and potential importance of cemented layers in inactive sulphide mine tailings. *Geochim. Cosmochim. Acta*, 55, 965–978, 1991.
20. Blowes, D.W., Ptacek, C.J., Jurovec, J., Mill Tailings: Hydrogeology and geochemistry, in: *Environmental Aspects of Mine Wastes*, vol. 31, J.L. Jambor, D.W. Blowes, A.I.M. Ritchie (Eds.), pp. 95–116, Mineral. Assoc. Can. Short Course Ser, Ottawa, Ontario, Canada, 2003.
21. Jambor, J.L., Mine waste mineralogy and mineralogical perspective of Acid-Base Accounting, in: *The Environmental Geochemistry of Sulphide Mine-Wastes*, vol. 31, J.L. Jambor and D.W. Blowes (Eds.), pp. 117–147, Mineral. Assoc. Can. Short Course Ser, Ontario, Canada, 2003.
22. Jurjovec, J., Ptacek, C.J., Blowes, D.W., Acid neutralisation mechanisms and metal release in mine tailings: A laboratory column experiment. *Geochim. Cosmochim. Acta*, 66, 1511–1523, 2002.
23. Rosner, T. and van Schalkwyk, A., The environmental impact of gold mine tailings footprints in the Johannesburg region, South Africa. *Bull Eng. Env.*, 59, 137–148, 2000.
24. Jambor, J.L. and Blowes, D.W., Theory and applications of mineralogy in environmental studies of sulphide-bearing mine tailings, in: *Modern Approaches to Ore and Environmental Mineralogy*, vol. 27, L.J. L.J Cabri and D.J. Vaughan (Eds.), pp. 367–401, Mineral Assoc. Can. Short Course Ser, Ottawa, Canada, 1998.

Part 2
AMD TREATMENT

5

Technologies that can be Used for the Treatment of Wastewater and Brine for the Recovery of Drinking Water and Saleable Products

Tumelo Monty Mogashane[1,2], Johannes Philippus Maree[1,3*], Munyaradzi Mujuru[1] and Mabel Mamasegare Mphahlele-Makgwane[1]

[1]*Department of Water and Sanitation, University of Limpopo, Private Bag X1106, Sovenga, South Africa*
[2]*Department of Chemistry, University of Limpopo, Private Bag X1106, Sovenga, South Africa*
[3]*ROC Water Technologies, Die Wilgers, Pretoria, South Africa*

Abstract

South Africa like many other countries has a water shortage. It is feared that the demand for water is already equal to the supply, and yet daily millions of liters of wastewater and brines are generated in various processes. The wastewater and brines can be a source of potable water and other valuable products like pigments and important salts. This chapter reviews technologies that have been used to treat various streams of wastewater such as neutralization technologies used in treating acidic mine water to new and emerging technologies which are now being commercialized to recover potable water and valuable products from various types of wastewater. The emerging technologies include the reverse osmosis (RO)/cooling process which uses RO and cooling techniques to treat brines and other wastewaters to recover drinking water and pigments. In the review, each of the technologies in the spectrum is reviewed with a discussion of the equipment and methods used in the process. The advantages and disadvantages of each technology are described. Also included in the discussion are case studies of where the technology has been applied. Concluding with a review of emerging water treatment technologies, emphatically illustrates that wastewater streams should be viewed as opportunities to recover saleable products and not as a threat to the environment.

Remark

This chapter is largely based on a report to South Africa's Water Research Commission (Dama-Fakir, P., Sithole, Z., van Niekerk, A. M., Dateling, J., Maree, J. P., Rukuni, T., Mtombeni, T., Ruto, S., Zikalala, N., Hughes, C., Wurster, A., Saunders, B., *Mine water treatment technology selection tool: Users' Guide,* Johannesburg: Water Research Commission, 2017). Permission has been granted by

**Corresponding author*: johannes.maree@ul.ac.za [ORCID: 0000-0002-8901-2512]
[ORCID: Mogashane, 0000-0002-9096-7877; Mujuru, 0000-0001-8833-1792; Mphahlele-Makgwane, 0000-0003-2585-788X]

Elvis Fosso-Kankeu, Christian Wolkersdorfer and Jo Burgess (eds.) Recovery of Byproducts from Acid Mine Drainage Treatment, (99–156)

the WRC on 2020-02-03 to use the text and images in this book chapter. It is also based on earlier publications by J. Maree, listed in the references.

Keywords: Acid mine drainage, neutralization, desalination, brine treatment, sludge processing

5.1 Introduction

5.1.1 Formation of Acid Mine Water

South Africa is known as a water scarce country. Various studies already indicate that the demand for water is already equal to the supply. These limited water resources are further polluted with mine water and poorly treated sewage. Mine water pollution is widely debated in the media locally and globally and mine water increases the mineralization of surface and groundwaters in some areas. It poses a health risk where water users are exposed to high metal concentrations over time. The environment becomes more polluted despite the fact that (i) South Africa has extensive legislation in place, (ii) large investments are made in Research and Development and construction of full-scale treatment plants, (iii) Environmental Impact Assessment studies are carried out, and (iv) despite mine water amounting to only 3.4% of the surface water.

Mine water typically contains the following components: acid, iron(II), several metals, and dissolved salts [27]. Acid mine drainage (AMD) formation is caused by the contribution of the following events: (i) oxidation of pyrite in the ore strata aided by bacterial action resulting from dissolved oxygen in the ingress water contacting exposed, pyrite containing rocks within the mining strata and producing acid, Fe(II), sulfate, and other dissolved salts; (ii) partial neutralization of acid because of natural alkalinity occurring in the mined and broken rock environments; and (iii) repeated contact of pyrite-rich rock with water and oxygen as the water level fluctuates on account of pumping at a constant rate and water being recharged in proportion to seasonal rainfall [27]. Depending on the amounts of exposed natural alkali present in the ore strata that contribute to full or partial neutralization, mine water can be classified into three groups: (i) acid mine water (pH < 5.6), largely rich in H^+ and Fe^{2+}; (ii) Ca^{2+}/Mg^{2+} rich mine water, where calcite or dolomite in the ore strata is sufficient to neutralize the protons; and (iii) Na^+-rich mine water. Reaction 5.1 shows the reaction for pyrite oxidation.

$$4\,FeS_2 + 15\,O_2 + 14\,H_2O \rightarrow 4\,Fe(OH)_3 + 8\,H_2SO_4 \tag{5.1}$$

5.1.2 Water Volumes

The users who are downstream of the pollution source are commonly affected by compromised water quality. For Mpumalanga, it was calculated that 44,000 m^3/d of mine water was discharged into the Upper Olifants River Catchment in 2000 [1]. It was estimated that it would increase to 131,000 m^3/d by 2020. The volume of AMD discharged in Gauteng amounts to 350,000 m^3/d. This is only 3.4% of the total volume of river water in the region (Table 5.1). The percentage will even be less if mine water is allowed to raise to higher

Table 5.1 Volume of mine water produced in Gauteng and Mpumalanga [2]. Flow in m^3/d.

River	Basin	Mine water	Mine water (sum)	River water	Mine water/river water (%)
Vaal	Central	82	242	10,800	2.24
	Eastern	110			
	Far Western	150			
Crocodile	Western	40	40	650	6.15
Olifants	Mpumalanga	200	200	2,728	7.33
Total			282	14,178	3.40

water levels before it is pumped to surface. The individual figures for mine water that is discharged into the two main rivers, the Vaal and Crocodile Rivers, amounts to 3.0% and 4.6%, respectively. Thus, since mine water amounts to only 3.4% of the surface water, a cost comparison is needed for desalination of the mine water *versus* complete prevention of mine water discharge into rivers. Methods such as forced evaporation and irrigation or one of these should be considered in combination with treatment of the residual leachate. In the controlled release of mine water, better quality water can be used to dilute the increased salinity of river water following mine water discharge. Up to seven times more good quality water, compared to the mine water, is used for dilution. It may be more cost-effective to keep surface water free from the mineralized mine water through forced evaporation or irrigation plus treatment of the residual water or leachate than to desalinate mine water.

In Mpumalanga, three mine water desalination plants are in operation [2]. The plant in eMalahleni was designed for treating 25,000 m^3/d of mine water and has been expanded to 50,000 m^3/d. The capital cost of the first phase amounted to R 400 million and the second phase to R 500 million with a water recovery of 99%. A second desalination plant was constructed at Optimum Colliery with a capacity of 15,000 m^3/d at a cost of R 800 million. Pilot studies in place are focused on treatment of the brine that is generated. A third plant was commissioned in 2015 at Middelburg Colliery with a capacity of 25,000 m^3/d. At a running cost of R 15/m^3, the annual running cost amounts R 493 million. A large portion of the 75,000 m^3/d treated mine water that is treated at eMalahleni and Optimum is sold as drinking water, and a smaller portion is used by the mine. The balance is discharged into the Olifants River.

5.1.3 Legislation

Mey and Van Niekerk have summarized the mining-related legislation in South Africa [3]. The Minerals Act (Act No. 50 of 1991) has specific stipulations for environmental management of mines as per the Environmental Management Program Reports (EMPR). In 2002, the Minerals and Petroleum Resources Development Act No. 28 (MPRDA) was introduced, which requires a comprehensive Environmental Management Plan (EMP) that includes public participation and financial allowances for post-closure environmental and water liabilities. In addition, the National Water Act (Act 36 of 1998) (NWA) makes provision for issuing of licenses by the regulator for the different water uses. Finally, the National Environmental Management Act (Act No 8 of 2004) (NEMA) stipulates the activities requiring a full Environmental Impact Assessment (EIA). The National Environmental Management Waste Act (Act 59 of 2008) addresses the fragmentation in waste legislation. Although the legislation in place is of world standard, certain adjustments need to be made for it to be more practical. EIA studies and the approval required for the construction of water treatment plants are time consuming. Each application needs to be considered by the authorities that lack manpower resources and experience. Delegation of responsibilities to the industry could result in low-cost solutions that can be implemented by each party.

5.1.4 Government Initiatives

The *Expert Team* of the *Inter-Ministerial Committee on Acid Mine Drainage* who, in 2010, investigated mine water issues, proposed, among other actions, that the AMD associated with mining on the Witwatersrand be managed as follows: (i) pumping facilities should be installed in each of the Gauteng mining basins to keep the rising water levels below the Environmental Critical Level; (ii) neutralize excess mine water in the short term; and (iii) desalinate excess mine water in the medium to long term [4].

The Trans-Caledon Tunnel Authority (TCTA) has been appointed by the Department of Water Affairs and Sanitation (DWAS) to implement neutralization and manage the processes needed for the neutralization of the water in the Western, Central, and Eastern Basins [5]. Treatment using limestone for neutralization of acid, followed by lime addition for the removal of iron(II) as $Fe(OH)_3$ and other metals, has been carried out. Detailed studies of the waters from the Witwatersrand mining basins have predicted that the costs of installing AMD neutralization plants in the three basins will be around R 924 million [6], based on the most cost-effective solution [7]. Using proven technology, the High-Density Sludge (HDS) process was recommended for neutralization and reverse osmosis (RO) for desalination. No attention was given to disposal of resulting brine and sludge. For treatment of 212,000 m^3/d of mine water in South Africa (Western Basin: 30,000 m^3/d; Central Basin: 72,000 m^3/d; Eastern Basin: 110,000 m^3/d), the capital cost will amount to R 10 billion and the annual running costs to R 2.7 billion [8]. The environment will benefit very little if only e.g., 50% of the mine water is desalinated. All mine water has to be desalinated to benefit the environment in the long term. It is therefore important that the most cost-effective technology or method be identified that will prevent mineralization of surface water as a result of the discharge of mine water.

Several locally developed and imported treatment technologies are available. The most appropriate technology will be project specific. Various factors, including life cycle costs, incoming water quality, waste generation and management, environmental aspects, project associated risks, regulatory aspects, interested and affected party buy-in among others will influence the selection of the most appropriate technology. All the precipitation treatment processes rely on chemical precipitation through the production of sparingly soluble salts to remove unwanted components from the mine water. The precipitation treatment processes are generally only effective at removing multivalent ions from water and elevated concentrations of sodium or chloride cannot be removed.

5.1.5 Required Criteria

The ideal treatment configuration for mine water treatment should meet the following criteria:

- Water quality: Water of drinking water quality needs to be produced. Even when treated water is discharged into public streams, legislations require that the Total Dissolved Solids (TDS) need to be less than 750 mg/L.
- Zero waste: Tenders from the mining industry requires that brine and sludge need to be removed from the mining site. New legislation requires that liquid that is disposed at waste disposal sites should have a maximum water content of 40%.
- Easy approval of EIA studies: If the two above criteria are met, the EIA approval process is much easier than when not met. If these criteria are not met, the EIA process can take more than a year.
- Minimum treatment cost: The mining industry deals with high production cost. Treatment cost needs to be minimized by selecting a technology with lowest cost, if possible, it should generate income to offset the cost completely or partially.
- Water utilization: Treated mine water should be utilized to justify the cost of treatment, rather than to discharge purified water into public streams with contaminated water.
- Job creation: South Africa has a high unemployment rate of 30%. Job losses in the mining industry have contributed to this. The mining industry has access to large areas of land, 400,000 m^3/d of excess water and labor. By using purified water for crop production, the need or job creation and food production will be met.
- It would be preferred that local labor and local technologies are used.
- If saleable products can be produced, this reduces costs and can address chemical imports/exports trade deficits.

5.2 Neutralization Technologies

5.2.1 Neutralization Using Lime

5.2.1.1 Conventional Treatment With Lime

Lime is used to treat AMD for the neutralization of acid and subsequent metal removal. Gypsum is formed in the process when the sulfate concentrations exceed the gypsum solubility product. Due to the low solubility of gypsum, partial sulfate removal is achieved between 1,500 and 2,500 mg/L. The solubility of the gypsum, apart from temperature, depends on the composition and ionic strength of the solution [9].

The low-density sludge (LDS) process (Figure 5.1) that allows for mixing of AMD and lime and has provision for functions such as mixing, aeration, and removal of solids, but the settling rate of the solids is slow, the sludge is bulky and problematic to dispose of [10].

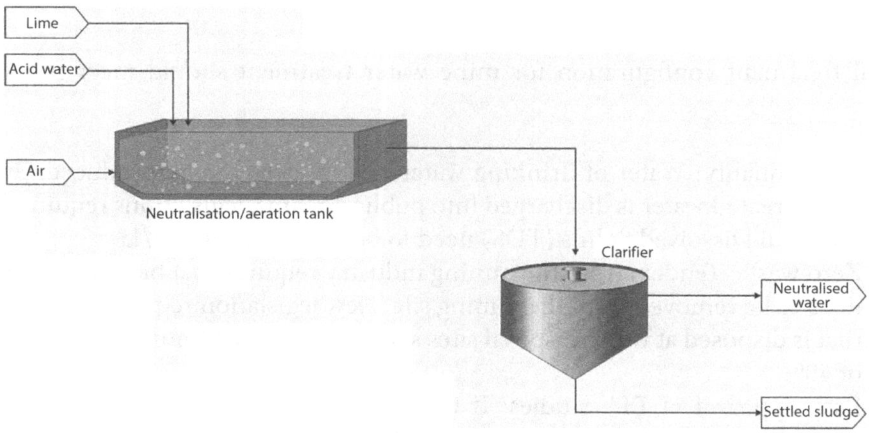

Figure 5.1 Low-density sludge (LDS) treatment process for acid water neutralization [9].

5.2.1.2 High-Density Sludge Process

The recycled sludge or high-density sludge (HDS) process was developed by the Bethlehem Steel Corporation (USA) and published in 1970 [11]. In this process (Figure 5.2), a part of the settled sludge is initially mixed with the AMD being treated, followed by the addition of lime. Due to the "seeding" effect, a denser sludge is produced [10] and the HDS process consists of the following stages [9].

- pH correction/sludge conditioning;
- Neutralization/aeration, and
- Solid/liquid separation.

The pH correction phase is done in a reaction tank by lime addition and a sludge conditioning tank where the recycled, settled sludge coming from the settling tank underflow

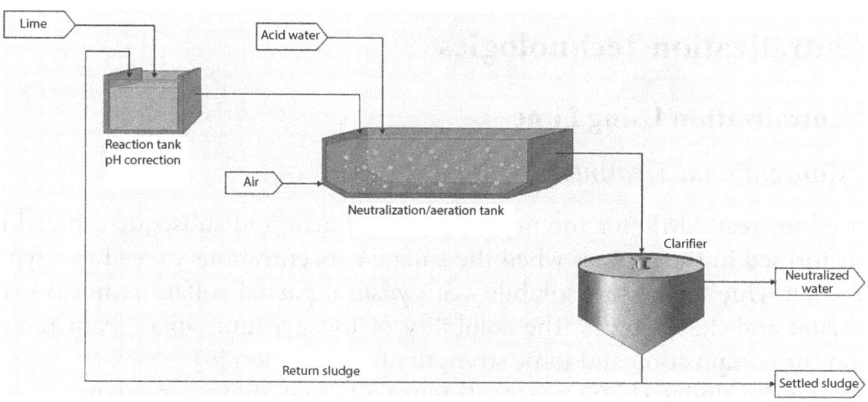

Figure 5.2 The high-density sludge neutralization process for AMD treatment [9, 11, 12].

and the lime solution addition flow. Addition of lime in the pH correction stage results in the final treated water having a pH of around 8. The conditioned sludge from the pH correction stage is allowed to flow into the neutralization and aeration tank, which also serves as a mixer that keeps the solids in suspension, mixing the conditioned sludge with the mine water entering the tank. It also serves to aerate the tank contents resulting in the oxidation of ferrous iron. After the neutralization and oxidation, the tank effluent overflows into the clarifier where sludge can settle and separated from the liquid. A poly-electrolyte solution can also be added to the clarifier to initiate and promote flocculation. The oxidation stage of the HDS process also serves to oxidize metals, including oxidation of Fe(II) to Fe(III) at around pH 7.2, Mn (II) to Mn(IV) at a pH > 9 or precipitate others such as manganese hydroxide (pH > 11) or aluminium hydroxide (pH 6.5) [13].

Precipitation of metals in the HDS process results in an excess of alkalinity (50–100 mg/L as $CaCO_3$), especially when there are high lime dosages. A gelatinous sludge is formed, which does not settle easily, even when polyelectrolyte polymers are added to aid precipitation. To produce large sludge particles, recirculation of a portion of the settled sludge is carried out [11] and this leads to improved sludge settling [14].

The HDS process has the following advantages over the conventional LDS neutralization process [11, 12]:

- Sludge, which is denser, in fact up to 10 times higher than that obtained in the LDS process is generated and consequently less demanding sludge drying operations are required.
- Because the sludge settles faster, a smaller clarifier can be used leading to a cost saving on the clarifier of approximately 38% [9].

5.2.2 Limestone Neutralization

Gunther *et al.* reported on the first full-scale implementation of the integrated limestone neutralization in South Africa [15] at the Landau Colliery, Anglo Coal in 2001, where lime neutralization with the HDS process was replaced with limestone with the aim to save on alkali cost [15]. Limestone was previously not considered as an effective alkali due to its limited solubility and armouring rock limestone particles, which resulted in poor utilization of the limestone [16].

In the limestone neutralization and iron(II) oxidation process, powdered limestone is used for neutralization [17], which is achieved by contacting the powdered $CaCO_3$, a high sludge addition and oxygen at pH 5.5 or higher. By this procedure, precipitation of, e.g., Fe^{3+} and Al^{3+}, and gypsum occur in one reactor. Additionally, in a processing facility, limestone slurry can be added to neutralize acid leachate from unwashed coal. In the Landau Colliery, lime was not used. Instead, powdered limestone was employed, that lowered the neutralization cost by 55% to R 4.5 million/a [15]. Mass balance modeling was used to optimize the treatment process. Other aspects that were investigated included: (i) a novel, robust limestone handling and dosing system was devised where powdered, waste limestone from the paper industry was slurried to a constant density, and (ii) a load cell–based density meter was used to control the limestone slurry density [17].

5.2.3 Limestone Handling and Dosing System

Landau Colliery uses two limestone make-up and dosing units (Figure 5.3). The first is at the Navigation Plant that contributes up to 50 t/d. The second is at the Kromdraai open-cast mine that supplies up to 30 t/d. The units consist of inclined concrete slabs, slurry tanks, and recycle pumps. Spray nozzles, load cell density control systems, and pumps make up the lime slurry transfer system (Figure 5.4). The flat concrete slab at an angle of 3° was replaced with a V-shape concrete structure. This offers the benefit that limestone only needs to be slurried at the bottom of the V and the limestone higher up in the dump will fall to the bottom of the V as limestone is slurried away [15].

The control philosophy is as follows: Limestone is loaded onto the inclined V-concrete slab and is slurried (5–15% solids) by a waterjet, the direction of which is determined by the density of the slurry which is monitored continuously using a density meter. When the slurry reaches a density less than a set value, the water-jet is directed onto the limestone-dump.

Figure 5.3 Limestone handling and dosing system at Navigation Section of Landau Colliery.

Figure 5.4 New V-shape limestone storage system at Sibanye Gold Mine.

Additionally, when the density of the slurry is at a further set value, the waterjet is directed onto the lower part of the slab where the slurry is allowed to enter the limestone-slurry make-up tank without slurrying additional limestone. An automatic valve, linked to a level, controls the level of lime slurry in the make-up tank. A stone trap between the slab and the slurry make-up tank is used to remove any stone and grit in the powdered $CaCO_3$. Limestone slurry from the make-up tank is transferred *via* pumps to the various points of application [15, 18].

The in-line density meter controls the density by using a load cell to measure the mass of a specific volume of slurry (side stream off the recycle line) that flows through the pipe. A pneumatically controlled spray system has a density meter, which, as required, directs the nozzles to spray onto the limestone. The load unit is equipped with a digital readout displaying the mass of the density cell that is able to be calibrated. It can be zeroed when the cell is empty and can be calibrated at a density of 1 kg/L when filled with clear water. During normal operation, it provides readouts of the density of the actual slurry flowing through the cell. The density meter is pre-set to activate or stop a control-unit at pre-determined high or low values [15].

Initially, limestone was dosed to replace lime, in June 2001, at the Navigation Plant. Thereafter, it was also implemented during June 2002 in the coal processing plant at Navigation and at the Kromdraai Liming facility, during January 2003, to replace lime completely [15]. The chief aim of the Navigation Liming Plant is to supply neutralized, make-up water to the coal processing facility. The feed water to the liming facility has a pH of ≈3, an Fe(II) concentration of about 50 mg/L as Fe^{2+}, an acidity of approximately 380 mg/L (as $CaCO_3$), and a flow rate of around 5,000 m^3/d. The average acid load amounted to ≈1.9 t/d (as $CaCO_3$). Over the 12-month trial period, the average limestone usage mounted to 2.2 t/d. The over dosage amounted to a low 0.3 t/d.

5.2.4 Utilization of Alkali in Mine Water for Removal of Iron(II)

Mine water with a low Fe^{2+} concentrations is normally treated with lime and aeration for Fe^{2+} removal, resulting in high sludge volumes due to $CaCO_3$ and ferric hydroxide precipitation. Mohajane showed that hydrogen peroxide can be used for removal of Fe^{2+} from alkaline mine water [19, 20]. It was found that by dosing hydrogen peroxide in a mole ratio of 1:1, Fe^{2+}-concentrations decreased from 120 mg/L to 0.1 mg/L, total acidity from 200 mg/L to zero, alkalinity from 300 mg/L to 50 mg/L, and the pH was raised from 5.6 to 7.6 by stripping. The drop in acidity was due to Fe^{2+} oxidation and precipitation and the drop in alkalinity due to $CaCO_3$ precipitation. The benefits associated with hydrogen peroxide treatment compared to the conventional lime/aeration treatment include less sludge production and lower chemical costs.

5.2.5 Modeling

After the limestone neutralization process was successfully implemented by Anglo Coal at Navigation Section of Landau Colliery, gypsum scaling in the coal washing plant was still a problem due to high sulfate concentrations of up 11,500 mg/L in selected streams prior to neutralization. Modeling was used to predict the effects on gypsum scaling of various treatment strategies [21].

The toe-seep water, with an acidity of 12,000 mg/L (as $CaCO_3$) and 12,500 mg/L sulfate, was found to be the next water to focus on as part of the overall water management strategy. A modeling procedure was employed to seek to estimate the following: (i) the degree of gypsum precipitation taking place within the prevailing systems in the Navigation liming and in the coal processing units; (ii) the effect of gypsum precipitation on the gypsum saturation index following separate and joint treatment of the waste coal dump toe-seep wastewater and the less polluted water-streams treated in the current Navigation liming unit; (iii) the effect of gypsum precipitation on the properties of the effluent from the coal processing facility, when a side-stream of the main stream from the lime-thickener to the coal processing unit, is treated; (iv) the additional sulfate concentration lowering required to ensure that the water in the coal processing facility is not over-saturated with gypsum, (v) the amount of sulfate that would precipitate by pre-washing the acid coal and (vi) the capital and running costs associated with different treatment alternatives [15].

A mass-balance model was developed with respect to water-flow and chemical species, and making provision for the following stages of the water network system which included: (i) the existing Navigation Liming Plant, (ii) the new gypsum precipitation facility located subsequent to the existing Navigation Liming Plant, (iii) the new toe-seep treatment facility that, includes or excludes gypsum precipitation, (iv) the biological sulfate removal plant, (v) the coal processing plant, (vi) the sludge disposal site, and (vii) the dumping site for discarded fine coal and coal discard [15].

The parameters into the modeling software included the following: (i) five flow-rates each from the liming unit, toe-seep to the discard leachate neutralization plant, from the sulfate removal unit, from the thickener underflow to remove dump and penstock return water; (ii) the chemical compositions of three feed waters including that to the liming unit, to the toe-seep unit and penstock; (iii) the composition of the treated water following sulfate removal; (iv) the amount of sulfate removal through gypsum precipitation where this occurs, i.e., at the discard dump and the coal processing unit and precipitation treatment facilities; and (v) the amounts of alkali used in the liming plant and the coal processing facility [15]. The output parameters were as follows: flow rates, chemical compositions, and gypsum saturation of all other streams, and the estimated capital and operating costs.

The model depended on the following parameters: (i) the steady-state equilibria at each point; (ii) ionic balance, i.e., the mole equivalents of the cation concentrations, acidity, iron(II), iron(III), calcium, and magnesium concentrations being the same as those of the anion present, i.e., sulfate; and (iii) the over saturation index (OSI = $[SO_4]_{solution}/[SO_4]_{equilibrium}$) where [15]:

$$[SO_4]_{equilibrium} = 1{,}500/48 + [Mg^{2+}] \text{ (determined empirically)} \tag{5.2}$$

The process flow-diagram used for the spreadsheet-based model included the following: Feed-waters and various plants/sites, i.e., limestone neutralization reactor, coal washing plant, coal discard dump, and desalination plant. The following treatment alternatives were evaluated: (i) the current situation, (ii) joint treatment of toe-seep coal mining water and less contaminated streams, (iii) separate treatment of toe-seep water and less contaminated streams, (iv) gypsum precipitation in water from the coal processing facility, (v) tertiary sulfate removal, and (vi) pre-washing water [15].

It was determined that gypsum precipitation in the liming and coal processing facilities amounted to 30% and 60%, respectively. During separate treatments of toe-seep, coal discard leachate water and the liming unit, the capital cost for a neutralization/gypsum precipitation unit was estimated to be R 3.0 million, relative to R 10.3 million when combined treatment is applied. Only slightly less gypsum precipitation takes place during separate treatment, i.e., 8.9 t/d against 9.5 t/d. Gypsum crystallization from the wastewater in the coal processing facility is a less effective way for controlling the SI that can be controlled at OSI = 1 by treating the feed wastewater to the coal processing unit for sulfate precipitation and subsequent removal. A flow rate of about 222 m^3/h requires treatment for precipitation and removal of sulfate to around 350 mg/L to yield an OSI of 0.98 (which is <1). The capital cost of a 222 m^3/h biological sulfate removal unit was estimated at about R 21.8 million and the running cost at R 4.10/m^3. Pre-washing of the coal will lower costs [15].

5.2.6 Lime/Limestone Neutralization

5.2.6.1 Description of the Process

In the integrated limestone and iron(II)-oxidation process, concomitant oxidation of iron(II) occurs when limestone is used for neutralization [22, 23] (Figure 5.5). Powdered limestone is used in this process to promote iron(II)-oxidation at a pH of about 5.5, neutralization of acid in the mine water, metal precipitation (e.g., Fe^{3+} and Al^{3+}), and gypsum precipitation in one reactor. The novelty of this system consists in the conditions that were identified where associated iron(II) oxidation is promoted at a pH of about 5.5, by the dosage of powdered limestone in the presence of air. Limestone is a comparably cheap alkali and is used for neutralization of the bulk of the acid, producing CO_2 which is stripped off by aeration and transported to the third stage. Lime is then used in the second stage for the precipitation of magnesium and other metals as their sulfates. The solubility product of gypsum governs the degree to which sulfate precipitates from solution. The CO_2 produced in the first phase leads to $CaCO_3$ precipitation in the third phase. This precipitation occurs at pH 8.3, and the $CaCO_3$ produced has high purity and can be sold as a by-product or recycled to the first phase of this process to augment the powdered limestone addition [23]. This process has several benefits including: (i) the treated effluent water is not saturated with gypsum and (ii) if the feed water contains Al, then sulfate precipitation

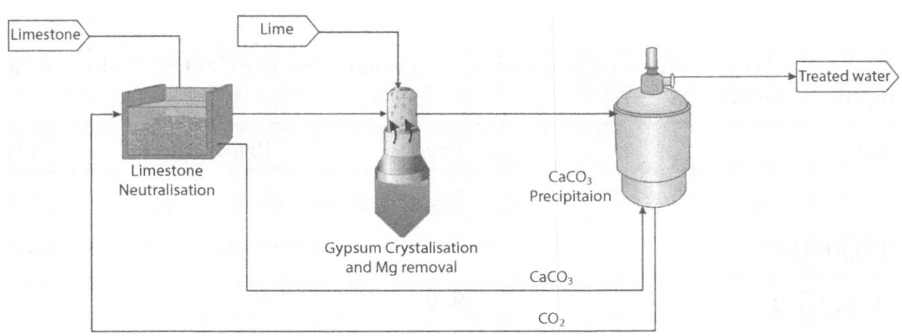

Figure 5.5 Setup for lime/limestone neutralization treatment process [9].

is not only achieved through gypsum crystallization, but also through ettringite formation ($3CaO.3CaSO_4.2Al_2O_3$) which precipitates in the pH range 11.3–11.4 [9].

The equipment used in this process consists of cheaper mixed or aerated reactors, followed by clarifiers. There are several system configurations, and each has advantages and disadvantages peculiar to that configuration. The limestone/lime process is a robust and proven process but usually fails to meet the legal treated water quality standards for discharge into a river or the re-use of the water. The treatment also produces large volumes of mixed precipitated sludge that requires long-term disposal. This treatment process can be used for the metal removal pre-treatment step in several desalination processes, such as RO or ion-exchange desalination [9].

5.2.6.2 *Removal of H_2SO_4, Fe^{3+}, and Al^{3+} with Limestone*

The cost of the three projects amounted to R 2.5 billion [24]. The construction cost of the plant in the Central Basin amounted to R 319 million and was operational by August 2012 [25]. The cost of the plant in the Eastern Basin was estimated at R 950 million and was operational by December 2015 [24]. In the Western Basin, it was decided to upgrade the existing neutralization facilities. The Trans Caledon Tunnel Authority (TCTA) elected to use proven technology that treats AMD with limestone for neutralization of acid, followed by the addition of a small amount of lime for removal of iron(II) and other metals (pers. comm. A. van Niekerk, 2011). The projects made provision for pumping of mine water to the environmental critical levels (ECL) of 165, 168, and 290 m for the Western, Central, and Eastern Basins, respectively [25, 26]. Ten specialized submersible pumps (double suction, 2400 kW, 6600 V, 50 Hz) were installed in the three Basins to pump water to the ECL [27]. Two were installed in the Central Basin, three in the Eastern Basin, and two in the Western Basin, while two were placed on standby. A tenth, smaller pump was initially installed in the Western Basin. The cost of two large pumps amounted to R 60 million [25].

Grab samples showed the water quality of the main parameters of the feed water, after pre-treatment with limestone and final treatment after lime treatment and gypsum crystallization (Table 5.2). As can be seen, the acidity was slightly lowered during the pre-neutralization with limestone (2,500 to 2,368 mg/L as $CaCO_3$), iron(II) is removed during lime addition and sulfate is removed from 3,630 mg/L to less than 1,900 mg/L owing to gypsum crystallization. This water has no metals remaining in solution at potentially toxic concentrations and can be discharged safely into the environment. The lime dosage amounts to

Table 5.2 Chemical composition of feed and treated water from the Central Basin neutralization plant (sampling date 2014-08-15).

Parameter	Feed	Pre-treated	Final
pH	5.2	5.3	9.3
Acidity (mg/L $CaCO_3$)	2,500	2,237	–
Iron(II) (mg/L Fe)	949	894	<1
Sulfate (mg/L SO_4)	3,630	3,559	1,867

1,391 mg/L and consumption for treatment of 83,000 m^3/d amounts to 4,485 t/month. At a price of R 2000/t, the lime cost amounts to R 9 million per month or R 3.55/m^3 (1 August 2015).

5.2.6.3 *Removal of H_2SO_4, Fe^{3+}, Al^{3+}, and Fe^{2+} with Limestone*

The research conducted by the Tshwane University of Technology (Department of Water, Environmental and Earth Sciences) revealed that limestone could be used for the total removal of iron(II) in about 90 min reaction time following which lime could be used for the complete removal of the remaining metals [28]. The conclusion reached was that the cost of alkali for AMD treatment in the Western Basin was estimated at R 5.83/m^3 (2002 cost figures), compared to an estimate of R 80/m^3 AMD if lime is used for both stages.

5.3 Chemical Desalination

5.3.1 SAVMIN

The SAVMIN™ process involves the removal of metals and sulfate at ambient conditions from contaminated mine water (pers. comm. M. van Rooyen, 2015). The process can produce a product stream that meets the standards for drinking water and comprises four stages (Figure 5.6): Stage 1: Metal precipitation, Stage 2: ettringite precipitation, Stage 3: Carbonation, and Stage 4: Recovery of aluminium hydroxide *via* ettringite decomposition.

In the first stage of the SAVMIN™ process (metal precipitation), AMD is treated with lime to raise the pH to a value between 10 and 11.5. Gypsum as well as (semi-)metals, as metal hydroxides, are precipitated during this stage. The combined metal hydroxides and gypsum solids are separated from solution and leave the process as waste. Further treatment of the metal hydroxides and gypsum solid waste material might be required; however, this does not form part of the SAVMIN™ process. In some cases, this waste might be dumped in a properly lined hazardous waste disposal site. In Stage 2, the pH is raised further and aluminium hydroxide is added to the gypsum-saturated solution from Stage 1, which results in the removal of calcium and sulfate by the formation of ettringite ($3CaO.3CaSO_4.Al_2O_3.31H_2O$). The overflow solution is treated with carbon dioxide in Stage 3 to lower the

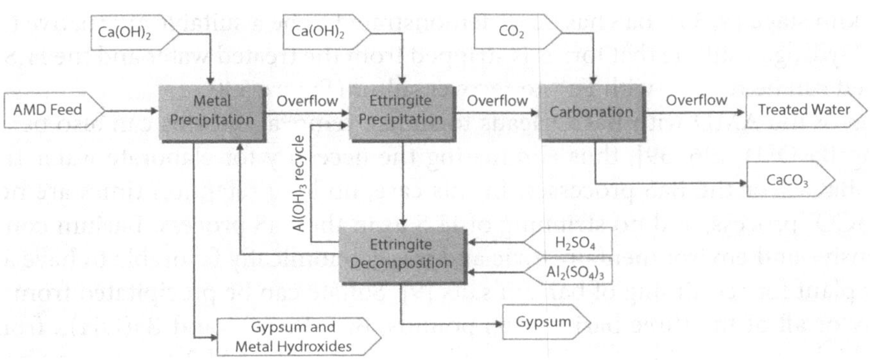

Figure 5.6 SAVMIN™ process block flow diagram [9].

pH and cause the precipitation of calcium carbonate, which is then separated from the treated water. The calcium carbonate can either be collected as a by-product or disposed of as waste. The solution from Stage 3 is the product: clean water obtained from the SAVMIN™ process. The final stage 4 of the process consists of the recovery of aluminium hydroxide *via* ettringite decomposition. Sulfuric acid and make-up aluminium sulfate are added to the ettringite slurry (underflow from Stage 2) to decrease the pH of the stream to a value between 6.0 and 9.0. Aluminium hydroxide precipitates, together with gypsum, which ultimately reports to the solids. The recovered aluminium hydroxide is separated from the gypsum solids and recycled to Stage 2. The "clean" gypsum is recovered as by-product or waste and the overflow, saturated in gypsum, is recycled to Stage 1. Trial studies at the Stilfontein mining facility in South Africa successfully treated 500 m^3 of water with a sulfate concentration of 800 mg/L to below 200 mg/L.

5.3.2 Barium Sulfate Treatment Process

The precipitation of sulfate using barium carbonate was demonstrated by Norris [29] and later by Trusler [30] and others [30–33] who showed that $BaCO_3$ can be used for sulfate removal. This process is based on the following, simplified reaction:

$$BaCO_3(s) + SO_4^{2-} \rightarrow BaSO_4(s) + CO_3^{2-} \quad (5.3)$$

Wilsenach demonstrated that the high cost of the method could be dealt with by reclaiming $BaCO_3$ through the roasting of barite ($BaSO_4$) to form BaS and purging an aqueous solution of BaS with CO_2 for $BaCO_3$ to be formed [31]. When magnesium is present in the water, it is removed by lime dosing. The high solubility of $MgCO_3$ lowers the extent of dissolution of $BaCO_3$ and lime dosing results in $CaCO_3$ precipitation which leads to higher Ba^{2+} concentrations in solution.

The slurry from the reactor is pumped to a thickener tank where treated water is removed and recycled in the mine and associated process operations (Figure 5.7). The thickened slurry is filtered, dried, and processed to reclaim barium carbonate and harvest the sulfur from the H_2S produced. In an improvement of this process by [32], a two-stage fluidized bed reactor system was designed to deal with the low reactivity of $BaCO_3$. However, a weakness of this process is that $CaCO_3$ and $BaSO_4$ are mixed and should be separated before or after the kiln stage [9, 34]. BaS has been demonstrated to be a suitable alternative to $BaCO_3$ [34, 35]. Hydrogen sulfide that forms is stripped from the treated water and the H_2S gas that is removed can be reacted with Fe^{3+} to recover sulfur (Figure 5.8).

Dosing of the AMD with $BaCO_3$ leads to sulfate removal but this can also be achieved by dosing $Ba(OH)_2$ [36–39], thus eliminating the necessity for elaborate water treatment with the $BaCO_3$ or the BaS processes. In this case, no long retention times are needed as in the $BaCO_3$ process, and no stripping of H_2S as in the BaS process. Barium compounds are expensive and environmentally toxic and it is economically favorable to have a barium recovery plant for reclaiming of barium salts [9]. Sulfate can be precipitated from solution using any or all of the three barium compounds, BaS, $BaCO_3$, and $Ba(OH)_2$, from unacceptably high concentrations to within regulatory standards. AMD streams can be directly treated using BaS and/or $Ba(OH)_2$, although in practice lime treatment is required for such

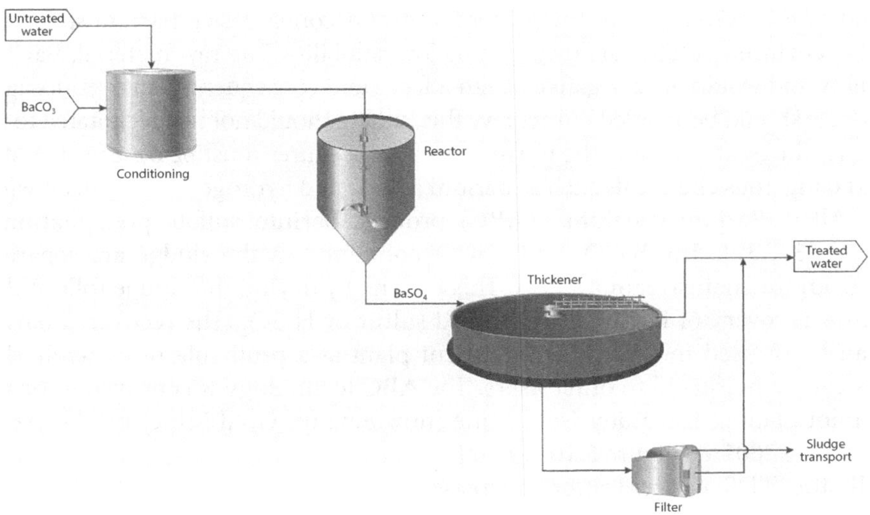

Figure 5.7 Process configuration of the barium carbonate process [9, 36].

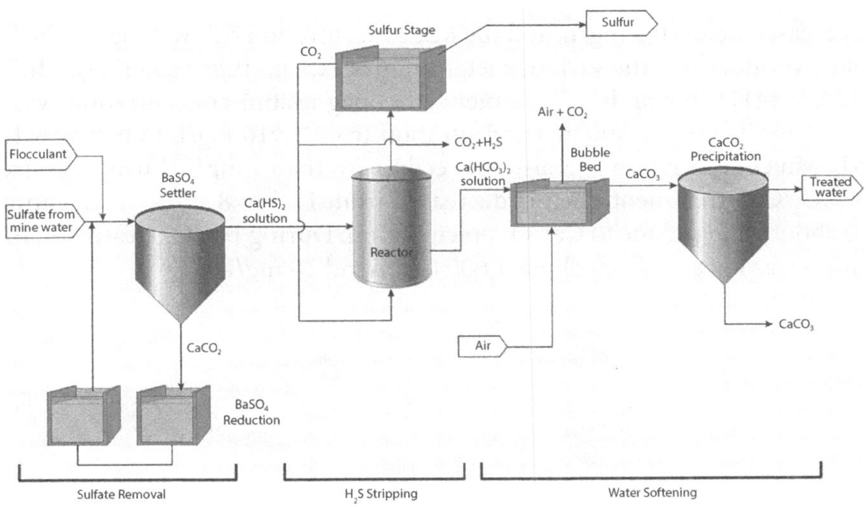

Figure 5.8 Schematic flow diagram of the BaS process [9, 36].

solutions to prevent metal hydroxide sludge being mixed with $BaSO_4$ sludge. The $Ba(OH)_2$ process also removes Mg and NH_3 and thus the TDS concentrations are lowered including concentrations of deleterious elements such as Mn, Al. A benefit of this treatment process is the recovery of saleable by-products and these products can be used to offset the costs of treatment. The $BaCO_3$ and BaS processes have the capacity to produce sulfur, NaHS, and $CaCO_3$ on a commercially viable scale [9].

The processes at the wastewater treatment site require a completely mixed reactor, clarifier, and sludge drier. Thus, the process required for desalination treatment with $BaCO_3$ bears similarities to the neutralization process. $BaCO_3$ dosage is stoichiometrically linked to the sulfate that should be removed and thus the process control becomes easier. Barium sulfate which is produced is environmentally safe owing to its low solubility. The raw material, $BaCO_3$, has a low solubility and would have negative health effects, and consequently, an over dosage of barium carbonate should be avoided. To achieve this, sulfate should not be precipitated to less than 100 mg/L. At the treatment site, the normal safety procedures must be observed as stipulated by law and dangerous chemicals such as barium sulfide and hydrogen sulfide dealt with [9].

In the Alkali-Barium Calcium™ (ABC) process, barium sulfate precipitation occurs (Figure 5.9) [32, 39]. The $BaSO_4$ and $CaCO_3$ contained in the sludge are separated and removed from the mainstream process. This is done by drying the sludge followed by processing it for recovery of $BaCO_3$, $CaCO_3$, and sulfur or H_2SO_4. The recovered barium carbonate can be re-used in the AMD treatment plant at a profitable price while the other chemicals can be marketed to other users. The ABC technology was proved to be viable in a 1 m^3/h pilot plant at Harmony Gold Mine (now Sibanye Gold Mine) on the West Rand, Gauteng during 2007/8 (Figure 5.10) [9, 32].

Overall, the TDS concentration decreased from 2,641 mg/L in the feed water to 360 mg/L in the treated water (Table 5.3) [38–40]. When treating with $CaCO_3$, the acidity of about 370 mg/L (as $CaCO_3$) was removed as shown by the pH increase from 3.3 to 5.8 and iron(III) and aluminium(III) concentrations decreased to below 1 mg/L. During pre-treatment with $CaCO_3$, $Ca(OH)_2$, or CaS, bivalent metal concentrations (e.g., Fe, Ni, and Co) decreased below 0.5 mg/L, and the Mn concentration below 4 mg/L. The following low solubility products for the various metal sulfides explain this: FeS (17.3), MnS (14.96), and NiS (20.7) [41]. During lime treatment, the magnesium concentration was lowered from 145 to 1 mg/L and the sulfate concentration from 1,910 mg/L to between 1,600 and 1,900 mg/L. Mn concentration was also reduced to less than 1 mg/L during lime treatment. During $BaCO_3/CO_2$ treatment, the pH decreased from 11.5 to 8.4. Ca^{2+} concentration was lowered to about 75 mg/L due to $CaCO_3$ precipitation. During $BaCO_3$ treatment, the sulfate concentration was lowered from about 1,600 to around 10 mg/L.

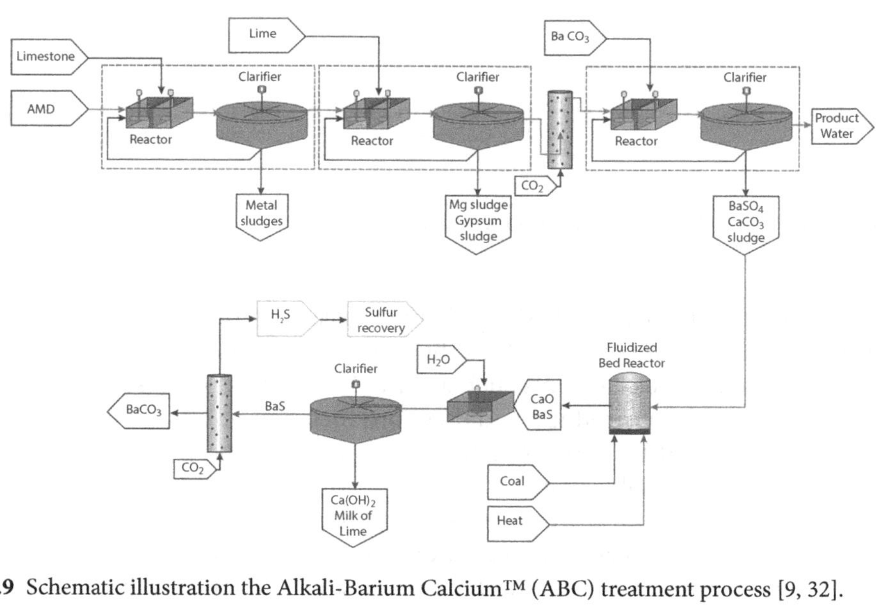

Figure 5.9 Schematic illustration the Alkali-Barium Calcium™ (ABC) treatment process [9, 32].

Figure 5.10 Chemical desalination pilot plant at Harmony Gold Mine based on the CSIR ABC™ process.

The low solubility of $BaCO_3$ allows the Ba^{2+} ions in solution to react with SO_4^{2-}. The dissolution of $BaCO_3$ can only continue to release Ba^{2+} if the CO_3^{2-} that is produced is precipitated from solution as $CaCO_3$. In order to obtain rapid removal of sulfate, it is necessary to dose excess $BaCO_3$. This was facilitated using a fluidized-bed reactor. Sulfate-rich water was pumped through a bed of $BaCO_3$ and in contact with high concentrations of excess $BaCO_3$, to form $BaSO_4$ and $CaCO_3$. This was continued until the $BaCO_3$ in the bed was exhausted followed by dosing with fresh $BaCO_3$. Further studies showed that a reactive $BaCO_3$ can be produced when $BaCO_3$ is forced to precipitate in micro fine crystals.

5.4 Membrane Processes

5.4.1 Reverse Osmosis

Membrane processes use a physical membrane barrier to help separate the charged ions from the clean water, producing two effluent streams, one with low dissolved solids concentration and the other with a high dissolved solids concentration. Pre-treatment for RO often involves limestone and/or lime dosing and aeration for the neutralization and oxidation of the mine water and the removal of metals. This is then followed by stringent filtration using either sand filtration and cartridge filters, or ultra-filtration, before the main RO step. The RO phase is a proven technology for the removal of dissolved solids from brackish or saline feed waters and it passes pressurized water in a parallel stream through a semi-permeable membrane, leaving dissolved ions behind the membrane. RO can retain bacteria, dissolved salts, sugars, proteins, particles, dyes, and other substances contained in the feed water with a molecular weight greater than 150–250 Daltons.

The potential use of RO is expensive since it can yield water of high quality. This reduction in TDS is costly and water recovery is less than ideal. The higher the required reduction in TDS concentrations, the more expensive a RO plant becomes. A recovery rate between 50% and 80% can be achieved with a single stage RO plant and can be increased up to 95% with multiple RO stages, thus reducing the waste brine volume and the cost of brine disposal.

Table 5.3 Chemical composition of AMD feed water before and after the various treatment stages [32]. Units in mg/L, pH without units, U in μg/L, cations and anions in meq/L, sample from 2008-04-09.

Parameter	Feed	Treated
pH	3.30	7.90
Alkalinity, $CaCO_3$	0	140
Sulfate	1,910	90
Chloride	44.46	49.52
Fluoride	5.40	0.07
Ammonia N	0.76	0.64
Nitrate N	0.41	0.46
Sodium	46.48	53.28
Potassium	4.38	4.80
Magnesium	124.61	0.98
Calcium	205	75
Barium	0.20	0.50
Silicon	11	0.45
Manganese	63.69	0.09
Total acidity, $CaCO_3$	714.16	0.59
Free acidity, $CaCO_3$	370	0.00
Iron(II)	180	0.30
Iron(III)	2	0.00
Aluminium	2.97	0.01
Chromium	0.26	0.00
Zinc	3.20	0.06
Copper	21	0.02
Cobalt	2.05	0.02
Nickel	5.60	0.01
Lead	0.03	0.01
Uranium	465	20
TDS	2,641	360
Σ Cations	41.84	6.35
Σ Anions	41.50	6.24

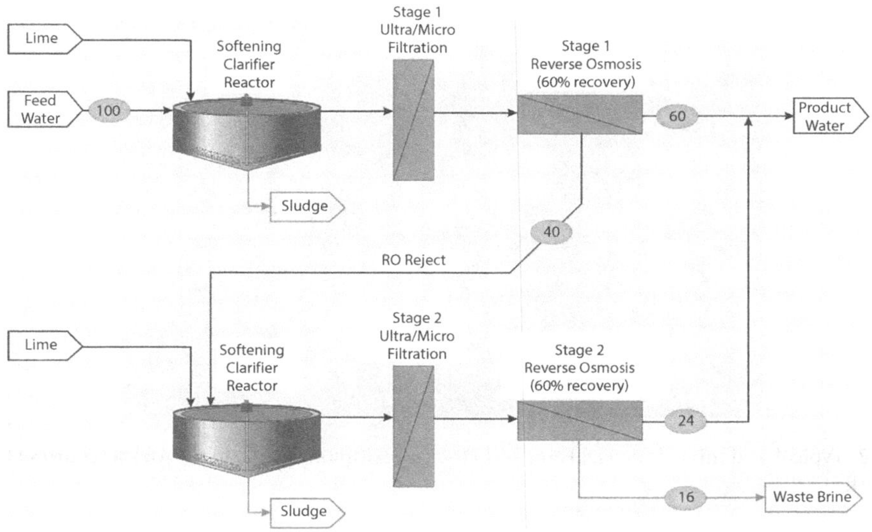

Figure 5.11 Illustration of a typical UF/RO membrane treatment process [9].

Multiple-stage RO can even achieve water recoveries of greater than 99%, depending on the feed water quality (Figure 5.11). These high-water recoveries are achieved where the feed water consists of predominantly divalent ions that can be precipitated from the preceding stage's brine before being treated in the next RO stage. Multiple-stage RO systems can also have some nanofiltration (NF) membranes to allow monovalent ions to pass through the membrane (see section on nanofiltration) and increase the overall water recovery by limiting or eliminating the production of brine.

Tubular RO systems, with regular flow direction reversal and cleaning using sponge balls, can be used for highly scaling water streams that cannot be treated in conventional spiral wound RO systems. The sludge and brine waste streams require long term disposal due to the hazardous nature and high concentration of dissolved salts.

5.4.2 NF Technologies

NF bears similarities to RO in terms of having the same pre-treatment expectations and waste disposal problems. However, in NF, the waste brine volume is reduced because NF effectively only separates the multivalent ions, while monovalent ions pass through the membrane. However, it is important to note that a solution with a mixture of monovalent and divalent ions such as sodium sulfate, will behave like a divalent salt in solution as the retained multivalent salt should remain charge neutral and will therefore not allow the sodium ions to go through the membrane. NF is a viable alternative where divalent ions like Mn^{2+} and Mg^{2+} are the main ionic constituents in the feed wastewater to be treated [9].

5.4.3 High Recovery Precipitating Reverse Osmosis (HiPRO®) Process

Since 2005, three mine water desalination plants were built in South Africa, where RO is the main section of the process configuration, to address AMD management needs in the Upper

Figure 5.12 Typical acid mine drainage; Fanie Nel Discharge, Mpumalanga, South Africa (courtesy Christian Wolkersdorfer).

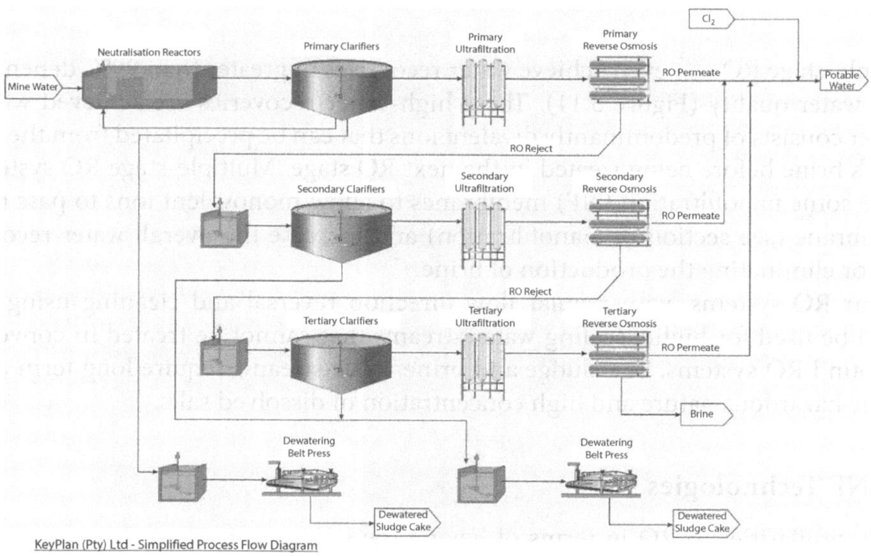

Figure 5.13 Process configuration for the HiPRO process of KeyPlan [9, 47].

Olifants River Catchment (Figure 5.12). Such needs include preventing pollution of the water in the Upper Olifants River and to supply drinking water to Municipalities of Witbank and Middleburg [42]. The demand for drinking water increased by 3.5% per annum [43]. The first plant was commissioned in September 2007 [44] and thereafter upgraded to treat 30,000 m^3/d [45]. The second plant was commissioned in August 2011 [46] at Optimum Colliery with a capacity of 15,000 m^3/d and a peak capacity of 18,750 m^3/d [47]. Water recovery amounted to 98%. The third plant is currently in the commissioning stage at Middleburg Colliery, South 32. The plant has a capacity of 25,000 m^3/d. The capital cost of the plant was R 750 million, fixed cost of R 8/m^3, and variable cost of R 4/m^3 (pers. comm. M. Koen, 2015).

Figure 5.14 Demonstration plant (4.5 m^3/h) that was used to demonstrate the HiPRO process at the Navigation Section of Landau Colliery during 2004 to 2005 [44].

Keyplan's High Recovery Precipitating Reverse Osmosis (HiPRO®) Process has three stages, each stage having pre-treatment, ultrafiltration (UF), and RO steps (Figures 5.13 and 5.14) [47, 48]. The treated water produced by each stage is of potable quality and the brine of each stage becomes the feed for the next stage. Partial sulfate removal is achieved through precipitation of gypsum during lime dosing in the pre-treatment stages. The RO stage is mainly used for removal of soluble sodium sulfate and sodium chloride, resulting in 98% water recovery. In the pre-treatment stage, limestone is used for the removal of H_2SO_4, Fe^{3+}, Al^{3+}, and Fe^{2+} (partially) when the feed water has a low pH (eMalahleni plant, Table 5.4). Lime is used for removal of remaining Fe^{2+}, manganese, and other metals present in low concentrations. Ozone is used for iron and manganese removal at Optimum Colliery, where only low concentrations of iron and manganese are present in the feed water. Solids are removed in Stage 1 clarifier. An anionic polymer flocculant is dosed to promote solids agglomeration. Clarification is followed by sand filtration prior to further treatment *via* UF. Pre-treatment Stages 2 and 3 differ from Stage 1 in the sense that pH is raised to a concentration high enough for precipitation of magnesium hydroxide, in addition to gypsum precipitation. In the reactors, the formed sludge is kept in suspension with large agitators. Hydrocyclones are used to draw off large sludge particles. Small particles in the overflow are introduced into the respective Stage 2 or Stage 3 clarifiers while the coarse sludge from the underflow of the hydrocyclones are subsequently dewatered by a vacuum belt filter.

UF is the final solids-removal process before RO treatment. Each UF Skid has 38 modules in the case of the Optimum plant, run in dead-end mode. In order to remove the entrained solids, each skid is regularly backwashed. The backwash water from the Sand Filters and the UF's are collected in the Plant's Drain Sump and further treated. Antiscalant and sulfuric acid are dosed upstream of the UF's in Stage 1 to prevent scaling of RO membranes with gypsum. In the case of Stages 2 and 3 where the water has an elevated pH after pre-treatment, antiscalant acid is dosed after the UF's, as well as sulfuric acid for pH correction.

RO is the final treatment stage, designed for the feed water quality and to meet the required water recovery. The $CaSO_4$ saturation of Stage 1 RO feed water amounts to around

Table 5.4 Chemical composition of the feed and treated mine water in the eMalahleni plant. Parameters in mg/L, electrical conductivity in mS/cm, and pH without unit [49].

Parameter	Feed	Treated
pH	2.7	6.0–9.0
Electrical conductivity	460	<70
Total dissolved solids	4,930	<450
Chloride	70	<100
Sulfate	3,090	<200
Acidity, $CaCO_3$	1,050	0
Sodium	130	<100
Potassium	13	<25
Magnesium	230	<30
Calcium	660	<80
Aluminium	40	<0.15
Iron	210	<0.01
Manganese	35	<0.05

90–95%, and the recovery is limited by the $CaCO_3$ scaling tendency. The maximum obtainable Stage 1 RO recovery was about 70%, and the design of the Stage 1 RO used NF membranes to allow the monovalent bicarbonate ions to flow into the permeate. The advantages of this were: (i) the low salinity permeate becomes stable, and (ii) the lowered concentrations of bicarbonate in the reject places less demand for the use of additional lime in the Stage 2 precipitation reactors. The Stage 2 and Stage 3 RO designs are aimed at 65% and 60% water yield, respectively, because of higher $CaSO_4$ saturation concentrations in the feed water. The permeate stream from each RO skid is harvested in a common water tank, while the reject from each stage is pumped to the next stage, with a small amount of Stage 3 reject being passed on to the brine pond. Mixed sludge, gypsum, and brine which are produced become waste streams. It was determined that: (i) 98% water recovery is achieved, (ii) excellent quality drinking water is produced; (iii) less waste is generated, and specifically, less than 12-L brine/m^3 feed, less than 6-L mixed sludge (dewatered)/m^3 feed, less than 15-L gypsum slurry (dewatered)/m^3 feed; (iv) potentially useful by-product gypsum; and (v) the variable cost, excluding power, amounted to R 5.00/m^3 and fixed cost to R 2.50/m^3 [47].

5.4.4 Electrodialysis

Electrodialysis (ED) requires the same sort of neutralization and metal removal pre-treatment as RO. The suspended solids removal requirements are, however, not as stringent, and normal filtration is sufficient. ED or ED Reversal (EDR) involves the transport of ions across water-tight

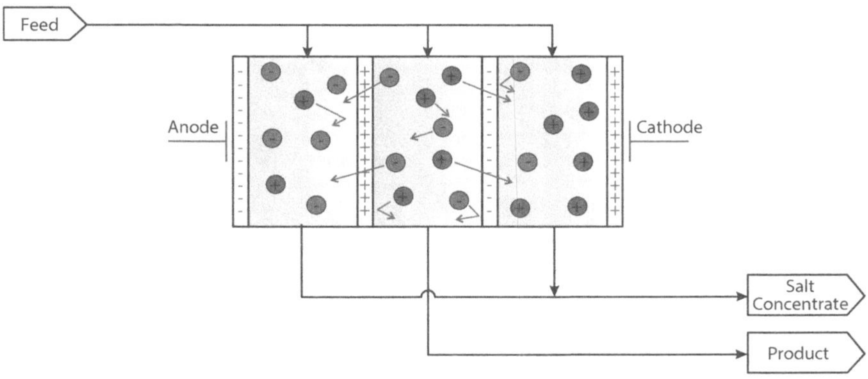

Figure 5.15 Illustration of the electro-dialysis process [9].

membranes that allow electrons to flow through them (Figure 5.15). A DC electrical current is applied in this process passing through a series of alternating cation and anion selective membranes. Anions migrate to the anode but cannot pass through anion-impermeable membranes and are concentrated, while cations migrate in the reverse direction and are impeded by cation-impermeable/anion-permeable barriers. The first container has thus had salts depleted and the clean water can be extracted. The process is greatly improved by current reversal by the anode and cathode being periodically reversed. Thereby the potential for membrane fouling is partially eliminated and membrane regeneration by self-cleaning is facilitated.

The current reversal used in the process is the reversal of the polarity of the plates and membranes. This results in a self-cleaning action by changing the direction in which ions flow through the membranes. The reversal of ion-flow prevents build-up of slime and lowers the requirement for pre-treatment chemicals. The main operating cost of EDR is the energy input for the transport of the ions across the membranes. The energy input required by EDR is almost directly proportional to the ionic concentration of the feed water. Therefore, EDR is typically more cost competitive, compared to conventional RO, at lower feed water TDS concentrations. EDR will selectively remove only ionic species and consequently if the feed water is high in organic compounds or non-ionic species, some mode of filtration prior to the EDR unit is required. Water of a very good quality can be produced by ED, similar in quality to ion-exchange or RO. As with conventional ion-exchange or RO, ED produces brine that requires long term disposal or thermal volume reduction followed by crystallization and disposal of the crystallized salts.

EDR does not respond to effluent temperature or pH. This is a major advantage over RO techniques. Capital costs of EDR are also less owing to lower working pressures. $CaSO_4$ scaling takes place owing to inadequate pre-treatment. At the Beatrix Gold Mine (South Africa), water recovery of about 80% was achieved during a pilot-plant demonstration. The raw water had high Fe, Mn, Na, and Cl concentrations, as well as high sulfate concentrations [50].

5.4.5 Vibration Shear Enhanced Process

The Vibration Shear Enhanced process (VSEP) is based on conventional RO processes but uses torsional vibration of the membrane surface, creating high shear energy at the membrane surface (Figure 5.16). This enhances throughput (98% water recovery) and reduces colloidal fouling, hence reducing the degree of pre-treatment required and possibly

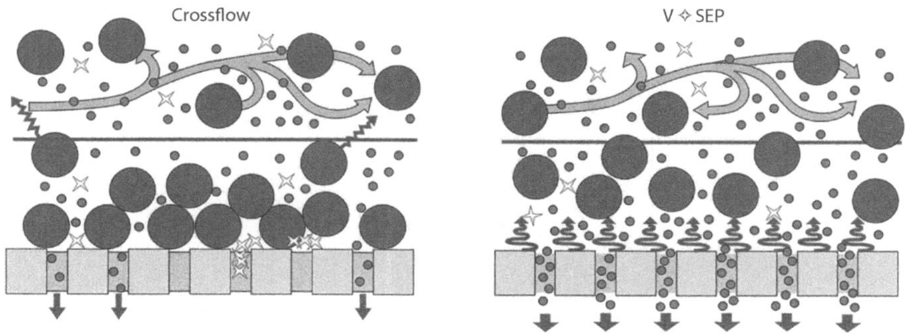

Figure 5.16 VSEP membrane (image courtesy New Logic Research; G. Johnson, 2020).

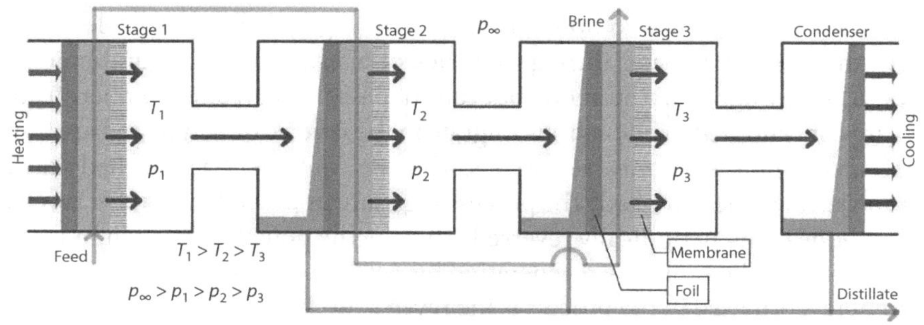

Figure 5.17 MEMSYS process (Image courtesy of *Quality Filters*) [9].

eliminating it. It achieves very high reductions of color, turbidity, or iron. VSEP has the potential to replace most of the RO pre-treatment requirements, resulting in lower operating costs and less brine production. Units are modular and portable. However, a small brine stream still remains that requires disposal.

5.4.6 Multi-Effect Membrane Distillation

The multi-effect membrane distillation (Memsys) system is similar to the multistage flash distillation process in that it makes use of waste heat and reduced pressures to vaporize the water (Figure 5.17). The process also makes use of hydrophobic membranes which allow the water vapor to pass keeping the concentrated brine behind. The water vapor, which passed through the membrane, condenses on a plastic surface and is recovered, while the concentrated brine is fed to the next stage. Studies indicate that concentrations of 15 g/L TDS can be achieved using this process.

5.4.7 Forward Osmosis Desalination

Forward Osmosis (FO), a recent alternative membrane process, is potentially capable of treating wastewater to produce high quality water. Technically, FO is the transport of water

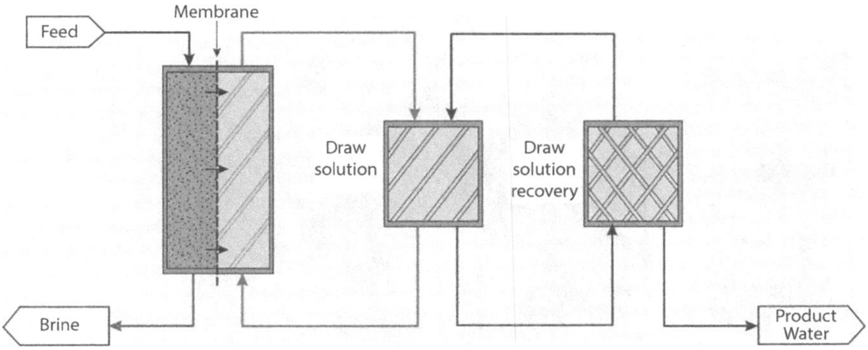

Figure 5.18 Forward Osmosis desalination [9].

molecules across a semi-permeable membrane [51]. What drives this separation is an osmotic pressure gradient, created by a "draw" solution of higher concentration relative to that of the feed solution, being used to give rise to a net flow of water through the membrane into the "draw" solution, so as to separate the feed water from its constituent solutes. Free from energy input other than from the ambient environment, this technology relies on water molecules passively diffusing by natural osmosis into a more concentrated "draw solution", whose volatile "draw salt" is then evaporated by low-grade heat (Figure 5.18). One such process applies the ammonia-carbon dioxide FO process [52, 53]. Since dissolved ammonium ions and carbonic acid readily become gaseous, ammonia, and carbon dioxide, upon heating, the draw solutes can be recovered and reused in a closed loop system.

FO is dependent on the osmotic pressure difference across the membrane to separate out clean water from the feed, however, the FO step is mostly perceived as a "pre-treatment" process. To increase FO-wastewater treatment feasibility, new membrane designs are required, and improved "draw" solutions are required to enhance wastewater treatment, energy recovery, and operating conditions.

5.4.8 Biomimetic Desalination—Aquaporin Proteins

Aquaporin is a protein found in most living cells. The unique structure of this protein allows water molecules to penetrate the cell wall, while excluding other molecules based on size, shape, and electrostatic charge (Figure 5.19). Investigators have mimicked this protein and are in the process of manufacturing membranes with similar characteristics as the proteins found in cell walls. The aquaporin membrane will enable the desalination of high TDS waters at lower pressures than required by current RO membranes.

5.4.9 Carbon Nanotube Distillation

Carbon nanotube (CNT) enhanced membrane distillation can be used for water desalination [54]. The immobilization of the CNTs in the pores of a hydrophobic membrane promotes vapor permeability while preventing liquid penetration into the membrane pores. This is due to the change in water-membrane interactions. It has been found that a

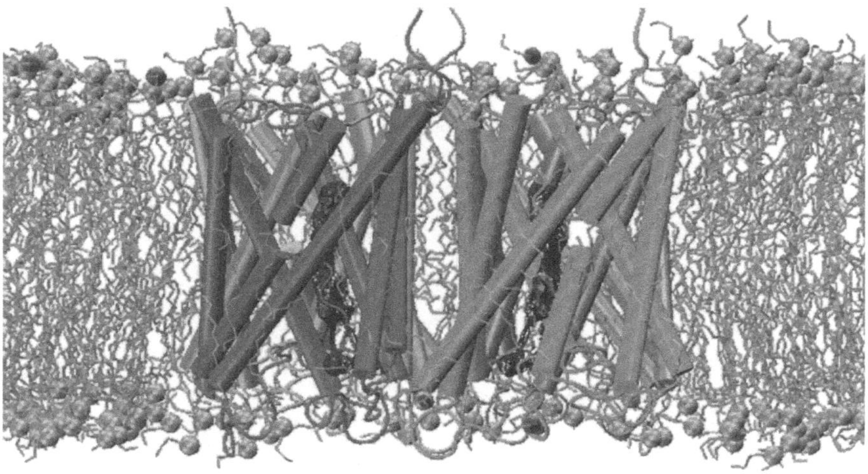

Figure 5.19 Illustration of lipid bilayer membrane containing aquaporin proteins [9] (pers. comm. H. Kandelia, 2020).

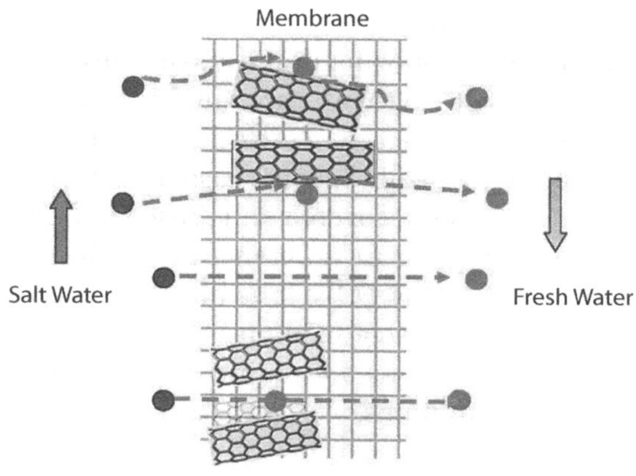

Figure 5.20 Illustration of the carbon nano-tube desalination process [54].

mineralization of 34,000 mg/L at 80 °C, the nanotube incorporation leads to a 1.85 increase in flux and a 15-fold reduction in TDS (Figure 5.20).

5.5 Ion-Exchange Technologies

5.5.1 Introduction

Ion-exchange technologies rely on the exchange of one charged ion with another to remove unwanted constituents from the water. When the exchanged cations are hydrogen and the anions, hydroxide, then the net effect is that the water is desalinated as the hydrogen and hydroxide ions combine to form water. The ion-exchange resins, whether natural or synthetic, are normally regenerated using acids and alkalis. The spent regenerant solutions contain the ions removed from the water plus some excess regenerant and typically require disposal if they are not beneficiated further. Ion-exchange technologies are generally efficient at removing all charged ions from the water being treated and as such can effectively desalinate water with high concentrations of sodium and chloride as well as removing some of the more difficult to remove metals like uranium.

5.5.2 Conventional Ion-Exchange

Conventional ion-exchange is a well proven process that still finds wide application in industry especially for boiler feed water demineralization or water softening. It is based on the use of well proven combinations of process equipment and the lowest cost regenerants that produce soluble spent regenerant solutions, typically hydrochloric acid and sodium hydroxide. The spent regenerant solutions require either disposal in brine ponds or further volume reduction treatment using thermal evaporation processes and disposal of the final salts as hazardous waste.

5.5.3 The GYP-CIX

The GYP-CIX process uses the same principles as conventional ion-exchange, except that the process configuration is based on reactors with fluidized resin beds and a fluidized bed resin regenerator. This process configuration renders the capital costs more expensive but allows for the use of low-cost sulfuric acid and lime as the regenerants while producing gypsum as by-products.

Previously, the limitation to the use of sulfuric acid and lime in the ion-exchange process was scaling or fouling, due to the calcium sulfate produced during the regeneration process. This fouling has been minimized by the use of a fluidized bed configuration as opposed to a packed bed or column and there the precipitation does not destroy or limit the performance of the resin.

5.5.4 KNeW

In the KNeW (Potassium Nitrate ex Waste) process, acid mine water is neutralized with Na_2CO_3 to precipitate calcium and magnesium followed by resin treatment for removal of dissolved solids (Figures 5.21 and 5.22) (pers. comm. J Bewsey, 2015). The remaining cations, mostly sodium, are removed on a cation resin and anions (SO_4 and Cl) on the anion resin (R-OH). The cation resin is regenerated with HNO_3 to produce a metal nitrate solution. Residual metals are removed as metal carbonates (mainly $FeCO_3$, $CaCO_3$, and $MgCO_3$) for use as soil ameliorants. The $NaNO_3$ solution is mixed with equimolar KCl, heated to evaporate water to exceed the solubility of NaCl (NaCl is less soluble than KNO_3), followed by cooling to 35 °C to crystallize KNO_3, which is a primary fertilizer. The anion resin is regenerated with ammonia to produce ammonium sulfate. Methanol addition is used to precipitate a pure $(NH_4)_2SO_4$ from the chloride contaminant. Ammonia is recovered by increasing the pH with NaOH to >12. The economic feasibility of the KNeW process depends on selling the products.

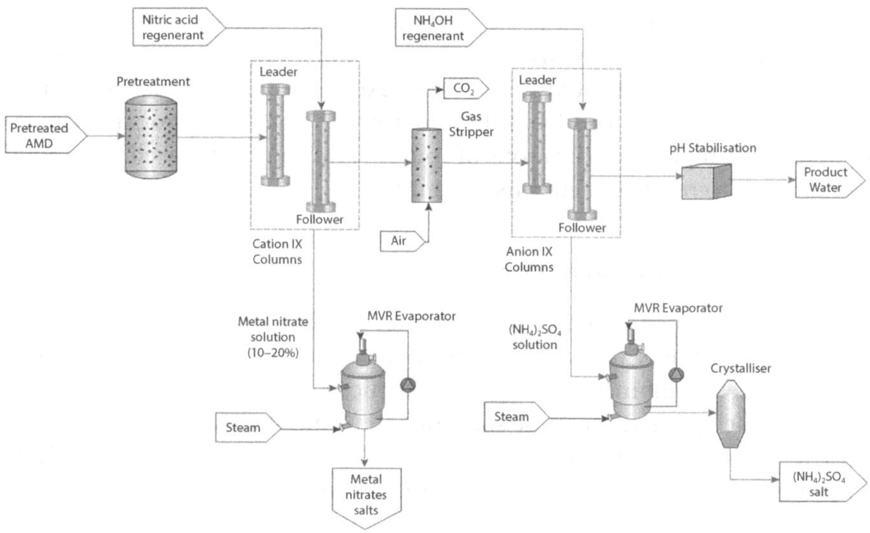

Figure 5.21 Example of a beneficial by-product recovery ion-exchange process [9].

Figure 5.22 KNeW pilot facility at 36 Steyn Road, Farm Rietvallei 180IQ, Krugersdorp (pers. comm. J. Bewsey, 2015).

5.6 Biological Processes

5.6.1 Background

Several biological processes can generate alkalinity. These include photosynthesis [55–57], denitrification [58, 59], ammonification, methanogenesis, and reduction of iron and sulfate [55, 58–60]. These processes offer the potential for neutralization of AMD [55]. Biological sulfate-reduction is one of the most promising bioprocesses for mine water treatment because of combined removal of acidity, metals, and sulfate together with metal recovery [61, 62]. The process mediated by sulfate-reducing bacteria (SRB) depends on biological sulfate reduction leading to hydrogen sulfide production and alkalinity (Reaction 5.4):

$$2\ CH_2O + SO_4^{2-} \rightarrow H_2S + 2\ HCO_3^- \tag{5.4}$$

where CH_2O denotes the electron donor. When hydrogen is used as an electron donor for sulfate reduction, the reaction yields hydroxide ions (Reaction 5.5):

$$8\ H_2 + 2\ SO_4^{2-} \rightarrow H_2S + HS^- + 5\ H_2O + 3\ OH^- \tag{5.5}$$

The biogenic hydrogen sulfide reacts with and causes precipitation of dissolved metals as low solubility sulfides (Reaction 5.6):

$$H_2S + M^{2+} \rightarrow MS(s) + 2\ H^+ \tag{5.6}$$

where M^{2+} denotes a metal ion, such as Zn^{2+}, Cu^{2+}, Ni^{2+}, Co^{2+}, Fe^{2+}, Hg^{2+}, Pb^{2+}, Cd^{2+}, or Ag^+.

The metal sulfide precipitation reaction releases protons, enhancing the acidity of the water. The acidity that is generated is neutralized by OH^- or HCO_3^- produced during sulfate reduction (Reaction 5.4). This results in precipitation of metals as hydroxides, e.g., $Fe(OH)_3$, or as carbonates, e.g., $CaCO_3$, and neutralization of H^+ (Reactions 5.7 and 5.8) [63, 64].

$$HCO_3^- + H^+ \rightarrow CO_2(g) + H_2O \tag{5.7}$$

$$OH^- + H^+ \rightarrow H_2O \tag{5.8}$$

Tuttle proposed the use of microbial sulfate reduction in mining applications as early as in the late 1960s [65]. Currently, much attention is being given to SRB-based passive treatment systems and active bioreactors for the treatment of AMD. Biologically produced H_2S is also used for selective recovery of metals from various bio-hydrometallurgical process streams.

5.6.2 Biological Sulfate Reduction

The five anaerobic, *Desulfovibrio* spp. and the three *Desulfotomaculum* spp. are responsible for the reduction of sulfate to hydrogen sulfide. These organisms possess a respiration metabolism, in which sulfate, sulfite, and other reducible sulfur compounds act as the final electron acceptors, with the formation of hydrogen sulfide [66]. The organic substrates for the bacterial action involved in these reactions are generally short-chain acidic compounds such as lactic and pyruvic acid. A wide variety of organic substrates have been studied, including molasses, sewage sludge, straw, newspaper, sawdust, manure, and short chain organic acids in wastes from the chemical industry. Nature provides these substrates through fermentative actions of other anaerobic bacteria on more complex organic substrates [67]. Sulfate reduction can be *in situ* for the treatment of mine water because of the natural occurrence of the sulfate reducing bacteria under anaerobic conditions. In anoxic conditions, sulfate may be reduced and precipitated from mine waters as stable sulfides. Under these conditions, sulfide minerals remain stable with low solubility [68]. Flooded underground mine workings, tunnels, and open pits can pre-exist in reduced states and provide suitable media for sulfate reduction in the presence of organic substrates. A dedicated reactor can be also be used for sulfate reduction or anaerobic conditions can be created in a passive system, such as the conditions prevailing in a constructed wetland. Sulfate reduction already occurs in mine workings as indicated by the presence of sulfide; hence, the H_2S odor from many mine water discharges [69].

Middleton and Lawrence determined the kinetics of microbial sulfate reduction in a completely mixed reactor using acetic acid as the carbon source and determined a sulfate reduction rate of about 0.29 g SO_4 L^{-1} d^{-1} (biomass content was not specified) [70]. Cork and Cusanovich employed a continuously purged system, using an inert carrier gas (75% Ar and 25% CO_2) for feeding sulfide removed from actively growing cultures of *Desulfovibrio desulfuricans* and *Chlorobium thiosulfatophilum* for oxidation to sulfur [71]. A sulfate reduction rate of *circa* 6.3 g SO_4 L^{-1} d^{-1} was observed in a completely mixed reactor with 12.6 g/L lactic acid as carbon source, at pH 6.5 and 30 °C (biomass content was not exactly specified). About 91% of the hydrogen sulfide produced was fed into a sulfide oxidizing reactor with *Chlorobium* spp., where about 88% was reduced to sulfur, leading to a sulfur yield of about 80%. In a process akin to the previous one, called the BIOSULFIX process, where hydrogen sulfide is removed in an external stripping chamber, a sulfate reduction of about 6.5 g SO_4 L^{-1} d^{-1} was observed by [72] and of similar order of magnitudes by [73].

A proposed biological sulfate removal process, for waste treatment containing sulfate, such as industrial wastewaters or waste gypsum is made up of primary anaerobic, aerobic, and secondary anaerobic stages [73, 74]. In the primary anaerobic step, sulfate is reduced to hydrogen sulfide in the presence of appropriate organic carbon sources, while metals are removed from solution as the corresponding insoluble sulfide precipitates. In the aerobic stage, soluble organic matter and calcium carbonate from the first stage are removed, while in the secondary anaerobic phase, barely biodegradable organic constituents, present in carbon sources such as molasses, are removed. The success of this process depends on the removal and downstream processing of the H_2S which is produced in the primary anaerobic stage and that carbon monoxide can be used as an energy source for biological sulfate removal [74, 75]. It was demonstrated to be cost-effective, as CO is often available as a waste gas [76]. The biological sulfate reduction processes in wastewater low in organic matter are largely dependent on the choice of external electron donor. A mixture of H_2 and CO (synthesis gas) has been suggested as a cheap alternative to pure H_2 [74, 77].

5.6.3 Constructed Bioreactors

An anaerobic bioreactor uses bacterial reduction of sulfate and iron to precipitate metals as metal sulfides (Figure 5.23). For effluent treatment to be effective, a uniform flow rate is required. The removal of sulfate was dependent on the energy source and residence time [78]. Producer gas was effectively used as a reliable energy source [79, 80]. Dill observed during pilot plant studies a sulfate reduction rate of 1.7–2.2 g L−1 d−1 with sulfate concentrations being lowered from 3,000 mg/L to less than 250 mg/L [81].

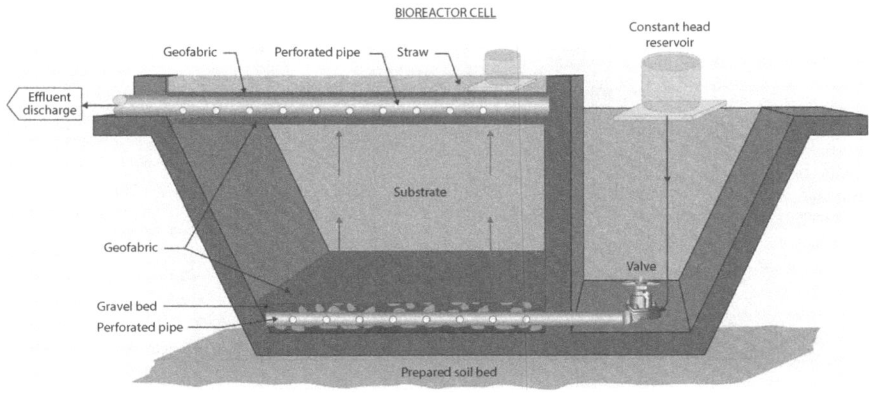

Figure 5.23 Schematic diagram of an upflow anaerobic bioreactor cell [8].

5.6.4 Paques Technologies

Paques Technology B.V. (The Netherlands) has commercialized the Upflow Anaerobic Sludge Blanket (UASB) reactor (Figure 5.24) technology for the anaerobic treatment of wastewater [82–84]. This treatment concept enables industrially used water to be purified and produces renewable energy, fertilizers, and soil conditioners and has been widely used by industry and municipalities. The SULFATEQ™ process removes sulfate to less than 300 mg/L and turns it into hydrophilic (non-clogging), elemental sulfur. It also allows recovery of valuable metals such as Co, Ni, and Zn as marketable metal sulfides. The typical range for this application is an influent sulfate concentration of 1,000–25,000 mg/L and a pH value of 2–8, and it can be used as a stand-alone installation or as the final treatment in a lime-gypsum plant [85, 86].

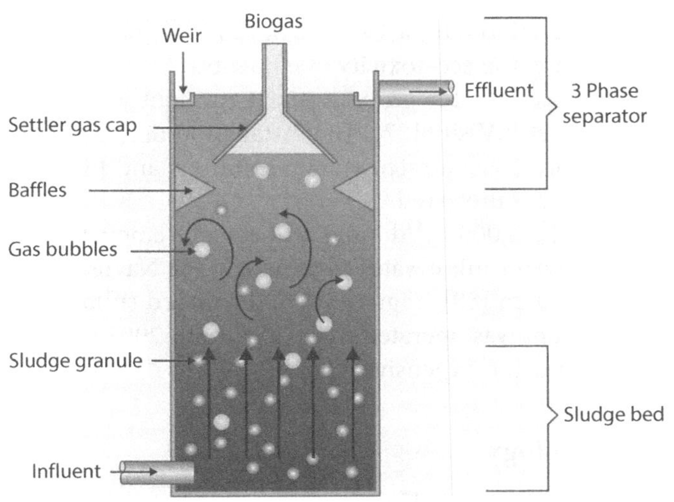

Figure 5.24 Up flow Anaerobic Sludge Blanket reactor [modified after 9, 82, 83].

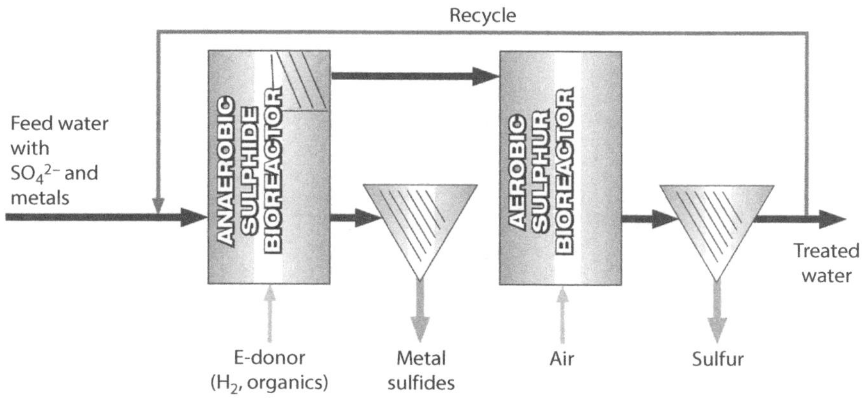

Figure 5.25 Process configuration of the SULFATEQ process [9, 85].

The process consists of two stages (Figure 5.25). In the first stage, sulfate is reduced biologically to dissolved sulfide in high-rate bioreactors. Ethanol or hydrogen can be used as energy source. In the second stage, sulfide is oxidized to sulfur with air and separates from the water. Valuable metals can be harvested when economically justified. The main advantages of this process over alternative ways for sulfate removal are: (i) effective removal of sulfate below 300 mg/L; (ii) use of the intermediate products, sulfide, and alkalinity, to harvest metals or produce elemental sulfur; (iii) removal of calcium and magnesium ions which are responsible for water-hardness to <100 mg/L and <4,000 mg/L, respectively; (iv) removal of acidity without diminishing of neutralization agents like caustic soda; (v) the process can be used for streams which are pH-neutral, e.g., sodium sulfate solutions; (vi) it is a safe process to employ under ambient conditions; and (vii) it has better selectivity for metals to produce high product quality. The process is also efficient for removal of "difficult" dissolved elements like manganese, selenium, and thallium and is a process which is flexible, robust, and proven that has been used for about 15 years in full-scale facilities with no perceptible release of odors and lowering the eco-toxicity of effluents.

The first full-scale applications for groundwater treatment with SRB were installed in 1992 by at the Nyrstar Budel B.V site [87]. The installation uses an organic energy source as electron donor. For more concentrated sulfate solutions and higher load applications, a system using hydrogen gas is preferred.

In 2002 Anglo Coal had a 3,000 m^3/d biological sulfate removal plant built, which was based on the Paques process for mine water treatment at the Navigation Section of Landau Colliery in the eMalahleni area [88] (Figure 5.26). It lowered sulfate concentrations from 2,300 to 180 mg/L. This plant was operated from 2002 until 2005 when the energy source, ethanol, or hydrogen became too expensive.

5.6.5 BioSURE Technology

A 10,000 m^3/d BioSURE biological sulfate removal process was operated at the Grootvlei Gold Mine. The process uses waste materials such as sewage sludge as an electron donor source. This technology is limited by the economic accessibility of sewage sludge or other organic wastes which can be used as carbon and energy sources. A benefit is that it provides

Figure 5.26 Biological water treatment plant at Landau Colliery [from 88].

an alternative co-disposal of sewage sludge, reducing the cost of the landfilling sewage sludge [89]. The running cost was calculated at R 3.30/m^3 of water, but the project was terminated when it became clear that the cost-effective availability of sewage sludge was the limiting factor at that site.

Greben *et al.* introduced an innovative biological sulfate reduction and sulfide oxidation to sulfur technology, in which the degradation products of grass cellulose from grass cuttings were used as the carbon and energy sources [90, 91]. Sulfate removal at an average of about 85% was demonstrated. A hybrid fermentation reactor was used that had been inoculated with bovine rumen fluid bacteria as source of the cellulose degrading and sulfate reducing bacteria, to reduce sulfate ions to sulfide. Biological sulfide oxidation was carried out with different air flow rates entering the sulfide oxidizing reactor. It was found that using 0.2 L/min of air most sulfide was oxidized to sulfur with the purity varying from 17 to 81%. It was shown that: (i) the acidic pH of the mine water was raised, (ii) the metals were removed as metal-sulfide precipitates, (iii) about 85% of the sulfate was removed, and (iv) the excess sulfide after metal-sulfide precipitation could be oxidized to almost pure sulfur.

5.6.6 The *VitaSOFT* Process

The *VitaSOFT* process was developed in response to additional development requirements found during the testing and full-scale demonstration of the BioSURE® process [92]. Further work was required to implement this process elsewhere, in particular where the mine water has a high acidity, low pH, high concentrations of dissolved metals, and a sulfate concentration in excess of 2,000 mg/L. One of the disadvantages is its reliance on primary sewage sludge, which may not always be available, as well as on a continuous supply of iron hydroxide, and availability of the associated disposal requirements for large amounts of iron sulfide sludge. VitaSOFT addressed these shortcomings and developed a more robust process with broader and more flexible application potential. Maize silage was used as an alternative carbon source, with advantages over primary sewage sludge, such as long shelf life, a higher percentage biodegradability, and lower nitrogen concentration. Enough alkalinity could be generated in the biological sulfate-reducing process to neutralize acidic mine water and precipitate contaminating metals as sulfides without the need for an upstream HDS process. Biological iron oxidation was used to regenerate the iron hydroxide required for sulfide removal, so that a constant supply would not be required. Finally, the effluent of the biological sulfate reducing process could be softened and stabilized by removal of calcium carbonate and magnesium hydroxide, to decrease the salinity of the water to meet the final effluent standards for discharge.

It was demonstrated that maize silage can be applied either as a supplementary carbon source where primary sewage sludge is available, or as a primary source where there are no alternatives. The lower nitrogen concentration of silage compared with primary sewage sludge resulted in a lower ammonia concentration in the effluent. The implication of this is that there is no requirement for integration of the process with a wastewater treatment plant, or for an alternative nitrification/denitrification step. An integrated biological sulfide oxidation reactor was introduced into the process. A substantial portion of the sulfide was removed biologically as elemental sulfur, minimizing the requirement for iron hydroxide. The generated alkalinity also precipitated all the calcium in the water as calcium carbonate without lime addition. Lime was only required for the removal of manganese and magnesium in a two-stage process. The cost saving using lime and limestone is therefore two-fold: there is no requirement for an upstream HDS process, and less lime is required for desalination than would typically be required without the contribution of the biogenic alkalinity. Biological iron oxidation was successfully proved as a practical means to regenerate iron hydroxide from iron sulfide.

5.6.7 *In Situ* Reactor

Waters rich in sulfate can be treated *in situ* because sulfate reducing bacteria occur naturally. Fixed bed reactors and in-pit reactors have been utilized but stirred reactors with a suspended solid media have also been suggested, to achieve higher reduction rates by improving operating conditions and reactor utilization [93].

The Summer Camp Pit Lake in Nevada, USA, was studied from a limnological-microbiological-geochemical viewpoint [94]. The upper (oxygen rich) oxic zone had high concentrations of total epifluorescent algae and heterotrophic aerobic bacteria and higher dissolved oxygen concentrations and this resulted in total sulfur being dominated by sulfate ions. This 6-m-deep oxic zone was followed by a transitional zone having an increase in heterotrophic anaerobic bacteria and having decreasing amounts of total epifluorescent algae and heterotrophic aerobic bacteria. In this latter transitional zone, sulfate and thiosulfate were slowly undergoing reduction to sulfide. Deeper this zone was followed by an anoxic zone with little or no dissolved oxygen and was dominated by heterotrophic anaerobic bacteria. In this zone, sulfate was reduced to sulfide and consequent metal sulfide precipitation. The addition of raw potatoes and cattle manure amended systems led to increased sulfate reduction [95]. The raw potato-stimulated sulfate reduction occurred best at low concentrations while manure stimulated sulfate reduction by bacteria proceeded best at mid to high concentrations. When using about 2,500 mg C/L raw potato, 80% sulfate reduction was achieved and additionally soluble arsenic reduction, following an initial increase, decreased to less than 1% of initial values in some cells. This was thought to be as a result of formation of arsenic sulfide.

5.6.8 Constructed Aerobic Wetlands

It is important to distinguish between aerobic and anaerobic constructed wetlands, as each of these systems treats a different type of mine water. While aerobic constructed wetlands treat net-alkaline mine water, anaerobic constructed wetlands treat net-acidic mine water. Constructed wetlands have found application at decommissioned mines where it is difficult to contain the effluent [96, 97]. In the Pelenna Valley pilot scale wetland, the designed discharge flow rate was 3 L/s at an average pH of 5.6 [98]. The total iron concentration was 21.7 mg/L, while the sulfate concentration was 459 mg/L. Sulfate concentrations gradually decreased over the 10-year period of the study.

5.6.9 Permeable Reactive Barriers

Contaminants can be eliminated from water in an aquifer by allowing the water to flow through a Permeable Reactive Barrier (PRB). The PRB is a reactive zone where substances are introduced to react with the contaminants. Examples of such substances are limestone for neutralization and organic material for biological sulfate removal. The lowering of high sulfate concentrations in groundwater at an Ontario, Canada, mine site have been reported [99]. A mixture of gravel and compost was used to form a sulfate reduction zone that precipitated iron sulfide equivalent to a rate of sulfate removal of about 14 mg/L/day over a 3-year period. A major disadvantage of such a system is the requirement for stoichiometrically equivalent amounts of metal ions and sulfide ions being present in the PRB in order to control sulfur dispersion.

5.6.10 General Aspects and Various Passive Technologies

Wastewater treatment systems are usually implemented at point sources that are commonly downstream of metal and acidity sources. At point sources, water flow is generally high and typically active processes are employed that require careful operation and diligent and timeous maintenance. This is a problem for the remote and diffuse sites such as those found in USEPA Region 8 (USA) where year-round access is difficult. By comparison, passive systems, such as passive bioreactors and permeable reactive barriers, can be used at the end of a tailing pile or at toe-seeps to treat a lower flow of water at the contaminant sources. Passive systems are also employed to implement both biologically and chemically mediated processes. Little operation and maintenance are required which is advantageous for remote sites.

Figueroa *et al.* have investigated a two-stage passive treatment system for effective removal of iron and aluminium [100]. Fe and Al present in mine water passing through passive bioreactor systems has reportedly caused clogging by $Fe(OH)_3$ and $Al(OH)_3$ precipitates [101]. This results in failure well before the organic substrates in the bioreactors have been consumed. To remedy this, Hedin *et al.* have suggested the implementation of pre-treatment for mine water having O_2, Fe, and Al concentrations above 1 mg/L. They suggested pre-treatment by wetlands or vertical flow reactors, similarly designed [102]. One method to prevent the clogging of bioreactors by $Fe(OH)_3$ and $Al(OH)_3$ is to pass the water through a limestone drain. The dissolution of limestone leads consumption of protons thereby increasing both alkalinity and pH of the water. Two major types of limestone systems that can be used have different chemistries and different mechanisms for metal removal. Anoxic limestone systems (ALSs) are employed when the mine water is in a chemically reduced state, with low dissolved O_2, or if the water is intercepted before exposure to atmospheric conditions [103]. The anoxic water conditions cause the metals to remain in reduced states. Most importantly, Fe will be predominantly in the Fe^{2+} state, and Mn in the Mn^{2+} state. Aluminium ions are not affected by the redox potential of the wastewater and are only affected by the pH change caused by the limestone. For wastewaters with pH levels greater than about 4.5, the Al precipitates.

The other main types of limestone systems for treating wastewaters with higher amounts of oxidized metal species are open to the atmosphere and are designated as oxic limestone systems (OLSs) [104–106]. In such systems, the increases in pH and alkalinity take place concomitantly with metal precipitation. As the pH in an OLS is increased, iron and aluminium will precipitate on the limestone surface, forming a passivating coating. Although the coating does not prevent complete limestone dissolution, it slows the process. This should be considered when designing the OLS [105, 107]. The accumulation of Fe^{3+} and Al^{3+} precipitates in the limestone drain will cause blocking, preferential flow-paths, and short-circuiting, and these problems should be considered in the design process to ensure that the OLS has the desired lifespan.

Anaerobic passive bioreactor systems have been installed at many sites in the United States and Canada. The success of these systems has been variable. The efficiencies of constructed anaerobic wetlands treating mine water vary with season and wetland age [108]. Ordóñez *et al.* were unable to achieve SO_4^{2-} reducing conditions in a pilot-scale wetland [109], while a field-scale, passive, bioreactor, constructed in West Fork, Missouri, effectively removed lead from the mine water. Periodic rotary tilling and back flushing were needed to prevent the cells from blocking [110]. In addition, work at sites in Colorado and Wyoming concluded that problems are encountered when wastewater at near freezing is treated in a passive treatment system [111]. It appeared as if the system would operate reasonably well for a certain period, but when stressed, the sulfate-reducing ability ceased. Subsequent attempts to restart SO_4^{2-} reduction proved difficult. Permeable, reactive barriers are should include a substantial fraction of gravel, to ensure good hydraulic conductivity.

Passive bioreactor treatment systems design has no consensus on design criteria. The absence of specifications for the organic substrate, inorganic material components, and physical, chemical, and biological characteristics is recognized examples. In one instance, the system design by Ordóñez *et al.* employed only undefined compost as the organic substrate, while Zaluski *et al.* recommended mixtures of different organic substrates [101, 109]. Consequently, there is a requirement for thorough physical, chemical, and biological characterization of the organic substrates to be used in passive bioreactor systems relative to SO_4^{2-} reducing activity and metal removal efficiency.

Table 5.5 Chemical composition of feed and treated water at the Sibanye Gold pilot plant using electrocoagulation. Parameters in mg/L, electrical conductivity in μS/cm, pH without units.

Parameter	Feed	Treated
Electrical conductivity	4,000	800
pH	6; 2.6	8.0
Sulfate	3,000; 2,500	100
Iron		0.05
Manganese		0.05

5.7 Electrochemical Processes

5.7.1 Electrocoagulation

Up to now, electrolytic wastewater treatment is more seldomly used compared to chemical treatment [112]. Electrolytic treatment is at times suitable and may be more effective in producing high quality water. Electrodes of Al, Fe, steel, and graphite are used. Applications such as electro-coagulation (EC), electro-flotation (EF), and electro-coagulation/flotation (ECF) are employed in the treatment of wastewater. The process operating parameters affecting these processes have been studied, i.e., reactor design, current density, time, and electrode type and arrangement. Among the electrochemical processes, the EC process is known to be the best choice, not solely because it can attain better removal but also because the process is affordable and technologically relatively simple.

5.7.2 Nanoelectrochemical Process for the Treatment of AMD

This sulfate reduction technology has been used successfully since 2008 for removal of metals, arsenic, and for cyanide destruction (pers. comm, P Zeevy, 2015). In this electrolytic process, sulfate and metals are removed through the formation of Crystalline Metallic Nano particles with the chemical formulas $M1M2SO_4$ or $M1M2M3SO_4$. The process occurs on the electrode surface in the reactor and within the flowing wastewater medium. The reactors have several individual electrolytic cells that combine modular cathodes and anodes of different types. These cathodes and anodes are chosen according to the types and contaminant concentrations in the wastewater and the pH values.

The electrolytic reactor includes a set of coated, carbon-carbon electrodes, and one or more sets made of pure metal. The electrodes are arranged in an alternating manner such that each cell is different from the one next to it. Parts of the inner electrolytic cells operate in series and other parts in parallel. When connected to an external power source, the anode material will be consumed electrochemically due to oxidation, while the cathode will undergo passivation. For this reason, bi-polar electrodes are used, that automatically change polarity and remove the passivation process. The current densities are controlled centrally by a system that monitors the current according to the influencing parameters of the wastewater and automatically allows the desired current to reach individual cells. The process stimulates a series of reduction and oxidation reactions that take place simultaneously within the reactor cell.

Mine water from the Western and Central Basins were treated successfully in a pilot plant for removal of metals, including uranium and sulfate, to produce water of drinking quality (Table 5.5). EDS measurements indicated the presence of two types of elemental compositions: (i) Fe-Al-S-O (according to different standard analyses, it was assumed that it belongs to the family of $Fe_aAl_b(SO_4)_c(OH)_d.x(H_2O)_e$) and (ii) Fe-Al-Ca-S-O, assumed to belong to the family of $Fe_aAl_bCa_c(SO_4)_d(OH)_e.x(H_2O)$. XRD analysis yielded spectra of different oxides of Fe, Al, and Mn in different oxidation states. This technology offers the following benefits: (i) no need for pre-treatment, (ii) little sludge production, (iii) no brine produced, and (iv) uranium removal to below 5 μg/L.

5.8 Freezing-Based Technologies

5.8.1 Basics

Freezing water involves the exchange of only about one sixth of the energy required to evaporate water, since the heat of fusion for ice is six times less than that evaporation. Approximately, 438.5 kJ of energy is required to freeze 1 kg of water, while 2,570.5 kJ is required to freeze 1 kg of water at 25 °C. Therefore, freezing based technologies can be more energy efficient than evaporation-based technologies.

5.8.2 Eutectic Freeze Crystallization

The eutectic temperature is the temperature at which a solute crystallizes out of solution when it is cooled to that temperature. Different dissolved salts have unique eutectic temperatures, and at a temperature of −25 °C, all dissolved salts within the solution will crystallize (Figure 5.27). This process enables recovery of relatively pure salts from a brine mixture. To improve efficiency, a fluidized bed–type configuration is used. This results in a carry-over of a part of salt, and thus, the water recovered in the form of ice may not meet drinking water standards. It will, however, be of a quality that can be returned to the feed stream of a typical mine water treatment plant.

5.8.3 HybridICE™ Technology

When water containing dissolved salts freezes, it preferentially forms ice crystals that are pure leaving behind a more concentrated salt solution. The ice can be separated and allowed to melt to produce a low dissolved solids product water. The removal of water from the resultant brine results in an equilibrium shift within the solution, resulting in salts crystallizing out of the solution. Since ice is less dense than water and the brine, it floats to the surface, while the denser salt crystals settle to the bottom. The pure water as ice and salt crystals can be easily separated. The waste heat produced in the cooling step is used in a

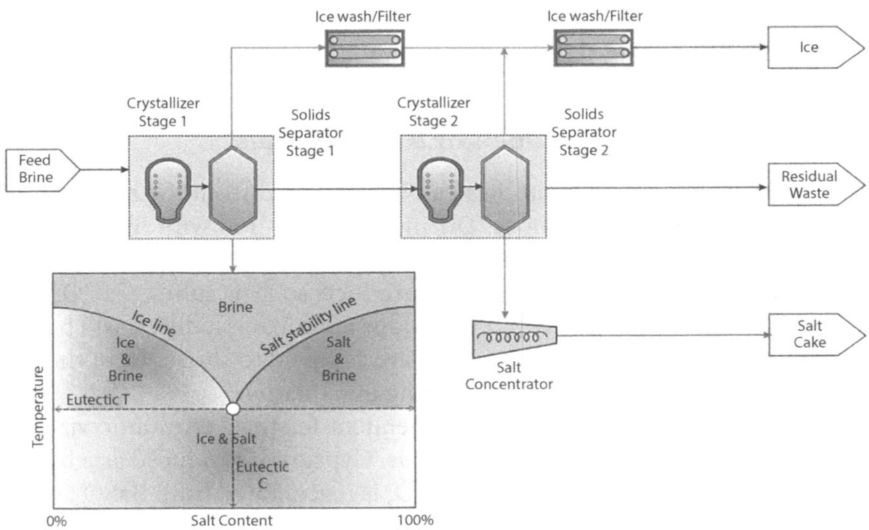

Figure 5.27 Eutectic freeze crystallization [9].

vacuum evaporator to evaporate the concentrate, leaving behind a concentrated, waste salt brine. With the HybridICE process, a mixture of salts is recovered. This is not saleable and would need to be disposed of at a hazardous waste facility in the case of mine water [9].

5.9 Sludge Processing

5.9.1 Background

Ideally, zero waste generation and disposal are required to protect the environment. Mine closure certificates are only granted upon acceptance of a mine closure plan. To ensure the continued existence of many mining operations, it is essential that low-cost solutions be sought and implemented. It is also important that legislation be realistic and in line with scientific principles. Where regulations are in place that do not lead to clear environmental benefits, they should be revised. If zero waste generation and disposal is required at the mine site, all sludges and brines have to be collected, transported, and dumped at existing toxic waste disposal sites.

The effects of gypsum- and $Fe(OH)_3$-rich sludge from an HDS plant, on the acid generation rate in a waste coal dump when pyrite was exposed to air, water, and iron-oxidizing bacteria has been studied [17, 22]. It was found that, when an $Fe(OH)_3$/gypsum mixture from a neutralization plant was mixed with coal discard, it reduced the acidity from 1,100 mg to 60 mg ($CaCO_3$/kg discard week). This was ascribed to the presence of some unused limestone in the sludge which resulted in a higher pH. It was further pointed out that the load of metals and acid in the sludge from the neutralization plant is not relevant when compared to the amounts of different metals present in the dump. Further, the sludge resulting from neutralization is inert. It does not contain pyrite which could generate acid when oxidized. Another argument is that the sludge from neutralization and desalination is not of a toxic nature. It contains mainly non-toxic compounds such as gypsum and $CaCO_3$.

The ideal way for waste minimization is to extract saleable compounds from the waste as discussed in the next section.

5.9.2 Recovery of Saleable Products or Raw Materials

Saleable products or process raw materials can be recovered from the sludge produced in several of the processes discussed above. During neutralization with limestone or lime, a sludge rich in gypsum and metal hydroxides is produced when the sulfate concentration is above the gypsum saturation [22, 36]. Saleable products such as sulfur, $CaCO_3$, and metals can be recovered by regeneration [110]. If the sludge is a mixed sludge, it will be considered as a waste that has to be disposed of. No values are assigned to sludge from the neutralization technologies, only to sludge produced during desalination.

Several of the desalination processes are dependent for their economic viability on the feasibility of the recovery of process raw materials. Gypsum is produced as a by-product in the HIPRO process. In the barium process, $BaCO_3$ is regenerated from $BaSO_4$, or the $BaSO_4$ sold as drilling mud. Other saleable products that can be recovered from the barium process sludge are sulfur and $CaCO_3$. In the case of the SAVMIN™ process, $Al(OH)_3$ should be recovered from ettringite ($3CaO.3CaSO_4.Al_2O_3$) while gypsum and $CaCO_3$ are produced as by-products. In the case of the KNeW process, the by-products, $NaNO_3$ and $(NH_4)_2SO_4$, can be sold to cover the cost of the raw materials, HNO_3 and NH_4OH. Where process raw materials are recycled, their costs are included in the cost of raw materials. These cost figures make provision for purity and utilization efficiency. In calculating the value of a recovered process raw material, the same price is used as the purchase price.

There are many different sludge disposal processes ranging from off-site transport to disposal in mine dumps [113]. The choice depends on many factors including: land availability, regulatory considerations, sludge production rate, sludge stability, available budget, and aesthetic considerations [114, 115]. The apparent disadvantages of these disposal options for large quantities pose an environmental problem due to leaching of rainwater or storage water producing an acidic effluent. Moreover, it results in loss of all commercial value. The effluent problems are mainly caused by the water solubility of gypsum (pers. comm. H. Kandelia, 2020).

Several ways of direct application and technologies to utilize waste gypsum have been developed in South Africa. These applications/technologies include the use of waste gypsum as: (1) fertilizer and soil conditioner, (2) cement and wallboard production, and (3) filler for road construction. However, only 5% to 10% of the total gypsum produced finds utility for such applications and the rest is stockpiled on land or dumped to sea.

5.10 Integrated Processes—ROC Process

5.10.1 Background

RO and Freeze Desalination (FD) are some of the main candidates for treatment of heavily polluted mine water. Although RO has been widely accepted as an efficient desalination process for both seawater and mine water, it suffers from predominantly two drawbacks, namely, membrane fouling/scaling and limited water recovery. In mine water, scaling is mainly driven by gypsum due to the use of lime/limestone pre-treatment, while water recovery as a function of osmotic pressure increase as water recovery increases. Limitations in water recovery lead to the generation of large volumes of brine which are currently disposed of in evaporation ponds. The use of lime/limestone also results in the generation of a mixed sludge which is costly to dispose of. On the other hand, FD processes are energy intensive and have been shown to be inadequate in producing the desired product water quality due to ice contamination. There is therefore need for energy-efficient, cost-effective, and environmentally friendly technology for treatment of brines, particularly those generated from mining affected waters.

5.10.2 Process Description

The ROC (Reverse Osmosis/Cooling) process was developed for the treatment of brines from desalination processes, such as reverse osmosis (Figure 5.28). The focus is to increase the water recovery from typically 85% to 99% and chemical recovery (e.g., calcium carbonate, magnesium hydroxide and sodium sulfate) to minimize sludge and brine disposal. In the ROC process, brine is treated with Sodium Carbonate (Na_2CO_3) and Sodium hydroxide (NaOH) in the pre-treatment stage to allow selective recovery of $Fe(OH)_3$, $Al(OH)_3$, $CaCO_3$, MnO_2, and $Mg(OH)_2$. Ammonia (NH_3) is removed with RO at pH 7. Magnesium in the brine is removed as $Mg(OH)_2$ at pH 11 by dosing NaOH. As the pH is already high, NH_3 can be stripped from the brine stream with air and absorbed in H_2SO_4. After pre-treatment, the sodium rich water is passed through a membrane stage to produce drinking water and brine. The brine has a concentration high enough to allow Na_2SO_4 crystallization upon cooling.

Use of Sodium salts enables the recovery of magnetite, pigments and other valuable products which is not possible with the current lime/limestone process. Sodium salts are used to manipulate stream chemistry and drive it toward more soluble Sodium Sulfate instead of Gypsum. This has an added advantage of reducing membrane scaling. Sodium Sulfate has a peculiar solubility: 4% at 0 °C and 20% at 20 °C. The ROC process exploits this peculiarity to treat the brine generated from RO by cooling it to 0 °C, thereby precipitating Sodium Sulfate, which, in turn, lowers the osmotic pressure. The lower osmotic pressure enables the use of RO to further treat the brine in a closed loop, resulting in a Zero Liquid Discharge process.

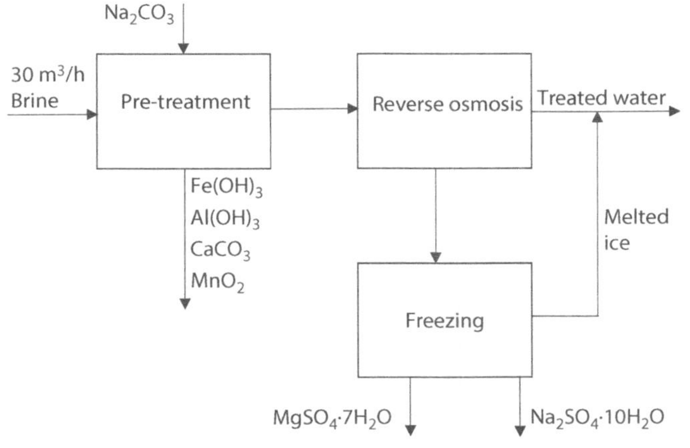

Figure 5.28 Process configuration of the ROC process.

Overall, combining the three stages, Sodium alkali pre-treatment, RO desalination, and Cooling crystallization, yields the following improvements to current technologies: (i) reduced RO membrane scaling—long membrane life and higher plant availability, (ii) increased water recovery—no brine generation, (iii) elimination of mixed sludge generation—no sludge disposal costs, and (iv) production of valuable products from polluted water.

The ROC process will solve one of the major environmental challenges of the worldwide industry. Waters laden with sulfate and other saline components will be treated to produce drinking water. Implementation of this cost-effective treatment technology will solve one of the major threats to food security which is freshwater salinization by producing water of good enough quality to discharge into public streams while valuable by-products could be recovered (e.g., gypsum and sodium sulfate). Lower freshwater demand will be one of the major benefits of this process which, in turn, will avail more water to other economic sectors.

It has been demonstrated that pigment can be recovered in the initial steps of treating mine water (Figure 5.29). Pre-treatment with UF and desalination with RO are proven technologies which have been implemented before (Figures 5.30 and 5.31). A commercial plant using the Freeze Crystallization (FC) process is being constructed for treatment of brine at A-Thermal Retort Technologies (Figures 5.32 and 5.33).

5.11 Feasibility Models

5.11.1 Introduction

A model was required to quantify the feasibility of competing process configurations that are considered for a specific application, where a process configuration is considered a combination of the technologies categorized as pre-treatment, desalination, brine treatment, and product recovery. The model was used to identify a cost-effective process configuration for the mine water from South Africa's Witwatersrand Western Basin, having the most polluted water of the three basins there, with a mine water make of 82,000 m^3/day. The input parameters of the model were used to quantify the output parameters.

Figure 5.29 Pigments produced from iron-rich mine water.

Figure 5.30 2.4 m^3/h RO stage of the R&D plant (Commissioned in 2018).

Figure 5.31 Completed desalination project (Namib Mills, Windhoek).

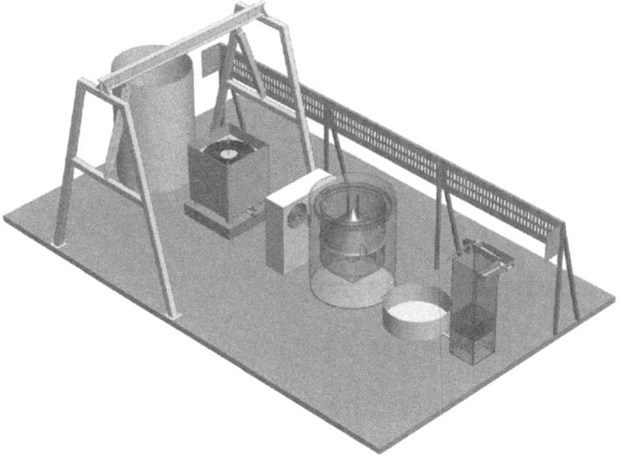

Figure 5.32 Integrated process for brine treatment.

Figure 5.33 Chiller and cooler reactor (a) and ice brine separation system (b).

Input parameters

- Application (e.g., Central Basin, Eastern Basin or Western Basin);
- Flowrate and quality of feed water;
- Price, purity, and utilization of raw materials (e.g., limestone, lime, sodium carbonate);
- Price of electricity;
- Staff and salaries/wages;
- Price and yield of products (e.g., water, by-products, crops to be produced from irrigation); and
- Transportation (e.g., R 1,000/t) and disposal cost (e.g., R 1,500/t) of waste (brine or sludge).

Output parameters

- Flow and water quality of treated water and brine;
- Dosage, usage, and cost of raw materials (e.g., limestone, lime, sodium carbonate), electricity, labor, and irrigation;
- Yield and value of water and by-products in R/m^3 (e.g., R 8.00/m^3 for water with TDS < 750 mg/L and R 0.00/m^3 for water with TDS > 750 mg/L);
- Mass and composition of sludge;
- Transport and disposal cost of waste (brine or sludge) in (R/m^3);
- Capital cost (in R) and capital redemption cost (in R/m^3); and
- Net cost/income of process stage or process configuration for a specific application (in R/m^3).

5.11.2 Feasibility of Individual Stages

5.11.2.1 Neutralization Technologies

Western Basin water was used to determine the main cost components when treated with the following stages: Pre-treatment with Na_2CO_3 for removal of metals, desalination, and

Table 5.6 Cost figures of neutralization technologies for Western Basin Water; if not stated otherwise, costs in R/m^3.

Cost item	Lime	$CaCO_3$/Lime	Na_2CO_3
Capital cost (million R/(1,000 m^3/d))	8	8	8
Capital Redemption Cost	3.79	3.79	3.79
Chemicals cost	5.57	3.61	11.97
Electricity (kWh/m^3 @ R 1.50/kWh)	1.50	1.50	1.50
Disposal cost	16.65	16.65	0.54
Labor cost	0.20	0.20	0.20
Maintenance (10% of Capital Redemption)	0.38	0.38	0.38
Value of products	0.00	0.00	−3.57
Gross cost (disposal cost excluded)	11.44	9.47	14.26
Gross cost (disposal cost included)	28.09	26.12	14.81

brine treatment. This was done for four selected technologies: ROC, HiPro, SAVMIN™, and CSIR ABC (Barium) processes.

Neutralization technologies need to be used as pre-treatment before desalination for removal of acid and metals. Due to the high TDS of the neutralized water, it cannot be released into public water bodies. A summary of the results shows the following (Table 5.6):

- Western Basin water can be neutralized with lime at a cost R 11.44/m^3 or with limestone/lime at a cost of R 9.47/m^3. As new projects require zero waste disposal, the disposal cost was calculated when sludge and brine streams need to be disposed at toxic waste disposal sites, as the sludge is a mixture of gypsum and metal hydroxides. Disposal cost will add R 16.65/m^3 when lime or lime/$CaCO_3$ is used for pre-treatment. The following assumptions were made for calculating the disposal cost: (i) transportation cost: R 1,000/t; (ii) disposal cost: R 1,500/t; and (iii) the sludge has a solids content of 50%.
- Western Basin water can be neutralized with Na_2CO_3 at a cost of R 14.26/m^3. This cost figure is less than the cost of lime neutralization. The higher price of Na_2CO_3 compared to lime (R 11.97/m^3 versus R 5.57/m^3) was offset by the saving in the cost of sludge disposal (R 0.54/m^3 versus R 16.65/m^3) and the value of pigments (R 3.57/m^3) that can be recovered.

5.11.2.2 Desalination Technologies

Using one of the desalination technologies, the TDS was reduced to less than 100 mg/L with the RO processes (HiPRO and ROC), while with the chemical processes (SAVMIN™ and CSIR ABC), it was greater than 400 mg/L (Table 5.7). This results from the chemical processes

Table 5.7 Cost figures of desalination technologies for western basin water; if not stated otherwise, costs in R/m³.

Cost item	SAVMIN™	Lime/$BaCO_3$ CSIR ABC	Lime/RO HiPRO	Na_2CO_3/ RO ROC
TDS of treated water (mg/L)	442.80	467.19	68.82	93.50
Capital cost (million R/(1,000 m³/d))	12	10	15	15
Capital redemption cost	5.68	4.74	7.10	7.10
Chemicals cost	11.44	29.23	0.35	0.20
Electricity (kWh/m³ @ R 1.50/kWh)	1.95	1.95	5.25	5.25
Disposal cost	23.46	21.78	169.80	130.00
Labor cost	0.33	0.33	0.33	0.33
Maintenance (10% of capital redemption)	0.57	0.47	0.71	0.71
Value of products		−4.79		−4.72
Gross cost (disposal cost excluded)	19.97	31.93	13.74	8.87
Gross cost (disposal cost included)	43.43	53.71	183.54	138.87

Table 5.8 Cost figures of Brine treatment technologies for Western Basin water; if not stated otherwise, costs in R/m³.

Cost item	Freeze crystallization	Evaporation
Capital cost (million R/(1,000 m³/d))	60	160
Capital redemption cost	1.42	3.79
Chemicals cost	0.00	0.00
Electricity (kWh/m³ @ R 1.50/kWh)	3.75	18.75
Disposal cost	13.00	6.50
Labor cost	0.10	0.10
Maintenance (10% of capital redemption)	0.14	0.38
Gross cost (disposal cost excluded)	3.99	19.23
Gross cost (disposal cost included)	16.99	25.73

not being able to remove monovalent ions (Na^+, K^+, and Cl^-). The chemical cost of chemical processes (SAVMIN™ and CSIR ABC) was higher than that of the RO processes (HiPRO and ROC) due to the need for dosing $Ca(OH)_2$ and $Al(OH)_3$ (assumed 30% losses) in the case of the SAVMIN™ process and $BaCO_3$ in the case of the ABC process. The cost of chemicals can be reduced through the recovery of $Ca(OH)_2$ from gypsum in the case of the SAVMIN™

Table 5.9 Cost figures for sludge treatment technologies for Western Basin water; if not stated otherwise, costs in R/m^3.

Cost item	ROC	SAVMIN	$BaCO_3$
Main compound in sludge	Na_2SO_4	$CaSO_4{\cdot}2H_2O$	$BaSO_4$
Capital cost (million R for a 75,000 m^3/d plant)	60	60	60
Capital cost (million R/1,000 m^3/d)	0.8	0.8	0.8
Product to produce	Na_2CO_3	$CaCO_3$	$BaCO_3$
Dosage required (mg/L)	2,632.62	2,591.59	4,050.00
Molar mass of main compound	142.04	172.17	233.39
Purity (%)	99.00	95.00	95.00
Carbon content in coal (%)	65.00	65.00	65.00
Coal required (mg/L Feed)	1,037.18	878.42	1,013.36
Coal waste (mg/L)	363.01	307.45	354.68
Sludge losses (20%) (mg/L)	705.34	891.51	958.02
Coal + sludge losses (mg/L)	1,068.36	1,198.96	1,312.70
Coal price (R/t)	600.00	600.00	600.00
Disposal cost (R/t)	2,600.00	2,600.00	2,600.00
Lime (replace 20% losses)	1.11		
$BaCO_3$ (replace 20% losses)			5.85
Electricity usage (kWh/m^3)	1.00	1.00	1.00
Electricity price (R/kWh)	1.50	1.50	1.50
Capital redemption cost (R/m^3 Feed)	1.57	1.57	1.57
Coal cost (R/m^3 Feed)	0.62	0.53	0.61
Electricity cost (R/m^3 Feed)	1.50	1.50	1.50
Labor (R/m^3 Feed)	1.00	1.00	1.00
Disposal cost (R/m^3 Feed)	1.83	2.32	2.49
Maintenance cost (10% of capital redemption)	0.16	0.16	0.16
Total (R/m^3 Feed)	9.30	8.57	
Product cost (R/t)	3,531.69	3,307.38	3,622.54
Product Purchase Price	4,500.00	2,500.00	7,000.00
Savings (Trade deficit offset)	968.31		3,377.46
Recovery through thermal treatment (%)	70.00	70.00	70.00
Product value	8.29	4.54	19.85

Table 5.10 Feasibility of various process configurations; if not stated otherwise, costs in R/m^3; [1]million R/(1,000 m^3/d), [2]kWh/m^3 @ R 1.50/kWh, [3]10% of Capital Redemption.

		P1.1	P1.2	P2.1	P2.2	P3.1	P3.2	P4.1	P4.2
Process configuration		**Lime | RO**	**Lime | RO**	**Na_2CO_3 | RO**	**Na_2CO_3 | RO**	**Lime | $BaCO_3$**	**Lime | $BaCO_3$**	**Lime | SAVMIN**	**Lime | SAVMIN**
Cost item		Sludge disposal	Freeze crystal	Freeze crystal	Freeze crystal		$BaCO_3$ recovery		$CaCO_3$ recovery
Capital cost[1]	Pre-treatment	8	8	8	8	8	8	8	8
	Desalination	15	15	15	15	10	10	12	12
	Brine treatment		60		60				
	Sub-total	23	83	23	83	18	18	20	20
Capital Redemption Cost	Pre-treatment	3.79	3.79	3.79	3.79	3.79	3.79	3.79	3.79
	Desalination	7.10	7.10	7.10	7.10	4.74	4.74	5.68	5.68
	Brine treatment	0.00	1.42	0.00	1.42				
	Sub-total	10.89	12.31	10.89	12.31	8.53	8.53	9.47	9.47
Chemicals cost	Pre-treatment	5.57	5.57	11.97	11.97	5.57	5.57	5.57	5.57
	Desalination	0.35	0.35	0.20	0.20	29.23	29.23	11.44	11.44
	Sub-total	5.92	5.92	12.17	12.17	34.80	34.80	17.01	17.01

(*Continued*)

Table 5.10 Feasibility of various process configurations; if not stated otherwise, costs in R/m^3; [1]million R/(1,000 m^3/d), [2]kWh/m^3 @ R 1.50/kWh, [3]10% of Capital Redemption. (*Continued*)

		P1.1	P1.2	P2.1	P2.2	P3.1	P3.2	P4.1	P4.2
Process configuration		**Lime | RO**	**Lime | RO**	**Na_2CO_3 | RO**	**Na_2CO_3 | RO**	**Lime | $BaCO_3$**	**Lime | $BaCO_3$**	**Lime | SAVMIN**	**Lime | SAVMIN**
Electricity[2]	Pre-treatment	1.50	1.50	1.50	1.50	1.50	1.50	1.50	1.50
	Desalination	5.25	5.25	5.25	5.25	1.95	1.95	1.95	1.95
	Brine treatment	0.00	3.75	3.75	3.75				
	Sub-total	6.75	10.50	10.50	10.50	3.45	3.45	3.45	3.45
Disposal cost	Pre-treatment	16.65	16.65	0.54	0.54	16.65	16.65	16.65	16.65
	Desalination	169.80		130.00		21.78	21.78	23.46	
	Brine treatment		13.00		13.00				
	Sub-total	186.45	29.65	130.54	13.54	38.43	38.43	40.11	16.65
Labor cost	Pre-treatment	0.20	0.20	0.20	0.20	0.20	0.20	0.20	0.20
	Desalination	0.33	0.33	0.33	0.33	0.33	0.33	0.33	0.33
	Brine treatment	0.00	0.10	0.00	0.10				
	Sub-total	0.53	0.63	0.53	0.63	0.53	0.53	0.53	0.53

(*Continued*)

Table 5.10 Feasibility of various process configurations; if not stated otherwise, costs in R/m^3; [1]million R/(1,000 m^3/d), [2]kWh/m^3 @ R 1.50/kWh, [3]10% of Capital Redemption. (*Continued*)

		P1.1	P1.2	P2.1	P2.2	P3.1	P3.2	P4.1	P4.2
Process configuration		**Lime | RO**	**Lime | RO**	**Na_2CO_3 | RO**	**Na_2CO_3 | RO**	**Lime | $BaCO_3$**	**Lime | $BaCO_3$**	**Lime | SAVMIN**	**Lime | SAVMIN**
Maintenance[3]	Pre-treatment	0.38	0.38	0.38	0.38	0.38	0.38	0.38	0.38
	Desalination	0.71	0.71	0.71	0.71	0.47	0.47	0.57	0.57
	Brine treatment	0.00	0.14	0.14	0.14				
	Sub-total	1.09	1.23	1.23	1.23	0.85	0.85	0.95	0.95
Products	Pre-treatment			3.06	3.06				
	Desalination					4.79	4.79		
	Brine treatment				4.72				
	Sub-total	0.00	0.00	3.06	7.78	4.79	4.79	0.00	0.00
Gross cost (Disposal excluded)		25.18	30.59	35.31	36.83	48.16	48.16	31.41	31.41
Disposal cost		186.45	29.65	130.54	13.54	38.43	38.43	40.11	16.65
Income		0.00	0.00	3.06	7.78	4.79	4.79	0.00	0.00
Gross cost—Income		25.18	30.59	32.25	29.05	43.37	43.37	31.41	31.41

process and $BaCO_3$ from $BaSO_4$ in the case of the ABC process. The transport and disposal costs for the sludges and non-saleable products typically amount to R 1,000/t and R 1,500/t, respectively. This serves as motivation why waste streams need to be processed, especially in the case of the RO technologies where the disposal cost exceeds R 100/m^3 feed water.

5.11.2.3 Brine Treatment

FC and evaporation can be sued for brine treatment (Table 5.8). As has been shown, the cost of FC is less than the cost of evaporation due to lower energy cost and lower cost for material of construction. It is further noted that treatment of the brine by FC results in large disposal saving costs due to the reduction in the brine volume. Without brine treatment with FC, the disposal cost of the ROC process (Na_2CO_3 pre-treatment/RO) amounts to R 130.00/m^3 feed (Table 5.7). By treating the brine with FC, the disposal cost is reduced to R 13.00/m^3. This cost can be further reduced by treating the brine/Na_2SO_4 sludge further to recover the raw material, Na_2CO_3.

5.11.2.4 Product Recovery

The previous sections showed that the cost of mine water treatment is dominated by the cost of brine and sludge disposal. Treating the sludge will not only reduce the cost of sludge disposal but also the cost of raw material when recovered from the sludge (Table 5.9). Na_2CO_3 can be recovered at a cost of R 3,532/t (compared with R 4,500 when purchased) and $BaCO_3$ can be recovered at a cost of R 3,623/t (compared with R 7,000/t when purchased). These figures were not included in the feasibility section.

5.11.3 Feasibility of Various Process Configurations

Cost of desalination differs when: (i) cost of waste disposal is excluded and (ii) cost of waste disposal is included (Table 5.10). The HiPRO process is currently used for desalination of 100,000 m^3/d of mine water in Mpumalanga, a coal-rich province in South Africa. The treatment cost was calculated to be R 25.18/m^3 (Column P1). This figure included capital redemption cost (R 3.79/m^3) but excluded waste disposal cost (R 186.45/m^3). This high cost for waste disposal shows the importance of processing of waste streams when zero waste disposal is implemented. Column P1.2 shows that disposal cost is substantially reduced when FC is used to reduce the volume of brine to be disposed by 90%. The total treatment cost increases from R 25.18/m^3 to only R 30.59. The disposal cost is reduced from R 186.45/m^3 to only R 29.65/m^3.

The ROC process was developed to deal with the shortcomings of the HiPRO process. By replacing lime with Na_2CO_3 in the pre-treatment stage, gypsum scaling of the RO membranes is avoided, metal compounds could be recovered selectively instead of as mixed gypsum/metal hydroxide sludges, and the brine concentration could be increased from 30 g/L TDS to 80 g/L TDS. This way, the treatment cost was reduced from R 30.59/m^3 to R 29.05/m^3 and the Waste Disposal cost from R 29.65/m^3 (Column P1.2) to R 13.54/m^3 (Column 2.2).

Chemical technologies competed with the RO technologies if the concentrations of monovalent ions (Na^+, K^+, and Cl^-) are low and the sludge is processed through thermal treatment for the recovery of raw materials. The total cost of the ABC process ($BaCO_3$) amounted to R 43.37/m^3 (column P3.2 in Table 5.10). This included the cost of $BaCO_3$ (R 29.23/m^3). By processing the $BaSO_4$ sludge to $BaCO_3$, the chemical cost of R 29.23/m^3 (column P3.2 in Table 5.10) can be reduced to R 14.67/m^3 (Table 5.9). The total cost of the SAVMIN™ process amounted to R 31.41/m^3. This included the cost of lime (R 11.44/m^3). By processing the gypsum sludge to produce Lime, the chemical cost of R 11.44/m^3 (column P4.1 in Table 5.10) can be lowered to R 9.51/m^3 (Table 5.9).

5.12 Conclusions

This chapter showed that a combination of technologies needs to be considered in order to achieve maximum environmental protection at affordable costs. For this reason, water treatment technologies were divided into the following categories: (i) neutralization, (ii) desalination, (iii) brine treatment, and (iv) sludge treatment. Affordability is a key factor, since water treatment needs to be applied by all mine water users. The environment does not benefit if only a portion of the mine water is treated to a good water quality and the balance is only neutralized or not treated at all.

Neutralization with $CaCO_3$, $Ca(OH)_2$, or Na_2CO_3 is essential, as it removes metals to below potentially toxic concentrations as well as achieving partial desalination due to gypsum crystallization. Desalination needs to be applied if treated water is to be used for drinking or industrial purposes. NF and RO are used for desalination in full-scale applications. Ion-exchange is also a promising technology to consider. Chemical desalination using processes such as SAVMIN™ or CSIR ABC can be used for treatment of water low in sodium. In the case of NaCl-rich waters, the chemical desalination processes need to be combined with ion-exchange or membrane processes for final "polishing". Desalination using biological processes will find application where biological energy sources are available at low cost. Passive treatment biological systems can be used in applications where small volumes of water, which are not heavily polluted, need to be treated in remote areas. With the ROC process, brine and sludge can be treated by combining FD and thermal treatment with chemical pre-treatment and RO. This allows for reduced treatment cost as saleable products can be recovered such as drinking water, pigment, aluminium sulfate, calcium carbonate, and sodium sulfate.

The feasibility section showed that the waste disposal cost is unaffordable if the concept of zero waste disposal at the treatment site is applied. Brine treatment with FC and thermal treatment of the sludge might reduce the cost of waste disposal to acceptable ranges.

Acknowledgements

The authors thank the South African Water Research Commission (WRC) for allowing us to write and reuse the WRC Report: *Mine Water Treatment Technology Selection Tool: Users' Guide* [8]. The technologies discussed in that report were summarized in this document. Fritz Carlsson provided proof-reading and editorial advice on an early version of this chapter.

References

1. Van Zyl, H.C., Maree, J.P., van Niekerk, A.M., van Tonder, G.J. and Naidoo, C., Collection, treatment and re-use of mine water in the Olifants River Catchment, *SAIMM*, 101, 1, 41–46, 2000.
2. Rukuni, T., Maree, J.P., Carlsson, F., An integrated low cost solution to the mine water problem in South Africa, in: *Proc. WISA Biennial Conf*, 3S Media, South Africa. 2016.
3. Mey, W.S. and Van Niekerk, A., Evolution of mine water management in the Highveld coalfields, in: *Water Institute of Southern Africa & Intern. Mine Water Assoc.: Proc. Intern. Mine Water Conf*, pp. 38–45, Pretoria, South Africa 2009.
4. Expert Team of the Inter-Ministerial Committee, *Mine Water Management in the Witwatersrand Gold Fields with Special Emphasis on Acid Mine Drainage - Report to the Inter-Ministerial Committee on Acid Mine Drainage, Council for Geoscience*, Pretoria 2010.
5. Creamer, T., TCTA again warns of funding shortfall for acid water projects. *Eng. News*, Creamer Media, 19 January 2012, https://www.miningweekly.com/article/tcta-again-warns-of-funding-shortfall-for-acid-water-projects-2012-01-19
6. Van Vuuren, L., Acid mine drainage raring to go. *Water Wheel*, 10, 6, 16–17, Dec 2011.
7. Department of Water Affairs (DWA), Feasibility study for a long-term solution to address the acid mine drainage associated with the East, Central and West Rand underground mining basins. Study Report No. 1: Inception Report—DWA Report No.: P RSA 000/00/16112. Pretoria: DWA, 2012. Available from http://www.dwaf.gov.za/Projects/AMDFSLTS/Documents/AMD%20FS%20LTS_Assessment%20of%20the%20Water%20Quantity%20&%20Quality%20of%20the%20Mine%20Voids%20(Draft%20Report).pdf. Accessed April 14, 2013.
8. Solomons, I., War on AMD. *Mining Weekly*, 3, July 2015. https://m.miningweekly.com/article/govt-private-sector-working-constructively-to-tackle-acid-mine-drainage-in-wits-basin-2015-07-03-1/rep_id:3861
9. Dama-Fakir, P., Sithole, Z., van Niekerk, A.M., Dateling, J., Maree, J.P., Rukuni, T., Mtombeni, T., Ruto, S., Zikalala, N., Hughes, C., Wurster, A., Saunders, B., *Mine water treatment technology selection tool: Users' Guide*, Water Research Commission, Pretoria, 2017.
10. Pretorius, C.F., Design, construction and operation of a high density sludge lime treatment plant, in: *Proc. WISA Biennial Conf*, Cape Town, WISA, 1998.
11. Kostenbader, P.D. and Haines, G.F., High density sludge treats acid mine drainage. *Coal Age*, 75, 90–97, 1970.
12. Osuchowski, R., Advanced treatment of acid mine water. *Technol. SA*, April, 9–11, 1992.
13. Van Tonder, G.J. and Gunther, P., Manganese removal from acid mine drainage by high pH lime treatment in a high density sludge process, in: *Proc. WISA Biennial Conf*, Sun City, South Africa, WISA, 1998.
14. Bosman, D.J., The improved densification of sludge from neutralised acid mine drainage. *J. S. Afr. Inst. Min. Metall.*, 74, 9, 340–348, 1974.
15. Günther, P., Maree, J.P., Strobos, G., Mtimikulu, J.S., Neutralisation of Acid Leachate in a Coal Processing Plant with Calcium Carbonate, in: *Proc. 8th Intern. Congr. on Mine Water & the Environment*, Johannesburg, South Africa, 20–25 Oct, pp. 367–382, 2003.
16. Renton, J.J. and Brown, H.E., An evaluation of fluidised bed combustor ash as a source of alkalinity to treat toxic rock materials. *Eng. Geol.*, 40, 3–4, 157–167, 1995.
17. Maree, J.P., *Underground neutralisation of mine water with limestone*, WRC Report, Pretoria, 1996, No 609/1/96.
18. Maree, J.P., Limestone handling and dosing system. S.A. Patent 2001/7086; USA patent 6,592,246 B2, assigned to CSIR. 2000.
19. Mohajane, G.B., Maree, J.P., Panichev, N., Treatment of iron(II)-rich acid mine water with limestone and oxygen. *Water Sci. Technol.*, 70, 2, 209–217, 2014.

20. Mohajane, G.B., Maree, J.P., Panichev, N., Louw, W.J., Treatment of alkaline mine water with hydrogen peroxide, in: *Proc. WISA Biennial Conf*, Mbombela, South Africa, 3S Media, 2014.
21. Van Tonder, G.J., Theron, D.J., Maree, J.P., Cost optimisation of the water management strategy by steady-state modelling of the water network of a copper/nickel mine and processing plant, in: *Proc. WISA Biennial Conf*, Sun City, South Africa, WISA, 2000.
22. Maree, J.P. and Du Plessis, P., Neutralisation of acid mine water with calcium carbonate. *Water Sci. Technol.*, 29, 9, 285–296, 1994.
23. Maree, J.P., Van Tonder, G.J., Millard, P., Erasmus, C., Pilot scale neutralisation of underground mine water. *Water Sci. Technol.*, 34, 10, 141–149, 1996.
24. Cornish, L., AMD in the Eastern Basin sorted by September 2015. *Mining Rev. Africa*, 2014, 9, 36–38, 2014.
25. Odendaal, N., Group Five starts to work on TCTA AMD contract. *Mining Weekly*, 22, February 2013. https://www.miningweekly.com/article/group-five-starts-work-on-tcta-amd-contract-2013-02-22
26. Van Vuuren, L., Acid Mine Drainage solutions rearing to go. *Water Wheel*, 10, 6, 16–17, November/December, 2011 https://www.miningweekly.com/article/group-five-starts-work-on-tcta-amd-contract-2013-02-22
27. Ritz, Ritz Pumps SA Africa's leading dewatering submercible pumps supplier, Mining Review Africa, 41–43, September 2014.
28. Maree, J.P., Mujuru, M., Bologo, V., Daniels, N., Mpholoane, D., Neutralisation treatment of AMD at affordable cost. *Water SA*, 39, 2, 245–250, 2013.
29. Norris, L. R. III, *Barium Carbonate Treatment of Sulfuric Acid Wastewater*, MSc Thesis, Virginia Polytechnic Institute and State University, Blacksburg, 1972.
30. Trusler, G.E., *The chemical removal of sulphates using barium salts, M.Eng. thesis*, University of Natal, Durban, 1988.
31. Wilsenach, J., *Cost estimate for barium sulfate reduction, Internal report of the Division of Water Technology*, Pretoria 1986, CSIR, 620/2616/6.
32. Motaung, S., Maree, J., de Beer, M., Bologo, L., Theron, D., Baloyi, J., Recovery of drinking water and by-products from gold mine effluents. *Water Res. Dev.*, 24, 3, 433–450, 2008.
33. Trusler, G.E., Edwards, R.I., Brouckaert, C.J., Buckley, C.A., The chemical removal of sulfates, in: *Proc. 5th National Meeting of the South African Institute of Chemical Engineers*, Pretoria, SAIChe, 1988.
34. Adlem, C.J.L., *Treatment of sulfate-rich effluents with the barium sulfide process, M. Tech. thesis*, Pretoria Technikon, Pretoria, South Africa, 1997.
35. Bosman, D.J., Clayton, J.A., Maree, J.P., Adlem, C.J.L., Removal of sulfate from mine water. *Int. J. Mine Water.*, 9, 1–4, 149–163, 1990.
36. Hlabela, P., Maree, J.P., Bruinsma, D., Barium carbonate process for sulphate and metal removal from mine water. *Mine Water Env.* 26, 1, 14–22, 2007.
37. Maree, J.P., Bosman, D.J., Jenkins, G.R., Chemical removal of sulfate, calcium and metals from mining and power station effluents. *Water Sewage Effluent*, 9, 3, 10–25, 1989.
38. Adlem, C.J.L., Maree, J.P., du Plessis, P., Treatment of sulfate-rich mining effluents with the barium hydroxide process and recovery of valuable by-products, in: *4th Intern. Mine Water Assoc. Congr*, pp. 211–222, Ljubljana–Pörtschach, Austria, 1991.
39. Bowell, R., *A review of sulphate removal options for mine waters*, A. Jarvis, B.A. Dudgeon and P.L. Younger (Eds.), Mine Water, Newcastle upon Tyne, 2004.
40. Maree, J., Wilsenach, I., de Beer, M., Motaung, S., Carlsson, F., Bologo, L., Redebe, V., Baloyi, J., *Drinking water and by-products from gold mine effluents – Water stage in pilot studies on CSIR ABC process*, Contract Report for Western Utilities Corporation, CSIR, Pretoria, 2008.

41. Sillén, L.G. and Martell, A.E., Stability Constants of Metal-Ligand Complexes, *Royal Soc. Che.*, London, 1964.
42. Tshwete, L., Gunther, P., Mey, W., van Niekerk, A., Emalahleni (Witbank) mine water reclamation project, in: *Proc. WISA Biennial Conf*, Durban, South Africa, 3S Media, 2006.
43. Gunther, P. and Mey, W., Selection of mine water treatment technologies for the eMalahleni (Witbank) water reclamation project, in: *Proc. WISA Biennial Conf*, Sun City, South Africa, 3S Media 2008.
44. Hutton, B., Kahan, I., Naidu, T., Gunther, P., Operating and maintenance experience at the eMalahleni water reclamation plant, in: *Water Institute of Southern Africa & Intern. Mine Water Assoc.: Proc. Intern. Mine Water Con.*, C. Wolkersdorfer (Ed.), pp. 415–430, Pretoria, South Africa, Document Transformation Technologies cc, 2009.
45. Aveng Water, eMalahleni Water Reclamation Plant, 2014. https://web.archive.org/web/20190731174235/http://www.avengwater.co.za/projects/emalahleni-water-reclamation-plant (orignal site not available any more).
46. Kolver, L., Kromdraai water treatment plant follows success of eMalahleni, *Mining Weekly*, 2012-05-04, 2012. http://www.miningweekly.com/article/contract-awarded-to-increase-capacity-of-emalahleni-mine-water-treatment-plant-2012-05-04
47. Karakatsanis, E. and Cogho, V., Drinking Water from Mine Water Using the HiPrO® Process – Optimum coal Mine Water reclamation Plant, in: *Mine Water and Innovative Thinking, Proc. Int. Mine Water Con.*, C. Wolkersdorfer and A. Freund (Eds.), pp. 135–138, Cape Breton University, Sydney, NSW, 2010.
48. Gunther, P., Naidu, T., Mey, W., eMalahleni mine water reclamation project – Key learnings, in: *Proc. WISA Biennial Conf*, Sun City, South Africa, 3S Media, 2008.
49. Gunther, P. and Naidu, T., Mine water reclamation – towards zero disposal, in: *Proc. WISA Biennial Conf*, South Africa, 3S Media, 2008.
50. Juby, G.J.G. and Pulles, W., Evaluation of electrodialysis reversal for desalination of brackish mine water. *WRC Rep.*, 1990. 179/1/90, Pretoria.
51. Lutchmiah, K., Verliefde, A.R.D., Roest, K., Rietveld, L.C., Cornelissen, E.R., Forward osmosis for application in wastewater treatment: A review. *Water Res.*, 58, 179–197, 2014.
52. McCutcheon, J.R., McGinnis, R.L., Elimelech, M., A novel ammonia-carbon dioxide forward (direct) osmosis desalination process. *Desalination*, 174, 1–11, 2005.
53. McGinnis, R., Osmotic Desalination Process, Desalination, 207, 370–382, USA Patent 7560029, 2009.
54. Gethard, K., Sea-Khow, O., Mitra, S., Water desalination using carbon-nanotube-enhanced membrane distillation. *ACS Appl. Mater. Interfaces*, 3, 2, 110–114, 2011.
55. Johnson, B., Biological removal of sulfurous compounds from inorganic wastewaters, in: *Environmental Technologies to Treat Sulfur Pollution: Principles and Engineering*, P. Lens and L. Hulshoff Pol (Eds.), pp. 175–205, IWA Publishing, London, 2000.
56. Robb, G.A. and Robinson, J.D.F., Acid drainage from mines. *Geogr. J.*, 161, 1, 47–54, 1995.
57. Van Hille, R.P., Boshoff, G.A., Rose, P.D., Duncan, J.R., A continuous process for the biological treatment of heavy metal contaminated acid mine water. *Resour. Conserv. Recy.*, 27, 157–167, 1999.
58. Kalin, M., Cairns, J., McCready, R., Ecological engineering methods for acid mine drainage treatment of coal wastes. *Resour. Conserv. Recy.*, 5, 265–275, 1991.
59. Johnson, D.B., Acidophilic microbial communities: Candidates for bioremediation of acidic mine effluents. *Int. J. Biodeter. Biodegr*, 35, 41–58, 1995.
60. White, C., Sayer, J.A., Gadd, G.M., Microbial solubilisation and immobilization of toxic metals: Key biogeochemical processes for treatment of contamination. *FEMS Microbiol. Rev.*, 20, 503–516, 1997.

61. Hulshoff Pol, L.W., Lens, P.N.L., Weijma, J., Stams, A.J.M., New developments in reactor and process technology for sulfate reduction. *Water Sci. Technol.*, 44, 67–76, 2001.
62. Lens, P., Vallero, M., Esposito, G., Zandvoort, M., Perspectives of sulfate reducing bioreactors in environmental biotechnology, Reviews. *Environ. Sci. Biotechnol.*, 1, 311–325, 2002.
63. Christensen, B., Laake, M., Lien, T., Treatment of acid mine water by sulfate-reducing bacteria: Results from a bench-scale experiment. *Water Res*, 30, 1617–1624, 1996.
64. Dvorak, D.H., Hedin, R.S., Edenborn, H.M., McIntire, P.E., Treatment of metal-contaminated water using bacterial sulphate reduction: Results from pilot-scale reactors. *Biotechnol. Bioeng.*, 40, 609–616, 1992.
65. Tuttle, J.H., Dugan, P.R., Randles, C.I., Microbial sulfate reduction and its potential utility as an acid mine water pollution abatement procedure. *Appl. Microbiol.*, 17, 297–302, 1969.
66. Chappelle, F.M., *Groundwater Microbiology and Geochemistry*, Wiley, Sussex, 1993.
67. Gould, W.D., Bechard, G., Lortie, L., The nature and role of microorganisms in mine drainage, in: *Short Course Handbook on Environmental geochemistry of Sulfide Mine-Wastes*, J.L. Jambor and D.W. Blowes (Eds.), pp. 103–132, Mineralogical Association of Canada, Waterloo, Ontario, 1994.
68. Arnesen, R.T., Monitoring water quality during filling of the Larkken mine: Role of sulfate reducing bacteria in metals removal, in: *2nd Int. Conf. Abatement Acid Mine Drainage*, pp. 16–18/9/91, Springer, Dordrecht, Montréal, Canada, 1991.
69. Huang, H.H. and Tahija, D., On the nature of Berkeley Pit Water, in: *Proc. Mineral and Hazardous Waste Processing Symposium*, Butte Montana, 1990, 10/01/90.
70. Middleton, A.C. and Lawrence, A.W., Kinetics of microbial sulphate reduction. *J. Wat. Pollut. Control Fed.*, 49, 7, 1659–1670, 1977.
71. Cork, D.J. and Cusanovich, M.A., Continuous disposal of sulphate by a bacterial mutualism. *Dev. Ind. Microbiol.*, 20, 591–602, 1979.
72. Hilton, B.S., Oleszkewicz, A., Oziemblo, Z.J., Sulphate reduction as an alternative to methane fermentation of industrial wastes, in: *Canadian Society for Civil Engineering Annual Conference*, pp. 243–261, Saskatoon, Canada, 1985.
73. Maree, J.P. and Strydom, W., Biological sulphate removal from a packed bed reactor. *Water Res.*, 19, 9, 1101–1106, 1985.
74. Maree, J.P., Gerber, A., Hill, E., An integrated process for biological treatment of sulphate containing industrial effluents. *J. Wat. Pollut. Contr. Fed.*, 59, 12, 1069–1074, 1987.
75. Du Preez., L.A. and Maree, J.P., Pilot-scale biological sulphate and nitrate removal utilizing producer gas as energy source. *Water Sci. Technol.*, 30, 12, 275–285, 1994.
76. Du Preez, L.A., Odendaal, J.P., Maree, J.P., Ponsonby, M., Biological removal of sulphate from industrial effluents using producer gas as energy source. *Env. Technol.*, 13, 875–882, 1992.
77. Sipma, J., Osuna, M.B., Lettinga, G., Stams, A.J.M., Lens, P.N.L., Effect of hydraulic retention time on sulfate reduction in a carbon monoxide fed thermophilic gas lift reactor. *Water Res.*, 41, 9, 1995–2003, 2007.
78. Maree, J.P., Hulse, G., Dods, D., Schutte, C.E., Pilot plant studies on biological sulfate removal. *Water Sci. Technol.*, 23, 1293–1300, 1991.
79. Du Preez, L.A., Maree, J.P., Jackson-Moss, C., Biological sulfate removal from industrial effluents. *Environ. Technol.*, 13, 875–882, 1992.
80. Van Houten, R. and Lettinga, G., Biological sulphate reduction with synthesis gas: microbiology and technology, in: *Progress in Biotechnology*, R.H. Wijffels, R.M. Buitelaar, C. Bucke, J. Tramper (Eds.), pp. 793–799, Elsevier, Amsterdam, 1996.
81. Dill, S., Du Preez, L.A., Graff, M., Maree, J.P., Biological sulfate removal from acid mine drainage utilizing producer gas as carbon and energy source, in: *Proc. Int. Mine Water Conf.*, J.D. Reddish (Ed.), pp. 631–641, International Mine Water Association, Nottingham, UK, 1994.

82. Lettinga, G., Roersma, R., Grin, P., Anaerobic Treatment of Raw Domestic Sewage at Ambient-Temperatures using a Granular Bed UASB Reactor. *Biotechnol. Bioeng.*, 25, 7, 1701–1723, 1983.
83. van Lier, J.B., Mahmoud, N., Zeeman, G., Anaerobic Wastewater Treatment, in: *Biological Wastewater Treatment: Principles, Modelling and Design*, M. Henze, M.C.M.V. Loosdrecht, G.A. Ekama, D. Brdjanovic (Eds.), pp. 415–456, IWA Publishing, London, 2008.
84. van Lier, J.B., van der Zee, F.P., Frijters, C.T.M.J., Ersahin, M.E., Celebrating 40 years anaerobic sludge bed reactors for industrial wastewater treatment. *Rev. Environ. Sci. Biotechnol.*, 14, 681–702, 2015.
85. Weghuis, M.O., High rate biotechnology for the metal and mining industry. *Environ. Sci. Eng. Mag.*, 26, 4, 28–32, 2013.
86. Paques Technology B.V., Sulfateq technology (The Netherlands), https://en.paques.nl/products/other/sulfateq, Accessed: 15 March 2020, s.a.
87. Dudeney, A.W.L., Chan, B.K.C., Bouzalakos, S., Huisman, J.L., Management of waste and wastewater from mineral industry processes, especially leaching of sulphide resources: State of the art. *Int. J. Min. Reclam. Environ*, 27, 1, 2–37, 2012.
88. Anglo Coal South Africa, *Towards Sustainability 2003*, 2003, https://www.angloamerican.com/~/media/Files/A/Anglo-American-Group/PLC/investors/annual-reporting/2004/br-2003-12-31b.pdf, Anglo Coal, Marshalltown.
89. Rose, P., Long-term sustainability in the management of acid mine drainage wastewaters – development of the Rhodes BioSURE Process. *Water SA*, 39, 582, 2013.
90. Greben, H.A., Sigama, N.J., Radebe, V., Wilsenach, J., An Environment-Friendly Technology for Biological Sulphate and Sulphide Removal from Acid Mine Drainage, in: *Water Institute of Southern Africa & Intern. Mine Water Assoc.: Proc. Intern. Mine Water Conf.*, C. Wolkersdorfer (Ed.), pp. 478–485, Water Institute of Southern Africa & Intern. Mine Water, Pretoria, South Africa, 2009.
91. Greben-Wiersema, H.A., *Biowaste as energy source for biological sulphate removal (PhD thesis, Water Resource Management), Univ*, Faculty of Natural & Agricultural Sciences, Pretoria, 2007.
92. Joubert, H. and Pocock, G., 2016. The VitaSOFT PROCESS: A sustainable, long-term treatment option for mining-impacted water, Water Research Commission Report No 2232/1/16, Pretoria.
93. Robins, R.G., Chemical, physical and biological interaction at the Berkerley Pit, Butte, Montana, in: *Tailings & Mine Waste*, pp. 529–541, Balkema, Rotterdam, 1997.
94. Bowell, R.J., Hydrogeochemical dynamics of Pit Lakes, in: *Mine Water Hydrogeology and Geochemistry - Spec. Publ. - Geol. Soc.*, vol. 198, P.L. Younger and N. Robins (Eds.), pp. 159–186, The Geological Society, London, 2002.
95. Gannon, J., Wielinga, B., Moore, J.M., Policastro, P., McAdoo, D., *Field investigations of the sulfate reducing potential in the Summer Camp Pit lake*, Bitterroot Consultants, Report to Gerchell Gold Corporation, 1996.
96. Machember, S.D., Reynolds, J.S., Laudon, L.S., Wildeman, T.R., Balance of sulfur in a constructed wetland built to treat acid mine drainage, Idaho Springs, Colorado. *Appl. Geochem.*, 8, 587–603, 1993.
97. Younger, P.L., Banwart, S.A., Hedin, R.S., *Mine water: Hydrology, Pollution, Remediation*, Springer, Dordrecht, 2002.
98. Rees, S.B. and Bowell, R.J., Stable Isotopic Modeling of the Longevity of Treatment Processes Operating in a Constructed Wetland for the Amelioration of Acid Mine Drainage, in: *Mine Water Env. Proc. Int. Mine Water Conf*, R. Fernándes Rubio (Ed.), pp. 585–589, International Mine Water Association, Seville, Spain, 1999.

99. Benner, S.G., Blowes, D.W., Gould, W.D., Herbert, R.B., Placek, C.J., Geochemistry of a permeable reactive barrier for metals and acid mine drainage. *Env. Sci. Technol.*, 33, 2793–2799, 1999.
100. Figueroa, L., Miller, A., Zaluski, M., Bless, D., Evaluation of a two-stage passive treatment approach for mining influenced waters, in: *Nation. Meeting Am. Soc. Mining Reclamation, 30 Years of SMCRA and Beyond*, 2007, Gillette.
101. Zaluski, M.H., Bless, D.R., Figueroa, L., Joyce, H.O., A Modular Field Bioreactor for Acid Rock Drainage Treatment, in: *Proc. 7th Intern. Conf. on Acid Rock Drainage*, SME, St Louis, 2006.
102. Hedin, R.S., Nairn, R.W., Kleinmann, R.L.P., Passive Treatment of Coal Mine Drainage. *Bur. Mines Inf. Circ.*, IC-9389, 1–35, 1994.
103. Skousen, J. and Ziemkiewicz, P.P., Performance of 116 Passive Treatment Systems for Acid Mine Drainage, in: *Proc. Nation. Meeting Am. Soc. Mining Reclamation*, Am. Soc. Mining Reclamation, Lexington, KY, 2005.
104. Ziemkiewicz, P.F., Skousen, J.G., Lovett, R., Open limestone channels for treating acid mine drainage: A new look at an old idea. *Green Lands*, 24, 36–41, 1994.
105. Ziemkiewicz, P.F., Skousen, J.G., Brant, D.L., Sterner, P.L., Lovett, R.J., Acid mine drainage treatment with armored limestone in open limestone channels. *J. Environ. Qual.*, 26, 1017–1024, 1997.
106. Cravotta, C.A. and Trahan, M., Limestone drains to increase pH and remove dissolved metals from acidic mine drainage. *Appl. Geochem.*, 14, 581–606, 1999.
107. Letterman, R.D., *Calcium carbonate dissolution rates in limestone contactors*, EPA Report/600/SR-95/068, Risk Reduction Engineering Laboratory, Cincinnati, OH May, 1995.
108. Weider, R.K., The Kentucky wetlands project: A field study to evaluate man-made wetlands for acid coal mine drainage treatment. Final report to the U.S. Office of Surface Mining, Villanova University, in: *Handbook of Technologies for Avoidance and Remediation of Acid Mine Drainage, ADTI, 1997*, Villanova, PA, Villanova University, The National Mine Land Reclamation Center, West Virginia University, Morgantown, West VA, 1992.
109. Ordóñez, A., Loredo, J., Pendás, F., Treatment of mine drainage water using a combined passive system, in: *Proc. 5th Intern. Conf. on Acid Rock*, SME, Littleton, 2000.
110. Gusek, J., Mann, C., Wildeman, T., Murphy, D., Operational results of a 1200 gpm passive bioreactor for metal mine drainage, in: *Proc. 5th Intern. Conf. on Acid Rock Drainage*, West Fork, Missouri, 2000.
111. Mbhele, N.R., Van der Merwe, W., Maree, J.P., Theron, D., Recovery of Sulphur from Waste Gypsum, in: *Water Institute of Southern Africa & Intern. Mine Water Assoc.: Proc. Intern. Mine Water Conf*, C. Wolkersdorfer (Ed.), pp. 622–630, Water Institute of Southern Africa & Intern. Mine Water Assoc., Pretoria, South Africa, 2009.
112. Liu, H., Zhao, X., Qu, J., Electrocoagulation in Water Treatment, in: *Electrochemistry for the Environment*, C. Comninellis and G. Chen (Eds.), pp. 245–262, Springer, Heidelberg, 2010.
113. Aubé, B.C., Clyburn, B., Zinck, J.M., Sludge Disposal in Mine Workings at Cape Breton Development Corporation, in: *Proc. Securing the Future 2005*, 2005, Skellefteå.
114. Vachon, D., Siwik, R.S., Schmidt, J., Wheeland, K., Treatment of Acid Mine Water and The Disposal of Lime Neutralization Sludge, in: *Acid mine drainage seminar/workshop*, Halifax, Nova Scotia, 1987.
115. Zinck, J., Disposal, reprocessing and reuse options for acidic drainage treatment sludge, in: *7th Intern. Conf. on Acid Rock Drainage*, St. Louis, Missouri, 2006.

Part 3

RECOVERY OF VALUES FROM AMD

6

Recovery of Ochers from Acid Mine Drainage Treatment: A Geochemical Modeling and Experimental Approach

Khathutshelo Netshiongolwe, Yongezile Mhlana, Alseno Mosai, Heidi Richards, Luke Chimuka, Ewa Cukrowska and Hlanganani Tutu*

Molecular Sciences Institute, School of Chemistry, University of the Witwatersrand, Johannesburg, South Africa

Abstract
Neutralization that results in the precipitation of high-density sludge (HDS) remains the preferred method for treating acid mine drainage (AMD). The HDS is usually discharged onto landfills, with little further exploitation and thus poses an environmental challenge. This study proposes aims at using various neutralizing agents (e.g., NaOH, $CaCO_3$, and $MgCO_3$) to obtain selective precipitation of ochers from AMD. The study approach involves establishing an experimental design and parameter optimization using geochemical modeling. Simulations of the neutralization process have been conducted using the PHREEQC geochemical modeling code. In the simulations, AMD and varying amounts of different neutralizing agents were used as input while the output was the different ochers and treated water. Selected simulations used to conduct the experimental studies were based on the yield of precipitated ochers and the chemistry of the treated water. Precipitates were generated using NaOH and $MgCO_3$ at different pH values (pH 3–9) at 25 °C. Precipitates varied from yellow, brown to black. Powder X-ray diffraction confirmed the mineralogy of these precipitates as goethite, hematite, and magnetite, respectively. Bluish and turquoise precipitates were obtained from solutions that contained ferrocyanide. Precipitates were used as simple art paints to demonstrate their potential.

Keywords: Ochers, geochemical modeling, high-density sludge, acid mine drainage treatment, neutralizing agents, selective precipitation

6.1 Introduction

Acid mine drainage (AMD) results from mines and contains various concentrations of metals and sulfate ions. (Di-)sulfides such as Pyrite (FeS_2) are the main starting point of AMD, due to the ease of oxidation when it is exposed to oxygenated water. Once discharged, AMD

**Corresponding author*: hlanganani.tutu@wits.ac.za [ORCID: 0000-0003-1248-130X]

[ORCID: Netshiongolwe, 0000-0001-6642-2941; Mhlana, 0000-0002-0451-4039; Mosai, 0000-0002-2762-9437; Richards, 0000-0003-1630-9400; Chimuka, 0000-0002-8552-2478; Cukrowska, 0000-0003-0777-0270]

Elvis Fosso-Kankeu, Christian Wolkersdorfer and Jo Burgess (eds.) Recovery of Byproducts from Acid Mine Drainage Treatment, (159–188)

can reach streams where it can negatively affect aquatic organisms [1]. Due to the potentially hazardous nature of the constituent elements (e.g., Hg, As, and U), there is usually a need for the affected water to be treated [2]. AMD formation can be summarized by the following reactions:

$$2\,FeS_2 + 7\,O_{2\,(g)} + 2\,H_2O_{(l)} \rightarrow 2\,Fe^{2+}_{(aq)} + 4\,SO^{2-}_{4\,(aq)} + 4\,H^{+}_{(aq)} \tag{6.1}$$

Ferrous ion (Fe^{2+}) is further oxidized to ferric ion (Fe^{3+}) in oxygenated water.

$$4\,Fe^{2+}_{(aq)} + O_{2\,(g)} + 4\,H^{+}_{(aq)} \rightarrow 4\,Fe^{3+}_{(aq)} + 2\,H_2O_{(l)} \tag{6.2}$$

Ferric ion produced depends on the pH of the solution. When the pH of the solution is >3.5, Fe^{3+} ion oxidizes pyrite further [3].

$$FeS_{2\,(s)} + 14\,Fe^{3+}_{(aq)} + 8\,H_2O_{(l)} \rightarrow 15\,Fe^{2+}_{(aq)} + 2\,SO^{2-}_{4\,(aq)} + 16\,H^{+}_{(aq)} \tag{6.3}$$

Finally, the ferric ion oxidizes the pyrite producing ferrous ion and protons (H^+), resulting in the formation of AMD. Surrounding compartments are exposed to high acidity, resulting in increased solubility, mobility, and bioavailability of potentially toxic trace elements [1]. Several methods have been used for the treatment of AMD, including: adsorption, ion exchange, membrane technology, and neutralization. Adsorption and ion exchange are effective but have shortcomings when it comes to selective removal of pollutants and show poor performance at high concentrations [1]. Membranes are also effective, but brines are produced as by-products which are deemed hazardous to the environment. Precipitation technologies using neutralizing agents have been the most widely used type of treatment for AMD. This results from their versatility, easiness to recover valuable minerals, and the ability to treat large volumes of water [3]. Using neutralizing agents such as lime results in solids with low densities that are usually difficult to precipitate. AMD treatment also involves the addition of limestone or quick lime to increase the pH and oxidizing ferrous ion to ferric ion [4]. Downside is that the resulting high-density sludge (HDS) has to be disposed of at landfill sites. In this study, the aim was to provide a simulated approach of neutralizing AMD while generating (precipitating) ocher during the neutralization process and to assess their potential use for pigment in paints. This is exacerbated by the fact that this sludge is heterogeneous and complex. Thus, beneficiation of sludge can produce material that can lead to the recovery of minerals such as hematite (Fe_3O_4) which can be used as the anticorrosive pigment [5]. Geochemical models assist in the characterization of contaminated environmental sites and in making predictions on the potentially negative effects caused by anthropogenic activities such as mining. PHREEQC [6] is a geochemical modeling code applicable in mine water treatment studies. It is a powerful tool to understand the chemical equilibria for water chemistry and includes acid-base reactions, reactions for precipitation/dissolution, aeration/gas transfer, and oxidation-reduction reactions, among others. Ochers are found in a variety of geological deposits (commonly in Fe ores and sulfide bearing ore deposits) [7]. Iron precipitates are defined as the natural clay earth pigments with a variety of colors produced depending on the geochemistry of the commodity in the surrounding environment [8]. These Fe precipitates are formed as oxides and hydroxides, together with some trace metals [9, 10]. Sedimentary ore deposits are the main source of iron-rich minerals (e.g., hematite and magnetite) which are formed from geochemical processes that involve the reaction of iron and oxygen in marine environments [11, 12]. Different types of iron ores (e.g., magnetite, hematite, and pisolitic ironstone deposits) are formed, influenced by the geological and mineralogical settings of the surrounding environment [13]. Some minerals (e.g., siderite) can result from igneous rocks formed from hydrothermal fluids, volcanic, and metamorphic ore deposits [14, 15]. These Fe ores are formed by hot solutions which migrate Fe and metamorphosed rocks of favorable chemical composition with Fe minerals combined to form irregular ore commodities. Commonly, the color of the Fe precipitates is determined by the chemistry of the elements in their medium (e.g., mafic and ultramafic rocks are enriched with minerals like magnetite) based on the geological setup of the area [16, 17]. Mineral alteration during physical and chemical weathering forms a variety of Fe ore deposits (ores containing high quantities of greater than 60% Fe) and other oxides such as silicates (with SiO_2 ranges from 0.6 to 5.7%) and aluminum oxide (Al_2O_3 ranges from 0.6 to 3.7%) [18, 19]. When Fe ore commodities are weathered, they form a variety of ochers or Fe precipitates due to geochemical reactions taking place in that environment, making them the main source of Fe precipitates [20]. Black precipitates are mostly formed from magnetite deposits, while red and yellow iron precipitates can form from the weathering of magnetite [20, 21]. Iron precipitates (hematite and magnetite) are generally higher in grade with Fe percentages from 64 to 67% [4]. With time, the chemical composition of Fe ores with fine textures continues to decline, as economic ore deposits (higher grade reserves) are replaced with low grade commodities for both delivering goods from the supplier and beneficiated products [22]. Low grade Fe commodities from the suppliers are likely to be replaced in the future with beneficiated fine Fe precipitates and some concentrates from banded iron formation (BIF) [23]. Global Fe ore trade (both imports and exports) released by the Australian Government Bureau of Resources and Energy Economics (BREE), Bloomberg, United National Conference on Trade and Development (UNCTAD) shows that the world trade in terms of exports in the year 2018 was 6,124 Mt and 9,463 Mt in 2019. This shows that here might be a high demand of Fe precipitates globally. In 2018, South Africa exported 75 Mt (import was 1,200 Mt) and 65 Mt (import was 1,190 Mt) in 2019, implying a decline that could be a worldwide trend. Iron ore imports in countries like China and Japan inclined in 2018 and 2019 (China in 2018 was 1.064 Gt and 1.069 Gt in 2019; Japan in 2018 was 100 Mt and 115 Mt in 2019). Natural ochers were used in ancient times in artistics and crafts for wall and body paintings [24]. Due to the beneficial use of ochers, they have been transformed into commercially acceptable pigments [25–27]. In general, iron precipitates contain different quantities of octahedral Fe oxides (e.g., hematite and goethite) [28, 29]. Reddish (hematite) and yellow (goethite) Fe oxides are commonly observed in the natural Fe oxides [30, 31]. An increased demand for Fe minerals or ochers for art and paints and the depletion of the naturally occurring resources will undoubtedly lead to the exploitation of alternatives. AMD is one such alternative that can be considered owing to its elevated concentrations of dissolved Fe. Current neutralization methods of AMD have yielded HDS which is not of much use as a result of its mixed nature, i.e., it contains various precipitates in a single mass.

6.2 Methodology

6.2.1 Simulation Studies—Model Setup as an Experimental Design Approach

Previous descriptions have shown that the simulated neutralization of AMD and subsequent precipitation of ochers can be used as an experimental design technique. It helps in gaining insights into the experimental setup to be used and what can be anticipated. A forward geochemical modeling approach was used in which two or more initial states were used to determine the resulting state. For instance, reacting the AMD (initial state 1) with a neutralizing agent (initial state 2) while changing the temperature of the reaction (initial state 3) will result in treated water (final state 1) and some precipitates (final state 2). AMD compositions in this study were taken from a previous study by [32] (Figure 6.1). Using PHREEQC (Phreeqc.dat database), the AMD's speciation and solubility was assessed and the neutralization and treatment of the affected water using a variety of neutralizing agents was simulated. In these simulations, different parameters with different neutralizing agents were assessed: varying pH values, equilibration with gases such as CO_2 and O_2 and varying temperatures. PHREEQC was also used for batch calculations that can be reacted or equilibrated with a gas (e.g., O_2 for fugacity). Our aim was to provide a holistic approach in combating the AMD problem and to generate ochers during AMD treatment using PHREEQC. These neutralizing agents were used for the treatment of the AMD: $MgCO_3$, MgO, $Mg(OH)_2$, $CaCO_3$, CaO, $CaHCO_3$, $Ca(OH)_2$, Na_2CO_3, Na_2O, NaOH, $NaHCO_3$, NH_3, and NH_4OH. In the input file, the chemical composition of the AMD was used as "SOLUTION 1", with the above neutralizing agents dosed as "REACTION 1" in 1 L AMD (Figure 6.1). Magnesium carbonate and sodium hydroxide were found to be the best neutralizing agents in improving the water quality and the precipitation of Fe oxides (including yield wise). Each of the reaction steps represented a different simulation, e.g., in the reaction with $CaCO_3$, 4 mol of the neutralizing agent were added in 10 steps. In a separate reaction step

```
TITLE RAW AMD WATER
SOLUTION 1
Temp       25
pH         2.8
pe         4
redox      pe
units      mg/l
density    1
Al         472.5
Ca         470
Cd         0.01
Fe         6051.3
K          1
Mg         0.03
Mn         125.9
S(6)       18000
Si         30
Zn         0.2
-water     1 # kg
```

```
Addition of neutralizing agent
REACTION 1
CaCO3      4
4 moles in 10 steps
END
```

Figure 6.1 PHREEQC script for forward modeling of AMD.

(Figure 6.2), $MgCO_3$ is reacted with the same initial water to obtain a different output. In another step, a mixture of $MgCO_3$ and NaOH was reacted with the initial water, giving another outcome. In yet another reaction step, $MgCO_3$ was reacted with the same initial water at a fixed pH of 9.0 (shown as "Fix H+ −9.0 $MgCO_3$ 10") to reach a different output. Notable changes in the treated water were the pH, pe (redox potential), and total alkalinity as the species change. Up to 1,000 simulations were conducted with about 200 proving to be satisfactory with respect to the desired predicted treated water quality and ochers precipitated. In addition, the solution was equilibrated with CO_2 (Figure 6.3) and with atmospheric O_2 (Figure 6.4).

```
Addition of neutralizing agent
REACTION 1
MgCO3        4
4 moles in 10 steps
END

Addition of neutralizing agent
REACTION 1
NaOH         4
4 moles in 10 steps
END

Combined neutralizing agent
REACTION 1
MgCO3        4
NaOH         4
4 moles in 10 steps
END

Fixed pH
PHASES
Fix_H+
H+=H+
log_k        0

EQUILIBRIUM_PHASES 1
Fix_H+       -9.0 MgCO3        10
END
```

Figure 6.2 PHREEQC script for forward modeling of AMD at fixed pH value.

```
Equilibration with CO2
REACTION 1
MgO          4
NH4OH        4
2 moles in 10 steps

EQUILIBRIUM_PHASES 1
CO2(g)       -2.0  8
END
```

Figure 6.3 PHREEQC script for forward modeling of AMD equilibrating with CO_2.

```
Equilibration with O2
REACTION 1
MgO          4
NH4OH        4
4 moles in 10 steps

EQUILIBRIUM_PHASES 1
O2(g)        -0.5  8
END
```

Figure 6.4 PHREEQC script for forward modeling of AMD equilibrating with atmospheric O_2.

6.2.2 Experimental Studies

6.2.2.1 Experiment 1

The aim of the experiments was to produce Fe precipitates using synthetic AMD (Table 6.1) based on [4] without heating (energy) throughout the experiments. A stepwise selective precipitation mechanism was applied to assess the effect of parameters such as oxidation, pH adjustment, settlement rates, and temperature (experiments conducted at room temperature). All the weighted chemicals were transferred into seven 600 mL beakers and dissolved in 500 mL of deionized water. A magnetic stirrer was used to homogenize the samples at 2 rpm (Figure 6.5).

Table 6.1 Chemicals (names, formula) and masses, *m*, used to synthesize AMD in Experiments 1 (m_1) and 2 (m_2).

Chemicals	Chemical formula	m_1, g	m_2, g
Aluminium potassium sulfate dodecahydrate	$AlK(SO_4)_2{\cdot}12H_2O$	0.50	0.50
Manganous sulfate monohydrate	$MnSO_4.H_2O$	0.30	0.30
Ferrous sulfate heptahydrate	$FeSO_4.7H_2O$	1.00	1.00
Copper sulfate anhydrous	$CuSO_4$	0.20	0.20
Calcium sulfate dihydrate	$CaSO_4.\ 2H_2O$	15.0	15.0
Ferric sulfate anhydrous	$Fe_2(SO_4)_3$	1.00	1.00
Nickel nitrate hexahydrate	$Ni(NO_3).6H_2O$	0.20	0.20
Cadmium nitrate tetrahydrate	$Cd(NO_3)_2.4H_2O$	0.20	0.20
Zinc nitrate tetrahydrate	$Zn(NO_3)_2.4H_2O$	0.30	0.30
Ferrocyanide	$[Fe(CN)_6]^{4-}$	–	1.00

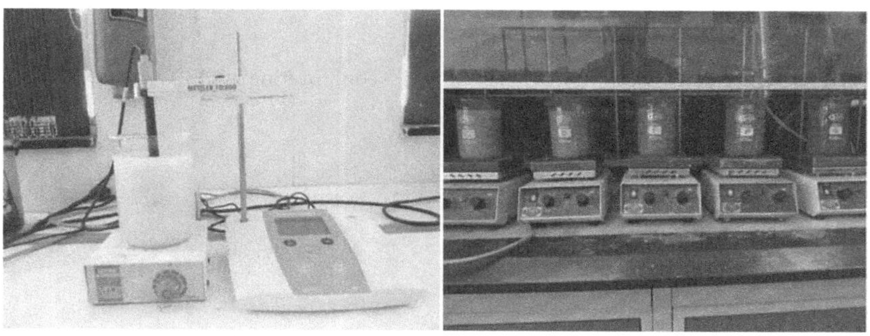

(a) (b)

Figure 6.5 (a) pH adjustment on each beaker; (b) Stirred solutions at 2 rpm at ambient temperature for 48 h.

6.2.2.2 *Using NaOH as a Neutralizing Agent*

pH values were measured (Figure 6.6) after homogenization and ranged between 2.76 and 2.80 in the seven beakers. To precipitate iron oxides from the synthesized AMD, different pH values were used, ranging from pH 3 in beaker A to pH 9 in beaker G. To reach the targeted pH, drops of 1 mol L^{-1} NaOH were added to raise the pH. Seven beakers were stirred at 2 rpm for 48 h at room temperature. After 48 h, the beakers were removed from the magnetic stirrer to allow the precipitates to settle through a vacuum suction using a 0.45 μm Whatman cellulose nitrate filter paper. After filtering the Fe precipitates, some of the filtrates were selected for retreatment with NaOH for further precipitation to yield more precipitates.

6.2.2.3 *Addition of Ferrocyanide to Mineral Salts Used to Simulate AMD (Experiment 2)*

To produce the turquoise Fe precipitates in Experiment 2, ferrocyanide ($[Fe(CN)_6]^{4-}$) was added to the chemicals used to synthesize AMD (Table 6.1). This was just a once-off experiment that was conducted to assess the possibility of obtaining such precipitates. Chemicals were transferred into four 600 mL beakers and 500 mL of deionized water was used to dissolve them. A magnetic stirrer was used to homogenize the samples at 2 rpm (Figure 6.7).

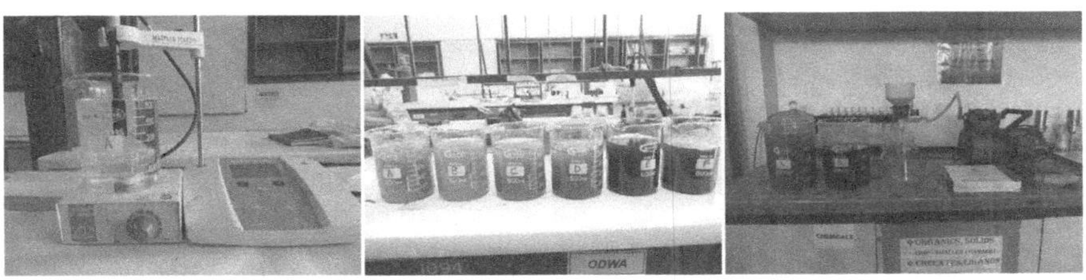

(a) (b) (c)

Figure 6.6 (a) pH adjustment with NaOH in initial solutions; (b) Precipitates formed in solutions after stirring for 48 h at 25 °C; (c) Vacuum system used to filter precipitates.

(a) (b)

Figure 6.7 (a) Mineral salts that were weighed out (600 mL beakers) in weighing balance and dissolved in deionized water; (b) Beakers stirred for 48 h to allow oxygen circulation in the reactions.

pH values in each beaker after homogenization ranged between 2.58 and 2.67. To precipitate the cyanide containing iron ochers from the synthesized AMD, different pH values were used by adding NaOH, ranging from pH 3 in beaker A to pH 9 in beaker D. After the pH adjustments, the precipitates were extracted through filtration.

6.2.2.4 *Using $MgCO_3$ as a Neutralizing Agent*

For neutralization with $MgCO_3$, the experimental setup was similar to that for NaOH, with the same initial pH values in the beakers as above. Adjustment of pH was conducted using $MgCO_3$, added as a powder followed by stirring at room temperature, (25 °C) as for the experiments using NaOH. Selected filtrates from the $MgCO_3$ neutralization were used for retreatment with $MgCO_3$ to assess the possibility of precipitating more ochers.

6.2.3 Characterization of Fe Oxides

Precipitated ochers characterization was conducted using powder X-ray diffraction (PXRD) (Bruker D2 Phaser, Germany). Main aim of this analysis was to confirm the Fe oxides precipitated using both $MgCO_3$ and NaOH. Inductively coupled plasma-optical emission spectrometry (ICP-OES) (Spectro Genesis, Spectro, Germany) was used to analyze different trace metal concentrations in the synthesized AMD and treated water. Filtrates were analyzed to verify if the water quality improved.

6.3 Results and Discussion

6.3.1 Simulation Studies

6.3.1.1 *Individual Neutralizing Agents*

Different neutralizing agents gave different pH values and varying amounts of precipitated iron minerals (Table 6.1). Magnesium-based neutralizing agents provided satisfactory results in terms of treatment of AMD, pH, and the iron precipitates produced. In particular $MgCO_3$, MgO, and $Mg(OH)_2$ proved to be the best neutralizing agents, as they raised the pH to above 9 which is in agreement with what the industries are currently allowing for waste disposal after treatment [1]. Advantage of using $MgCO_3$ is that gypsum does not precipitate with the iron minerals, allowing their easy separation. This is a challenge with the current HDS process, where co-precipitation with iron minerals is unavoidable, resulting in a heterogeneous and complex sludge as indicated previously. Calcium carbonate ($CaCO_3$) and sodium hydroxide (NaOH) also produced acceptable treated AMD and iron precipitates at pH > 9. Though these individual neutralizing agents yielded acceptable pH values, poor overall water treatment results were obtained (e.g., ineffective removal of potentially toxic elements and sulfate). Other neutralizing agents such as NH_3, NH_4OH, and $NaHCO_3$ gave poor results with both, the precipitated minerals and the treated water. Maximum pH after the addition of these agents was less than 7, resulting in ineffectively treated AMD. Iron precipitates were very few and had low saturation indices (SI), implying that they may not be precipitating effectively. Therefore, these neutralizing agents were not used for subsequent simulations for the AMD treatment.

At lower concentrations (0.2–1 mol) of the neutralizing agents, goethite (FeOOH), hematite (Fe_2O_3), and siderite ($FeCO_3$) could be separated at acid pH values (5–6.9) for all the neutralizing agents used. Hematite was observed to be a derivative of goethite as in the treated water a small amount of hematite was observed. Essentially, a loss of water by goethite will result in the formation of hematite. At higher concentrations (1.2–2 mol), the following iron precipitates were observed: $Fe(OH)_3$, FeS, goethite (FeOOH), hematite (Fe_2O_3), mackinawite (FeS), pyrite (FeS_2), and siderite ($FeCO_3$) at varying pH values for the effective neutralizing agents. They were chosen based on the acceptable pH reached and better quality of the water treated (pH values in brackets): $MgCO_3$ (9.44), MgO (12.15), $Mg(OH)_2$ (11.34), Na_2O (12.77), $CaCO_3$ (9.43), and NaOH (9.65).

6.3.1.2 *Combined Neutralizing Agents*

Main aim of combining the neutralizing agents was to identify changes in pH, treated water composition, and iron precipitates produced. This was done by combining the individual neutralizing agents that were considered effective for AMD treatment and those that were deemed to be poor as individual neutralizing agents. It should be noted here that the reasoning was premised on that a mixture of good and bad neutralizing agents would produce acceptable effects. pH range that was of interest was from 7 to 9 which is an acceptable pH range for discharging to streams. At this pH range, the concentration of trace metals in the water is low, reducing their potential environmental threat. Masses of iron minerals that precipitated were observed to be higher compared to those produced by the individual neutralizing agents. Combinations that yielded good precipitates at pH values between 7 and 9 and treated water were: $MgCO_3$+NH_3, MgO+NaOH, and $Mg(OH)_2$+NH_4OH (Table 6.2). At higher masses of the combinations, the phases $Fe(OH)_3$, FeS, goethite (FeOOH), hematite (Fe_2O_3), mackinawite (FeS), pyrite (FeS_2), and siderite ($FeCO_3$) precipitated. A benefit of using $MgCO_3$, MgO, $Mg(OH)_2$, $CaCO_3$, and NaOH as neutralizing agents is that the obtained minerals are not precipitated simultaneously with gypsum [33].

Table 6.2 Effect of the combined neutralizing agents on pH of the treated AMD.

Combined neutralizing agents	pH
$Na_2O + NH_3$	9.17
$CaO + CaHCO_3$	9.47
$MgO + NaOH$	9.84
$Mg(OH)_2 + NH_4OH$	9.19
$MgCO_3 + NH_3$	8.04
$MgCO_3 + NaOH$	7.23
$NH_3 + NH_4OH$	6.50

In other instances, depending on the individual and combined neutralizing agents used, the iron minerals precipitated were observed to decrease. For instance, the combination of $CaCO_3$+Na_2CO_3 decreased the iron precipitates by 30% compared to $CaCO_3$ individually. $Mg(OH)_2$+NH_4OH combination decreased the iron precipitates by 40% compared to $Mg(OH)_2$ individually. Likewise, the $MgCO_3$+NH_3 combination decreased the iron precipitates by 25% compared to the $MgCO_3$ individually. MgO had similar observations as $MgCO_3$. When the Na_2O+NH_3 combination was used, the iron precipitates were found to be 30% less compared to Na_2O individually.

6.3.1.3 *Equilibrating with CO_2*

To assess the effect of CO_2 on AMD, degassing of CO_2 was conducted with individual and combined neutralizing agents (Table 6.3) with the highest pH reached being 9. Most individual neutralizing agents gave satisfactory results when CO_2 was introduced. Gypsum was observed to precipitate before the introduction of CO_2, meaning that it would be possible to separate gypsum from other minerals by controlling equilibration with CO_2. This treated water quality showed greater improvement and more iron precipitates were produced with the most common iron precipitate being siderite which varies with the type of neutralizing agent used. Comparing individual $MgCO_3$ which had a molality of 0.1091 and the combined $MgCO_3$+NaOH which had a molality of 0.1041, the concentration does not vary much as it changes by only 0.44%. This implies that $MgCO_3$ is a good neutralizing agent when equilibrated with CO_2 as both individual and combined neutralizing agent.

6.3.1.4 *Equilibrating with O_2*

AMD was then equilibrated with atmospheric oxygen, where a pH of 8 was reached with most neutralization agents (Table 6.4). Iron precipitates and the resulting treated were not satisfactory for these neutralizing agents even though Jarosite-K ($KFe_3(SO_4)_2(OH)_6$) was observed to precipitate. Jarosite-K was predicted to precipitate for both the individual and the combined neutralizing agents when the system was equilibrated with oxygen. A less complex sludge would likely be obtained as the predictions showed that fewer precipitates would be produced.

Table 6.3 Final pH-values in AMD for individual neutralizing agents equilibrated with CO_2.

Neutralizing agent	**$MgCO_3$**	**MgO**	**$Mg(OH)_2$**	**Na_2O**	**$CaCO_3$**	**NaOH**
pH	9.58	9.58	9.58	9.76	9.51	9.64

Table 6.4 Final pH-values in AMD for individual neutralizing agents equilibrated with O_2.

Neutralizing agent	**$MgCO_3$**	**MgO**	**$Mg(OH)_2$**	**Na_2O**	**$CaCO_3$**	**NaOH**
pH	8.73	8.80	8.71	8.37	8.73	8.43

6.3.1.5 *Fixed pH*

pH was fixed to assess the changes in iron precipitates and the treated water (Table 6.5). Nitric acid was used to fix the pH at 3.0 resulting in fewer precipitated iron minerals, including Fe_2O_3, $Fe(OH)_3$ and FeOOH being common in the treated water of most of the neutralizing agents. As was expected, since the pH was low, the composition of the treated water did not change much. However, the precipitation of these minerals means that the pH must be lowered to 3.0 with nitric acid if the interest is in their precipitation. In the alkaline range, pH 9.5 was the maximum pH where the treated water was satisfactory, and the iron mineral precipitates were substantial (judging from the molar yields predicted). A pH of 9.5 proved to work better than pH values above 10. This, of course, depends on the mineral of interest for a study. Although most iron minerals precipitated at pH $\geq$ 9, using $MgCO_3$, MgO and NaOH improved water treatment. Consequently, these neutralizing agents can be used for AMD treatment.

6.3.1.6 *Varying Temperature*

Temperature was varied from 25 to 100 °C to assess the effect of heating the water before treatment. A temperature of 33 °C, with other factors held constant, was found to yield satisfactory. This was ideal as this temperature is quite ambient and easily attainable.

6.3.1.7 *Varying Concentrations of Neutralizing Agents*

Varying the dosages of neutralizing agents resulted in non-convergence of the simulation iterations as the highest pH reached was 14 at higher molarities (10 mol). Calcite was observed to precipitate gypsum at higher amounts and pH. At higher concentrations, there were fewer precipitates predicted and the treated water quality was poor as most metals dissolved in the solution. To be certain that the model works, a different AMD was used with the same neutralizing agents at a different temperature. It was observed that the identified effective neutralizing agents worked best (Table 6.5). Precipitated iron minerals are as follows: $Fe(OH)_3$, FeS, goethite (FeOOH), hematite (Fe_2O_3), mackinawite (FeS), pyrite (FeS_2), and siderite ($FeCO_3$) which were common in all the identified effective neutralizing agents.

6.3.2 Characterization of HDS

6.3.2.1 *Aims and Dry Matter*

Both fresh and aged (3 months) HDS were characterized to assess their physical properties as well as their chemical properties (through microwave digestion and leaching tests). Main aim was to set a benchmark for determining if either sludge is usable for further adsorption of elements from AMD (e.g., when pumped into mine voids) and to adsorb phosphate in

Table 6.5 pH values produced by the effective individual neutralizing agents.

Neutralizing agent	$MgCO_3$	MgO	$Mg(OH)_2$	Na_2O	$CaCO_3$	NaOH
pH	9.02	9.51	12.21	9.57	8.89	9.05

agricultural-type water. Dry solids in the HDS ranged from 2.5% to 37.2% and an increase from 20% to 25% was observed from fresh to aged HDS. Metal and sulfate concentrations in the aged sludge were found to be higher than for fresh sludge.

6.3.2.2 Physical Characterization of HDS

HDS samples from the Germiston mine water treatment plant were characterized to determine their mineralogy. Fresh sludge was divided into portions with some being aged through air drying in the laboratory. Mineralogical composition of HDS was determined using PXRD and X-ray fluorescence (XRF).

6.3.2.3 Chemical Characterization of HDS

To determine the release of metals from the dried sludge, leaching, and microwave digestion were done followed by analysis using ICP-OES for metals and IC for anions, mainly sulfate. For the metals, microwave digestion of the sludge was conducted using a 1:3 mixture of nitric and hydrochloric acids (0.1 g of sludge to 10 ml of the *aqua regia* mixture) followed by analysis using ICP-OES. For the sulfate content, 1 g of sludge:20 mL of deionized water (at pH 7) was used over a shaker for 24 h followed by filtration with a 0.45-μm cellulose nitrate filter followed by analysis using IC.

6.3.2.4 Mineralogy and Chemical Composition of HDS

Results for the mineralogy determined in the HDS showed that the dominant minerals include: magnetite, hematite, goethite, maghemite, pyrophyllite, chloritoid, mica, chlorite, jarosite, pyrite, gypsum, copiapite, and clay minerals, mostly kaolinite and montmorillonite. Observations of elevated concentrations of iron oxides in the HDS were also confirmed by XRF analysis (Figure 6.8). Presence of the iron minerals is apparent from the reddish-brown coloration of the sludge. In addition, the fresh HDS was leached in deionized water (Figure 6.9).

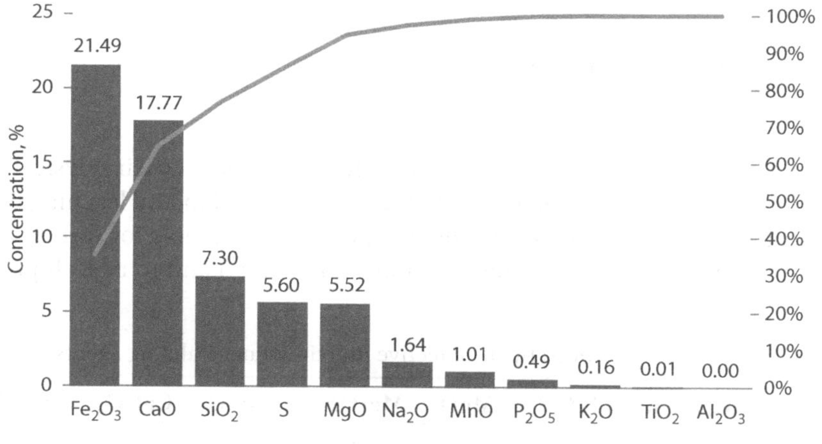

Figure 6.8 X-ray fluorescence analysis of HDS.

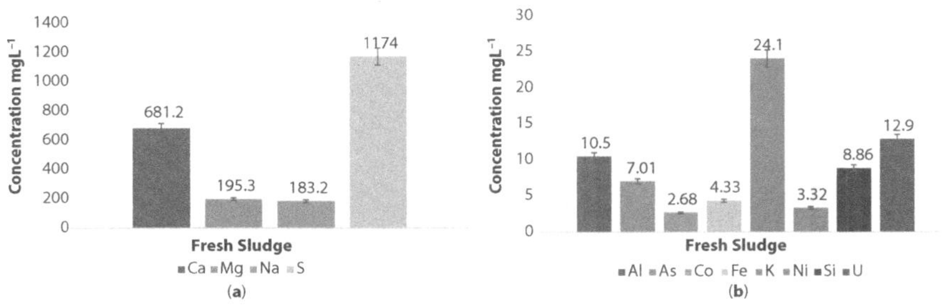

Figure 6.9 Major (a) and minor (b) elements in fresh HDS (sulfate reported as sulfur).

As can be seen, the results show elevated concentrations of sulfur and trace elements such as U. These are accumulated concentrations following precipitation and co-precipitation with iron precipitates during the neutralization process. Use of limestone and lime for neutralization in these plants can be seen by the elevated concentration of Ca resulting in large amounts of gypsum being precipitated (as substantiated by the PXRD results). Results of the aged HDS show elevated concentrations of elements compared to the fresh sludge (Figure 6.10). This can be explained by the fact that, in the fresh sludge, most of the minerals are likely to be in soluble forms, while in aged sludge, these have been allowed to precipitate onto the bulk sludge, resulting in the observed increased concentrations. A high sulfate concentration of 3,356 mg L^{-1} corroborates the observations for the fresh sludge. Three times higher concentrations were observed compared to that of fresh sludge. Low concentrations of nitrite, bromide, and nitrate show (Figure 6.11) that these ions did not play a substantial role in the chemistry of the sludge. Physical and chemical properties of the HDS showed differences between the fresh and aged sludge, with the aged sludge showing an increase in stability compared to the fresh sludge. Low concentrations of metals and high alkalinity in the fresh sludge may reduce metals leachability. Aged HDS properties may transform based on its mineral composition into carbonate rock enriched with iron oxide and metals may be accumulated in the matrix of solid sludge and might be a secondary source of pollution. Further characterization of both the fresh and aged sludge would shed some light as to the possibility of further adsorption that can occur on their surfaces. This will make it possible to deduce if use in conditioning AMD or adsorption of phosphate is feasible.

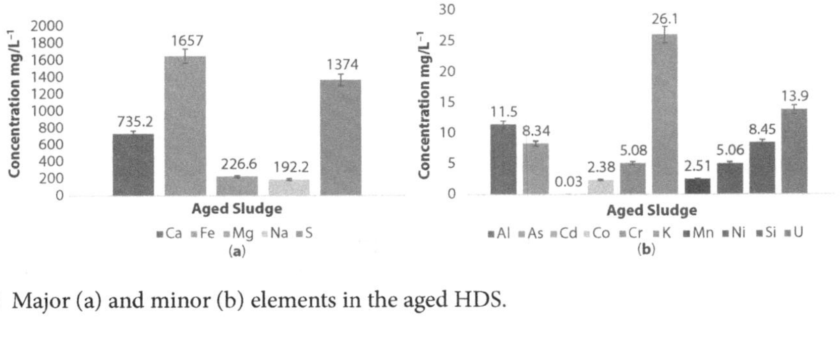

Figure 6.10 Major (a) and minor (b) elements in the aged HDS.

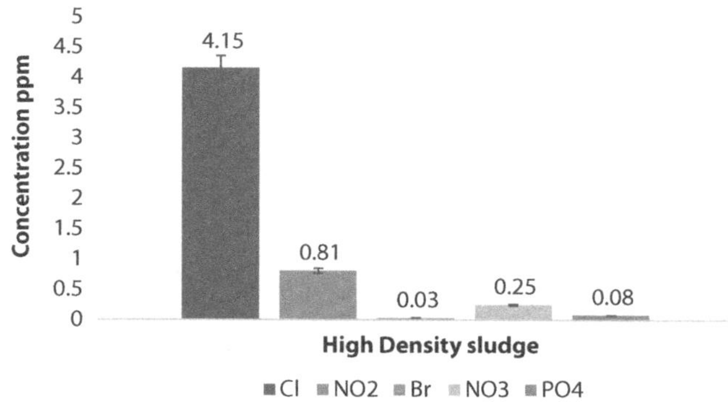

Figure 6.11 Minor anions in the high-density sludge.

6.3.3 Experimental Studies

6.3.3.1 Procedure Description

In this section, the formation of precipitates is presented and described. This includes the initial formation of the precipitates from the initial AMD solutions whose pH was adjusted with NaOH and $MgCO_3$. From those, some filtrates were selected for further treatment with each neutralizing agent to precipitate more ochers. Characterization of the ochers is also presented as well as some illustrations of the applications of the ochers to paint and artwork. Lastly, the changes in water chemistry are presented, that is, the composition of the initial AMD and the composition after treatment with NaOH.

6.3.3.2 Formation of Precipitates

6.3.3.2.1 Using NaOH as a Neutralizing Agent

Precipitates obtained at the seven pH values of precipitation (pH 3, 4, 5, 6, 7, 8, and 9) ranged from yellowish to brown (Figures 6.12 to 6.19) and are shown below (Figures 6.13 to 6.19, starting from the left beaker to the right one). Precipitates were dried at 180 °C in the

Figure 6.12 (a) Seven beakers stirred for 48 h to allow oxygen circulation in the reactions. The pH values for precipitation increase from the left to the right beaker (i.e., pH 3 to 9); (b) Vacuum system used to filter the samples.

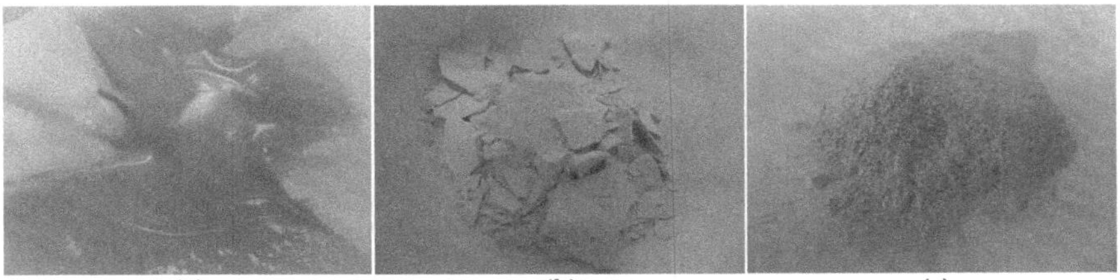

(a) (b) (c)

Figure 6.13 Yellow precipitates (pH 3): (a) Filtered (25 °C); (b) Dried (180 °C for 1 h); (c) Ground with pestle and mortar after drying.

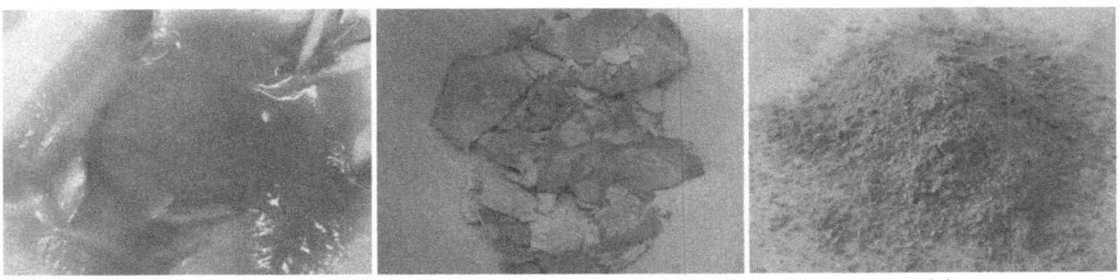

(a) (b) (c)

Figure 6.14 Yellow-brown precipitates (pH 4): (a) Filtered (25 °C); (b) Dried (180 °C for 1 h); (c) Ground with pestle and mortar after drying.

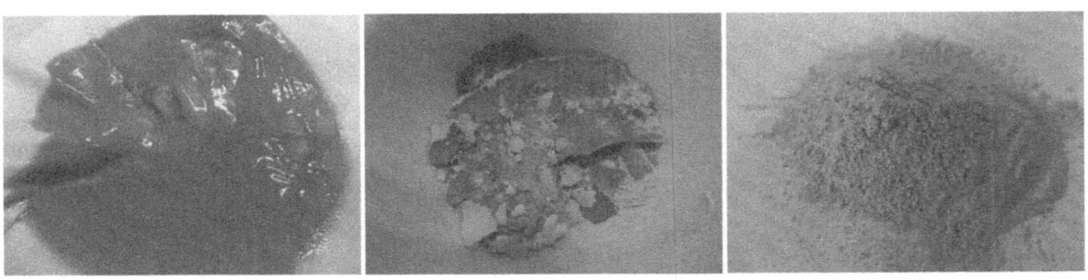

(a) (b) (c)

Figure 6.15 Yellow-brown precipitate (pH 5): (a) Filtered (25 °C); (b) Dried (180 °C for 1 h); (c) Ground with pestle and mortar after drying.

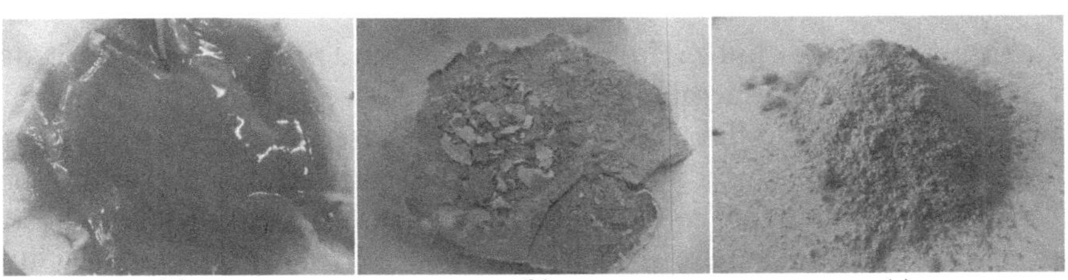

(a) (b) (c)

Figure 6.16 Brown precipitate (pH 6): (a) Filtered (25 °C); (b) Dried (180 °C for 1 h); (c) Ground with pestle and mortar after drying.

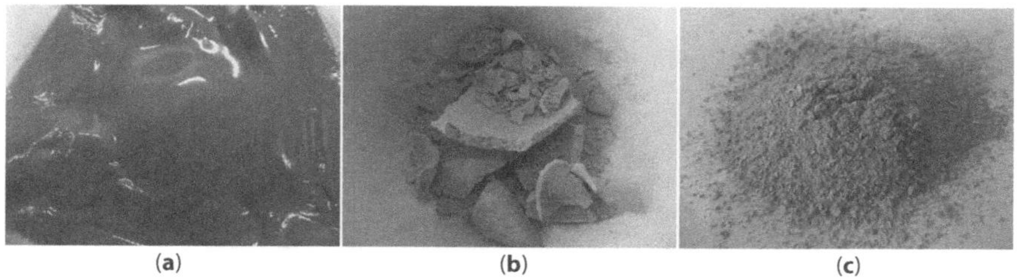

Figure 6.17 Light brown precipitate (pH 7): (a) Filtered (25 °C); (b) Dried (180 °C for 1 h); (c) Ground with pestle and mortar after drying.

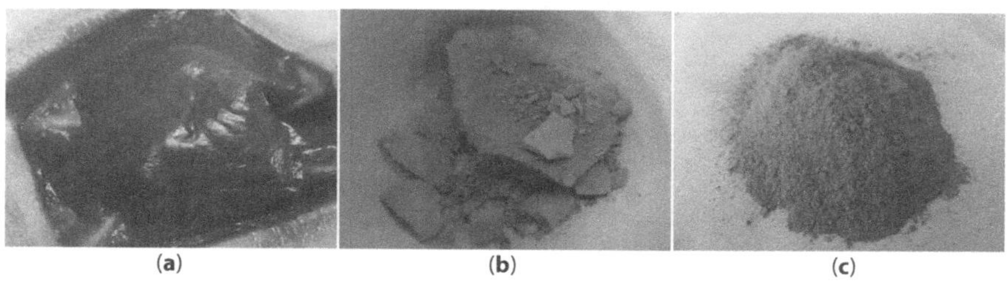

Figure 6.18 Reddish-brown precipitate (pH 8): (a) Filtered (25 °C); (b) Dried (180 °C for 1 h); (c) Ground with pestle and mortar after drying.

Figure 6.19 Brown precipitate (pH 9): (a) Filtered (25 °C); (b) Dried (180 °C for 1 h); (c) Ground with pestle and mortar after drying.

oven for an hour followed by pulverizing with a mortar and pestle to yield a powder. As no difference in coloration and mineralogy was observed between these precipitates and those that had been air dried over 5 days, oven drying was chosen as a faster method. Grinding with a mortar and pestle produced ochers that can be workable for art paintings as will be discussed later. Brown to reddish color was due to hematite (Fe_2O_3) and the yellowish color to goethite (FeOOH) as confirmed by PXRD analysis. Further details are provided in the characterization section. The recovered precipitates ranged from 22.2 to 27.8 g L^{-1} of treated AMD (Table 6.6).

Table 6.6 Mass (g L^{-1}) of precipitates recovered at different pH values using NaOH.

pH	Color of precipitate	*m* precipitate
3	Yellow	22.22
4	Yellow-brown	22.00
5	Yellow-brown	26.16
6	Brown	27.82
7	Brown	24.14
8	Reddish-brown	27.36
9	Brown	26.46

6.3.3.2.2 Retreatment of Filtrates with NaOH

After filtering the Fe precipitates above, three selected filtrates (i.e., those from pH of precipitation of 3, 5, and 7) were "retreated" (further neutralized) with NaOH followed by stirring for 48 h at 2 rpm and filtration. This was done to assess if further precipitates could be recovered following the initial precipitation. Reddish-brown to black iron precipitates were obtained (Figures 6.20 to 6.22), implying that the filtrates had potential to yield more precipitates. As mentioned before, the filtrate of pH 3 was neutralized to a new pH of 5; the

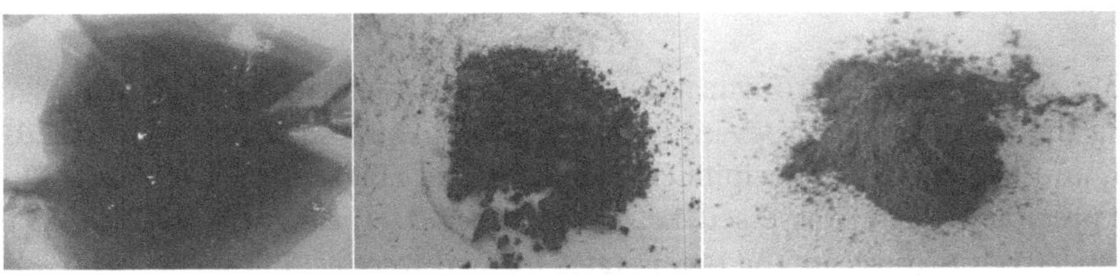

(a) (b) (c)

Figure 6.20 Reddish-brown precipitate (pH 5): (a) Filtered (25 °C); (b) Dried (180 °C for 1 h); (c) Ground with pestle and mortar after drying.

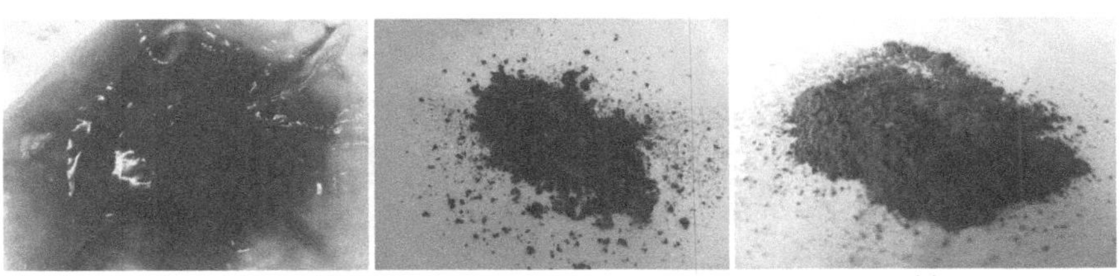

(a) (b) (c)

Figure 6.21 Brown precipitate (pH 7): (a) Filtered (25 °C); (b) Dried (180 °C for 1 h); (c) Ground with pestle and mortar after drying.

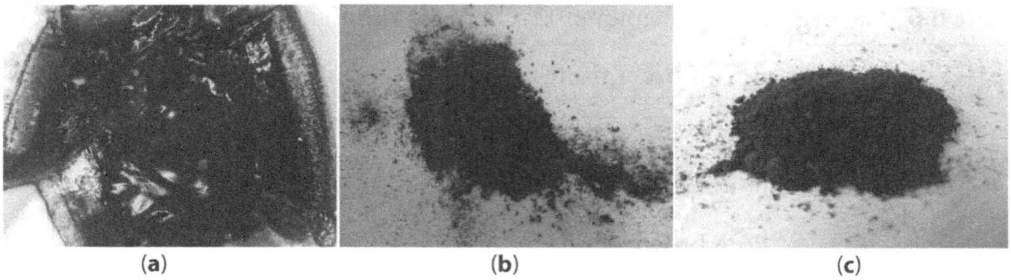

Figure 6.22 Black precipitate (pH 9): (a) Filtered precipitates (25 °C); (b) Dried (180 °C for 1 h); (c) Ground with pestle and mortar after drying.

Table 6.7 Mass (g L^{-1}) of precipitates recovered at different pH regimes following retreatment of selected initial filtrates with NaOH.

pH	Color of precipitate	*m* precipitate
5	Reddish-brown	2.672
7	Brown	2.242
9	Black	1.896

filtrate of pH 5 to pH 7; and that of pH 7 to pH of 9. Results show that close to 10% additional yield was obtained in the retreatment (Table 6.7).

6.3.3.2.3 Addition of Ferrocyanide and Precipitation with NaOH

Following the addition of ferrocyanide to the initial AMD solution (Figure 6.23), bluish, greenish, and brownish Fe precipitates were obtained (Figures 6.24 to 6.27). Only 4 pH values of precipitation were assessed, i.e., pH 3, 5, 7, and 9 (beakers A, B, C, and D in Figure 6.24a, respectively). To recap, the precipitates formed were as follows: at pH 3 a light blue precipitate, at pH 5 a dark blue precipitate, at pH 7 a green precipitate and at pH 9 a dark

Figure 6.23 (a) Bluish, green, and brown precipitates forming at various pH regimes during neutralization; (b) Sample filtration with a vacuum system.

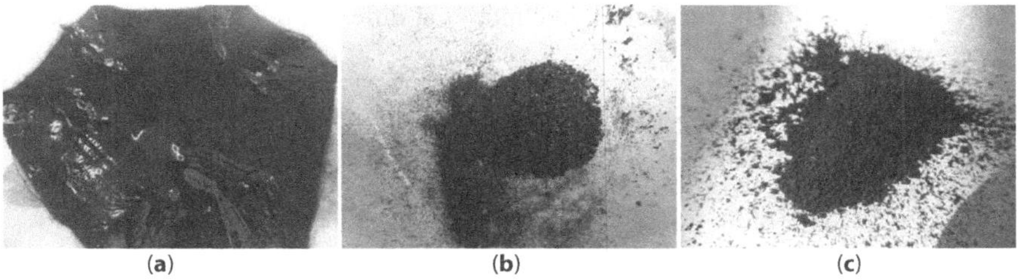

Figure 6.24 Light blue precipitates (pH 3): (a) Filtered (25 °C); (b) Dried (180 °C for 1 h); (c) Ground with pestle and mortar after drying.

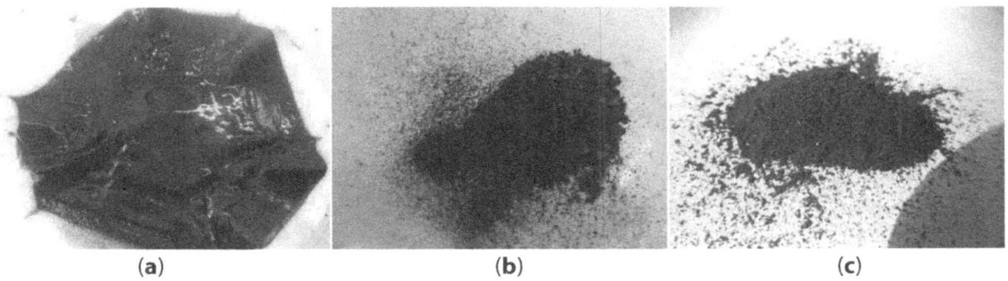

Figure 6.25 Dark blue precipitates (pH 5): (a) Filtered (25 °C); (b) Dried (180 °C for 1 h); (c) Ground with pestle and mortar after drying.

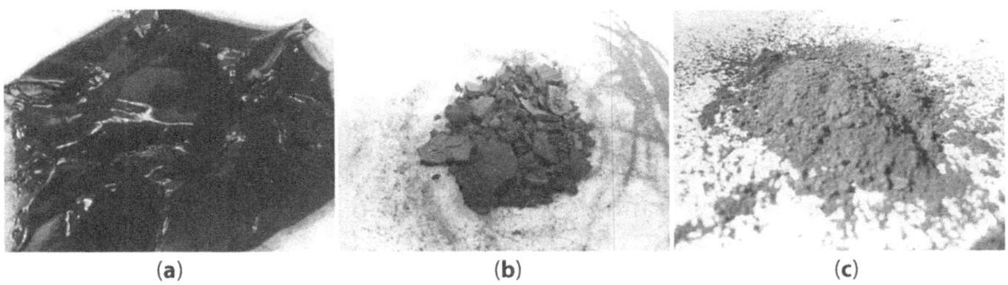

Figure 6.26 Green precipitate (pH 7): (a) Filtered (25 °C); (b) Dried (180 °C for 1 h); (c) Ground with pestle and mortar after drying.

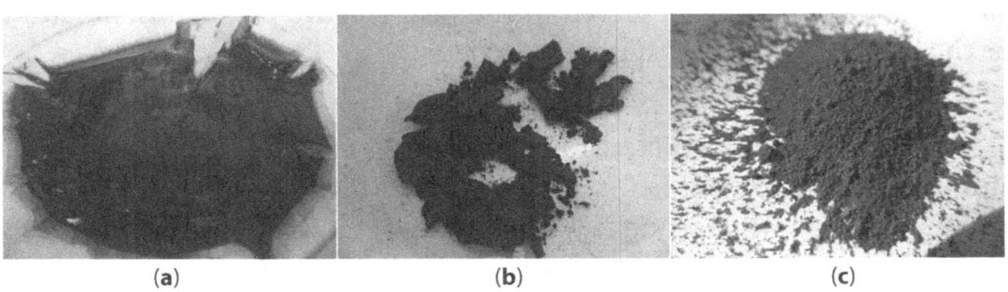

Figure 6.27 Brown precipitate (pH 9): (a) Filtered (25 °C); (b) Dried (180 °C for 1 h); (c) Ground with pestle and mortar after drying.

Table 6.8 Mass (g L^{-1}) of the recovered precipitates at different pH values (of ferrocyanide containing solutions) using NaOH.

pH	Color of precipitate	*m* precipitate
3	Light blue	1.644
5	Dark blue	2.194
7	Green	3.030
9	Dark brown	3.514

Figure 6.28 Prussian blue coloration: (a) in the foreground at an AMD affected stream, (b) in a water retain dam of a slimes dam (photo courtesy T. S. McCarthy), and (c) in a solution trench of a slimes dam (the pipe was discharging ferrocyanide containing water at pH 9 that mixed with AMD in the trench at pH 3, resulting in the formation of Prussian blue).

brown precipitate. Yield for the recovered precipitates ranged from 1.644 to 3.514 g L^{-1} of the initial synthetic AMD solution (Table 6.8). Precipitates involving ferrocyanide largely consist of Prussian blue ($Fe^{III}[Fe^{III}Fe^{II}(CN)_6]_3$) and have been observed in previous studies [34] in the vicinity of active gold tailings dams (Figure 6.28). These Prussian blue precipitates are formed *via* the following reaction:

$$4\ Fe^{3+} + 3[Fe(CN)_6]^{4-} \rightarrow Fe^{III}[Fe^{III}Fe^{II}(CN)_6]_3 \qquad (6.4)$$

Precipitates obtained experimentally (Figures 6.13 to 6.15) show that they are quite persistent and tend to overshadow the other precipitates, e.g., the yellow and brown ones. Distribution of the colorations observed in the field (Figure 6.17) substantiates this persistence as the bluish coloration is quite elaborate.

6.3.3.2.4 Using $MgCO_3$ as a Neutralizing Agent

Precipitates obtained following neutralization with $MgCO_3$ (Figure 6.29) show similar colors to those obtained when using NaOH (Figures 6.13, 6.16, and 6.19). These similarities are due to the same experimental conditions for both alkaline materials. Same chemicals were dissolved at room temperature and had similar fixed neutralization pH values. Yellowish to brown Fe precipitates were obtained (Figures 6.29 to 6.36). In addition, the

(a) (b)

Figure 6.29 (a) Beakers stirred for 48 h to allow oxygen circulation in the reactions; (b) Vacuum system used to filter the samples.

yields of the recovered precipitates are generally comparable to those obtained using NaOH (Table 6.9). Some of the filtrates from the initial neutralization experiments were used for further treatment with $MgCO_3$ to assess any potential precipitates that could be obtained (Table 6.10). Filtrates that were retreated were from beakers A, B, and C (with initial neutralization pH values of 3, 4, and 5, respectively). Targeted new pH values were 5 (for the filtrate that was at pH 3); 7 for the filtrate at pH 4 and 9 for the filtrate at pH 9 (Figures 6.37 to 6.39).

(a) (b) (c)

Figure 6.30 Yellow precipitates (pH 3): (a) Filtered (25 °C); (b) Dried in oven (180 °C); (c) Sample grinded with pestle and mortar after it was dried.

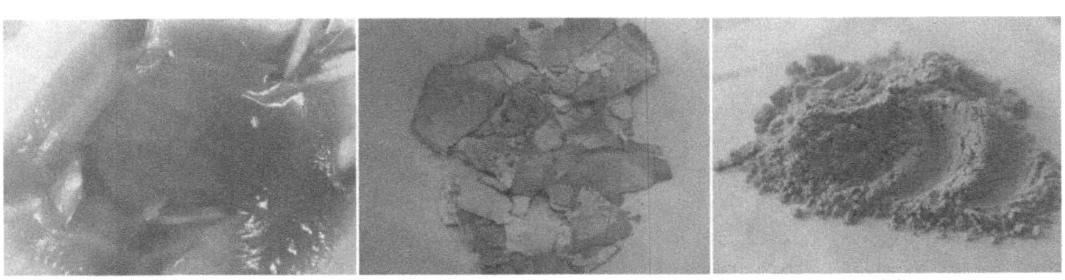

(a) (b) (c)

Figure 6.31 Yellow-brown precipitates (pH 4): (a) Filtered (25 °C); (b) Dried (180 °C for 1 h); (c) Ground with pestle and mortar after drying

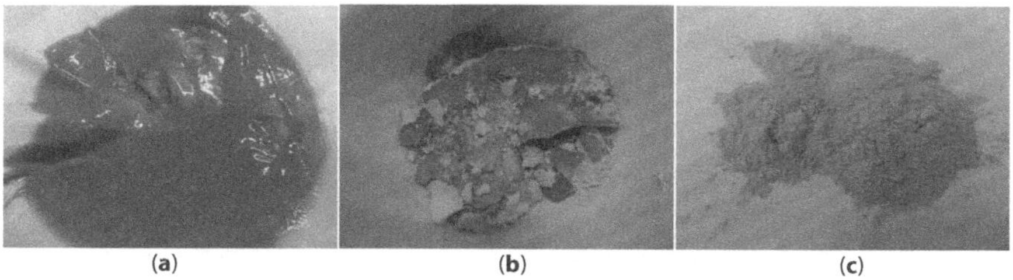

Figure 6.32 Yellow-brown precipitate (pH 5): (a) Filtered (25 °C); (b) Dried (180 °C for 1 h); (c) Ground with pestle and mortar after drying.

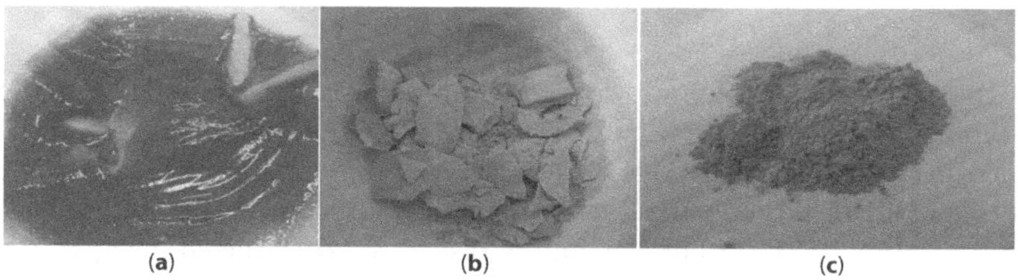

Figure 6.33 Yellow-brown precipitate (pH 6): (a) Filtered (25 °C); (b) Dried (180 °C for 1 h); (c) Ground with pestle and mortar after drying.

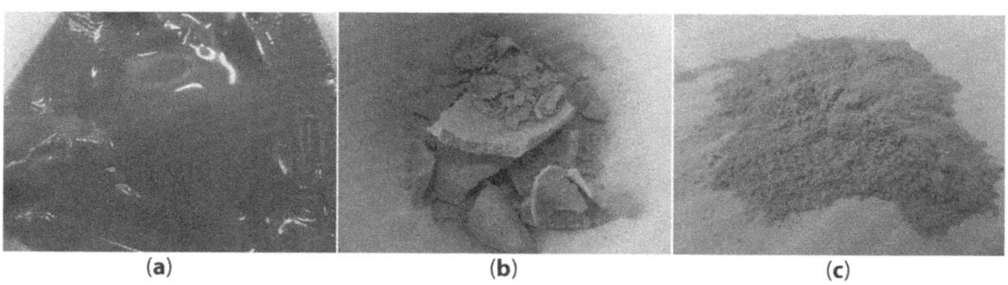

Figure 6.34 Light brown precipitate (pH 7): (a) Filtered (25 °C); (b) Dried (180 °C for 1 h); (c) Ground with pestle and mortar after drying.

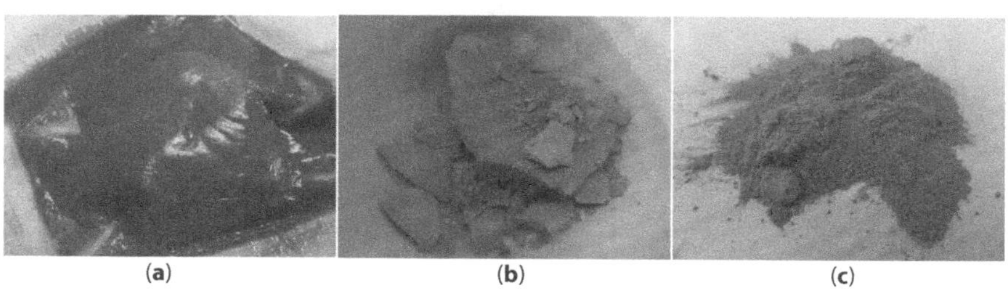

Figure 6.35 Reddish-brown precipitate (pH 8): (a) Filtered (25 °C); (b) Dried (180 °C for 1 h); (c) Ground with pestle and mortar after drying.

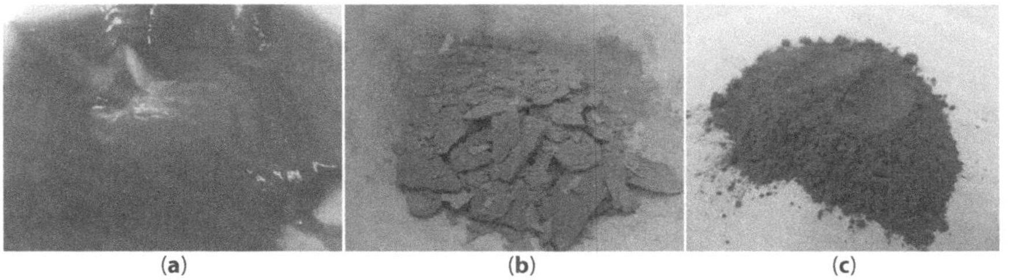

Figure 6.36 Brown precipitate (pH 9): (a) Filtered (25 °C); (b) Dried (180 °C for 1 h); (c) Ground with pestle and mortar after drying.

Table 6.9 Mass (g L^{-1}) of precipitates recovered at different pH values using $MgCO_3$.

pH		Precipitate
3	Yellow	17.0
4	Yellow-brown	19.1
5	Yellow-brown	27.1
6	Yellow-brown	28.5
7	Brown	25.0
8	Reddish-brown	29.4
9	Brown	31.6

Table 6.10 Mass (g L^{-1}) recovered from retreatment of selected filtrates with $MgCO_3$ at different pH values.

pH	Color of precipitate	*m* precipitate
5	Reddish-brown	1.82
7	Brown	2.04
9	Black	1.96

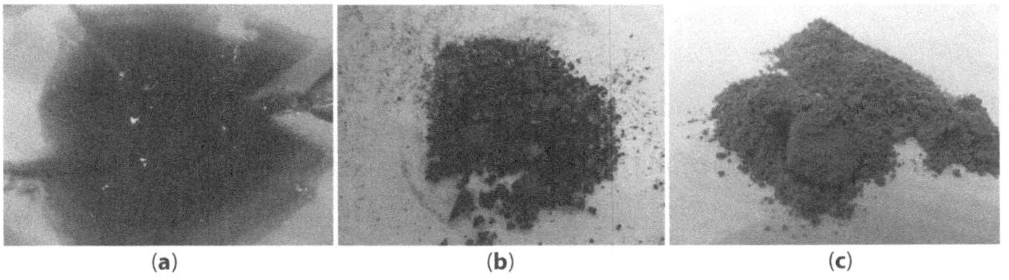

Figure 6.37 Reddish-brown precipitate (pH 5): (a) Filtered (25 °C); (b) Dried (180 °C for 1 h); (c) Ground with pestle and mortar after drying.

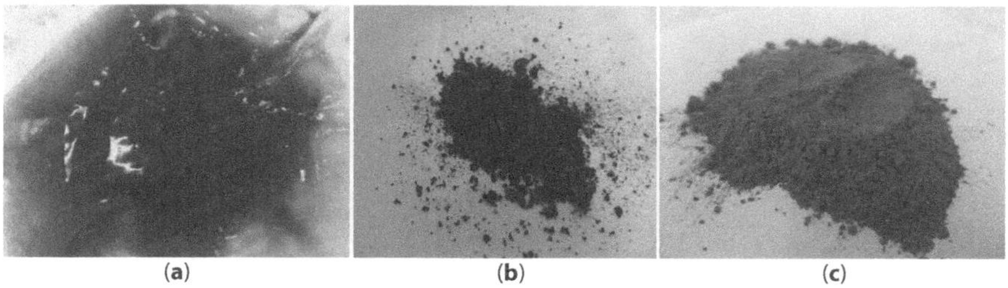

Figure 6.38 Brown precipitate (pH 7): (a) Filtered (25 °C); (b) Dried (180 °C for 1 h); (c) Ground with pestle and mortar after drying.

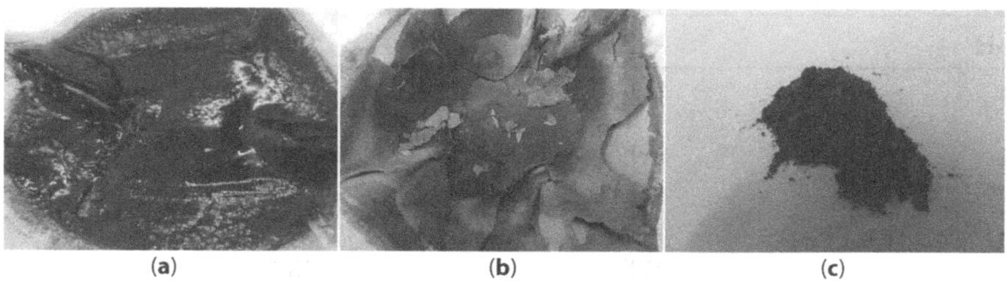

Figure 6.39 Brown-black precipitate (pH 9): (a) Filtered precipitates (25 °C); (b) Dried (180 °C for 1 h); (c) Ground with pestle and mortar after drying.

6.3.3.3 Characterization of Fe Precipitates

Iron precipitates were characterized using PXRD to determine their mineralogy, and the results showed that the dominant Fe minerals were goethite, hematite and magnetite (Table 6.11). Mineralogical analyses of the Fe precipitates showed elevated concentrations of Fe in the form of oxides and oxyhydroxides which was confirmed by PXRD analysis (Figure 6.40). pH in the experiments was found to be more influential in determining the colors of the precipitates produced than the temperature of reaction. This was proven during experimental work in that even at room temperature, it was possible to produce the desired Fe precipitates and it implies that less energy is required as no heating had to be done to enhance the production of the precipitates. Potential lower costs for recovering Fe precipitates from AMD makes it an attractive alternative to replace Fe oxides (ocher minerals) found in weathered natural iron ores [35].

Table 6.11 Major Fe precipitates and their composition.

Fe precipitate	Chemical formula	Chemical composition (wt %)	Density (g cm^{-3})
Goethite	FeOOH	Fe: 62.85; O: 36.01; H: 1.13	3.8
Hematite	Fe_2O_3	Fe: 69.94; O: 30.06	5.3
Magnetite	Fe_3O_4	Fe: 72.36; O: 27.64	5.15

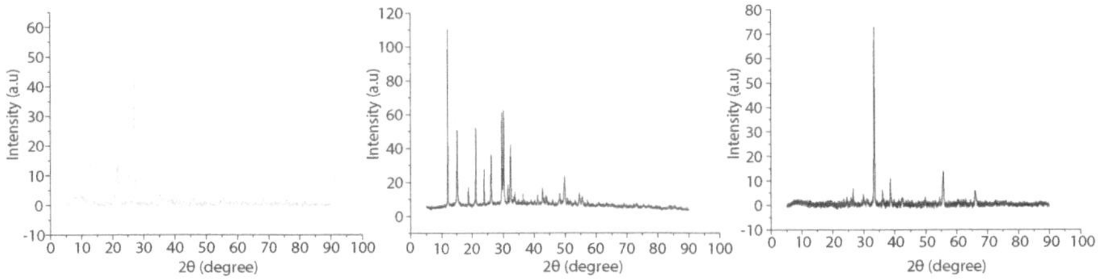

Figure 6.40 X-ray diffractograms (from left to right) of the yellow (goethite), reddish-brown (hematite), and black precipitate (magnetite).

6.3.3.4 *Application in Paintings and Artwork*

For the produced precipitates, their potential to be used in paintings and artwork was assessed (Figure 6.41). As pointed out earlier, the precipitates maintained their colors following addition of water to make painting pastes.

6.3.3.5 *Water Chemistry*

As expected, the nature of the synthesized AMD was influenced by the combination of chemicals dissolved in the solutions. Measured pH in the AMD solutions ranged from 2 to 3, with a high electrical conductivity (EC) and elevated concentrations of Fe and sulfate. Addition of NaOH to one of the chosen solutions (of pH 2.76), for instance, increased the pH and precipitated some metals out of the solution as observed by a substantial decreases in the concentrations of metals in the solution (Table 6.12). More than 95% of the Fe was precipitated in this case. Sulfate concentration also decreased by about 50%, suggesting that sulfate phases are formed in the precipitates. These were not observed in the PXRD

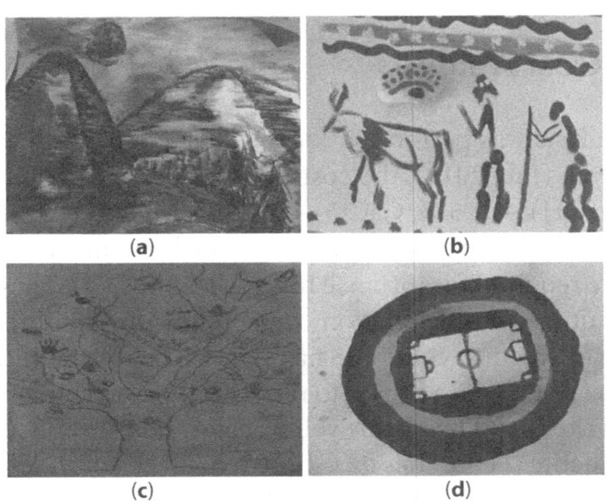

Figure 6.41 Art paintings from Fe precipitates recovered from synthetic AMD: (a) environmental terrains; (b) shepherd and flock; (c) painted tree; (d) soccer field (paintings by Khathutshelo Netshiongolwe).

Table 6.12 Analytical results for synthetic AMD and treated water (mg L^{-1}) for one of the solutions that was neutralized using NaOH; EC in μS/cm.

	pH	EC	Fe	Al	Ca	Cu	Zn	Mn	Cd	Ni	S
Synthetic AMD	2.04	4680	369	1.06	15.6	0.05	0.10	0.28	0.05	0.36	188
Treated AMD	6.34	363	7.57	0.06	0.29	<0.02	0.01	0.03	<0.02	0.01	92

*Sulfate reported as sulfur.

diffractograms, most likely due to the dominance of Fe minerals that tend to shield other minerals. Equation for neutralization by NaOH can be simplified as:

$$4\ NaOH(aq) + 2\ FeSO_4(aq) \rightarrow 2\ Fe(OH)_2(s) + 2\ Na_2SO_4(aq) \quad (6.5)$$

Oxidation of Fe^{2+} to Fe^{3+}, results in the formation of $Fe(OH)_3$. Fe^{2+} precipitates are yellowish in color while those for Fe^{3+} are reddish to brownish. Drying process results in the loss of the water component of the precipitated and formation of minerals such as FeOOH and Fe_2O_3. Magnetite (Fe_3O_4) forms as a result of the combination of oxides of Fe^{2+} and Fe^{3+}.

For neutralization with $MgCO_3$, the reaction can be simplified as follows:

$$MgCO_3(s) + FeSO_4(aq) \rightarrow FeCO_3(s) + MgSO_4(aq) \quad (6.6)$$

It should be noted that the stability and existence of the carbonate depends largely on the pH. At low pH, it is easily destroyed. While not shown in the results above, elevated concentrations of Na were analyzed (as well as Mg in instances where $MgCO_3$ was used). It is possible to recover salts such as epsomite ($MgSO_4$) from these additional ions.

6.4 Indicative Cost Analysis

An indicative cost analysis was performed for the effective neutralizing agents. According to Sigma-Aldrich/Merck (July 2019), the costs for the following chemicals (per 500 g) are: $MgCO_3$ – R 3350, $Mg(OH)_2$ – R 523, $CaCO_3$ – R 2332, MgO – R 553, Na_2O – R 1078, and NaOH – R 9568. Actual experiments would determine how much exactly of the respective neutralizing agents are required for instance to treat 1 L of AMD. These neutralizing agents were identified to result in the precipitation of the following iron precipitates: ferrihydrite ($Fe(OH)_3$) has a white to greenish color, FeS has a brown color, goethite (FeOOH) has a yellowish color, hematite (Fe_2O_3) has a brown to reddish color, mackinawite (FeS) has a bronze to white grey color, pyrite (FeS_2) has a pale yellow color, siderite ($FeCO_3$) has light brown to yellow color, and magnetite (Fe_3O_4) has dark brown to black color. A few grams (25 g) of an artist's paint pigment can cost anywhere above R 1500. Thus, precipitation of such ochers may provide an economically promising way of producing these pigments.

6.5 Conclusion

An experimental design using PHREEQC was established. Simulations successfully determined the resulting treated water chemistry as well as the ocher minerals of interest and the trace metals that would be removed from the water. Water was treated to acceptable quality for discharge, giving a pH between 7 and 9. Magnesium-based neutralizing agents resulted in treated water of acceptable quality and gave some important iron precipitates. Some of the combinations of neutralizing agents also showed that it was possible to achieve satisfactory results, an important factor when considering the costs of individual neutralizing agents. Also, the effects of varying some of the parameters were assessed. PHREEQC is a fast experimental design tool for establishing the neutralization of AMD and the precipitation of ochers. It saves time, is cheap, and showed reliable results that would otherwise be tedious to achieve through conventional trial-and-error methods. Yet, the economic value that could be derived from the precipitation of ochers would require further detailed determination. This study explored the possibility of beneficiating the AMD treatment process by forming Fe precipitates or ochers that can be useful for paint and artwork applications. Using the conditions established in computational simulations, NaOH and $MgCO_3$ were found to be the best choices for the neutralization of AMD with the aim of precipitating the desired ochers. Precipitates range in color from yellow, brown to black and were obtained by varying the pH of neutralization from 3 to 9 at room temperature. Bluish and turquoise precipitates were obtained from solutions that contained ferrocyanide. This meant that recoveries could be done without any extra energy applied to the reactions. Enough yields of the precipitates were obtained, e.g., above 30 g L^{-1} of AMD in some instances and further neutralization of some of the filtrates showed that more precipitates could be precipitated. Different ochers produced were used for some artwork and proved viable, suggesting that this route of their production may offer some potential. Water after treatment with the neutralizing agents showed an elevated pH and substantial decrease in elemental concentrations. Thus, the study has demonstrated that the treatment of AMD can enable more value to be derived in the form of treated water and precipitation of useful ochers. Study underscored the importance of computational simulations as an experimental design tool that can be used to save on time and experimental costs.

Acknowledgements

The authors thank the Water Research Commission (WRC) and the National Research Foundation (NRF) for financial support.

References

1. Masindi, V., Akinwekomi, V., Maree, J.P., Muedi, K.L., Comparison of mine water neutralization efficiencies of different alkaline generating agents. *J. Environ. Chem. Eng.*, 5, 3903–3913, 2017.

2. Masindi, V., Gitari, M.W., Tutu, H., De Beer, M., Fate of inorganic contaminants post treatment of acid mine drainage by cryptocrystalline magnesite: Complimenting experimental results with a geochemical model. *J. Environ. Chem. Eng.*, 4, 4846–4856, 2016.
3. Masindi, V., Recovery of drinking water and valuable minerals from acid mine drainage using an integration of magnesite, lime, soda ash, CO_2 and reverse osmosis treatment processes. *J. Environ. Chem. Eng.*, 5, 3136–3142, 2017.
4. Grover, B.P.C., *Geochemical modelling of the speciation, transport, dispersal and fate of metal contaminants in water systems in the vicinity of tailings storage facilities*, PhD thesis, University of the Witwatersrand, 2016.
5. Hedin, R., Recovery of a Marketable Ion Product from Coal Mine Drainage. *Proceedings of the West Virginia Surface Mine Drainage Task Force Symposium*, Morgantown, WV, pp. 7–8, 1998.
6. Parkhurst, D.L. and Appelo, C.A.J., Description of Input and Examples for PHREEQC Version 3 – A Computer Program for Speciation, Batch-Reaction, One-Dimensional Transport, and Inverse Geochemical Calculations. *U.S. Geol. Surv. Tech. Methods*, 6, 1–497, 2013.
7. Kingery-Schwartz, A., Popelka-Filcoff, R., Lopez, D., Pottier, F., Hill, P., Glascock, M., Analysis of geological ochre: Its geochemistry, use, and exchange in the US Northern Great Plains. *Open J. Archaeom.*, 1, 1, 15, 2013.
8. Kreissl, S., Bolanz, R., Göttlicher, J., Steininger, R., Mark, G., Structural incorporation of W^{6+} into hematite and goethite: A combined study of natural and synthetic iron oxides developed from precursor ferrihydrite and the preservation of ancient fluid compositions in hematite. *Am. Mineral.*, 101, 2701–2715, 2016.
9. Bernatchez, J.A., Geochemical Characterization of Archeological Ochre at Nelson Bay Cave (Western Cape Province), South Africa. *S. Afr. Archaeol. Bull.*, 63, 3–11, 2008.
10. Cornell, R.M. and Schwertmann, U. (Eds.), *The iron oxides: Structure, properties, reactions, occurrences and uses*, p. 664, Wiley-VCH, New York, 2003.
11. Will, R., Global markets for natural and synthetic iron oxide pigments-Proceedings of Iron oxides for colorant and chemical applications, Amsterdam, Netherlands, March 24-26: Portland, ME. *Intertech Corp.*, 2004, 25, 2004.
12. Woodland, A.B., Frost, D.J., Trots, D.M., Klimm, K., Mezouar, M., *In situ* observation of the breakdown of magnetite (Fe_3O_4) to Fe_4O_5 and hematite at high pressures and temperatures. *Am. Mineral.*, 97, 1808–1811, 2012.
13. Jambor, J.L. and Blowes, D.W., Theory and applications of mineralogy in environmental studies of sulfide bearing mine wastes., in: *Modern approaches to ore and environmental mineralogy*, Cabri, L.J. and Vaughan, D.J. (Eds.), pp. 367–401, 1998.
14. Popelka-Filcoff, R.S., Claire, E., Lenehan, K.W., Bennett, J.W., Stopic, A., Jones, P., Pring, A., Quinton, J.S., Durham, A., Characterization of Ochre Sources in South Australia by Neutron Activation Analysis (NAA). *J. Anthropol. Soc. South Aust.*, 35, 81–90, 2012.
15. Sobron, P., Bishop, J.L., Blake, D.F., Chen, B., Rull, F., Natural Fe-bearing oxides and sulfate from the Rio Tinto Mars analog site: Critical assessment of VNIR reflectance spectroscopy, laser Raman spectroscopy, and XRD as mineral identification tools. *Am. Mineral.*, 99, 1199–1205, 2014.
16. Glazyrin, K., McCammon, C., Dubrovinsky, L., Merlini, M., Schollenbruch, K., Woodland, A., Hanfland, M., Effect of high pressure on the crystal structure and electronic properties of magnetite below 25 GPa. *Am. Mineral.*, 97, 128–133, 2012.
17. Montalto, N.A., The Characterization and Provenancing of Ancient Ochres., in: *Cranfield Health, Translational Medicine*, PhD Thesis, Cranfield University, 2010.
18. Froment, F., Tournié, A., Colomban, P., Raman identification of natural red to yellow pigments: Ochre and iron-containing ores. *J. Raman Spectrosc.*, 39, 5, 560–568, 2008.

19. Murad, E. and Rojík, P., Iron-rich precipitates in a mine drainage environment. *Influence of pH on mineralogy, 3rd Australian New Zealand Soils Conference*, pp. 5–9, December 2004, University of Sydney, Australia. http://www.regional.org.au/au/asssi.
20. Hedin, R., Recovery of marketable iron oxide from mine drainage in the USA. *Land Contam. Reclamat.*, 11, 93–97, 2003.
21. Faria, D.L.A. and Lopes, F.N., Heated goethite and natural hematite: Can Raman spectroscopy be used to differentiate them. *Vib. Spectrosc.*, 45, 117–121, 2007.
22. Lopes, F.A., Menezes, J.C.S.S., Schneider, I.A.H., Acid Mine Drainage as Source of Iron for the Treatment of Sewage by Coagulation and Fenton's Reaction. *International Mine Water Association Conference*, p. 237–240, Aachen, Germany, 2011.
23. Guignard, J. and Crichton, W.A., Synthesis and recovery of bulk Fe_4O_5 from magnetite, Fe_3O_4. A member of a self-similar series of structures for the lower mantle and transition zone. *Mineralog. Mag.*, 78, 361–371, 2014.
24. Davidson, E.E., Mohd, T.L., Nanfe, R.P., Napoleon, W., Shukla, A., *Genesis, Uses and Environment Implications of Iron Oxides and Ores, Iron Ores and Iron Oxide Materials*, p. 75776, IntechOpen, London, UK, 2018.
25. Bikiaris, D., Sister Daniilia, S., Sotiropoulou, O., Katsimbiri, E., Pavlidou, A., Moutsatsou, P., Chryssoulaki, Y., Ochre-differentiation through micro-Raman and micro-FTIR spectroscopies: Application on wall paintings at Meteora and Mount Athos, Greece. *Spectrochim. Acta A Mol. Biomol. Spectrosc.*, 56, 1, 3–18, 2000.
26. Tiffany, R., Popelka-Filcoff, R.S., Lenehan, C.E., Towards identification of traditional European and indigenous Australian paint binders using pyrolysis gas chromatography mass spectrometry. *Anal. Chim. Acta*, 803, 194–203, 2013.
27. Valente, T. and Leal Gomes, C., Occurrence, properties and pollution potential of environmental minerals in acid mine drainage. *Sci. Total Environ.*, 407, 1135–1152, 2009.
28. Bibi, I., Singh, B., Silvester, E., Akaganeite (β-FeOOH) precipitation in inland acid sulfate soils of south-western New South Wales (NSW), Australia. *Geochim. Cosmochim. Acta*, 75, 6429, 2011.
29. Wang, M., Chou, I., Lu, W., de Vivo, B., Effects of CH_4 and CO_2 on the sulfidization of goethite and magnetite: An *in-situ* Raman spectroscopic study in high-pressure capillary optical cells at room temperature. *Eur. J. Mineral.*, 27, 193–201, 2015.
30. Brok, E., Frandsen, C., Lefmann, K., McEnroe, S., Robinson, P., Burton, B.P., Hansen, T.C., Harrison, R., Spin orientation in solid solution hematite-ilmenite. *Am. Mineral.*, 102, 1234–1243, 2017.
31. Manickavasagam, S., Saltiel, C., Giesche, H., Characterization of colloidal hematite particle shape and dispersion behavior. *J. Colloid Interface Sci.*, 280, 417, 2004.
32. Akinwekomi, V., *Formation of magnetite during desalination of Acid Mine Drainage*, D. Tech thesis, Tshwane University of Technology, 2017.
33. Bologo, V., Maree, J.P., Zvinowanda, C.M., Treatment of acid mine drainage using magnesium hydroxide. *Proceedings of the International Mine Water Association Conference*, Pretoria, South Africa, pp. 19–23, 2009.
34. Bakatula, E. and Tutu, H., Characterization and speciation modelling of cyanide in effluent from an active slimes dam. *S. Afr. J. Chem.*, 69, 140–147, 2016.
35. Lottermoser, B., Recycling, reuse and rehabilitation of mine wastes. *Elements*, 7, 405–410, 2011.

7

Innovative Routes for Acid Mine Drainage (AMD) Valorization: Advocating for a Circular Economy

Vhahangwele Masindi[1,2*] and Memory Tekere[2]

[1]Magalies Water, Scientific Services, Research & Development Division, Brits, South Africa
[2]Department of Environmental Sciences, College of Agriculture and Environmental Sciences, University of South Africa (UNISA), Florida, South Africa

Abstract

In the aftermath of intensive mining, the generation of acid mine drainage (AMD) has been an issue of prime concern to national, regional, and international scientific and water care communities. Elevated concentrations of Al, Fe, Mn, and sulfate, among other contaminants, dominate AMD. This makes AMD a viable source of valuable metals and minerals. Lately, recovery of metals from AMD has been intensively pursue, albeit, with minimal success. Ongoing research has been widespread due to the need for effective, efficient, and working solutions that aims to responds to increasingly stringent environmental regulations. Drainages from mining activities are being considered in a fresh light with a prime attempt of them being beneficiated through the recovery and synthesis of valuable minerals. To achieve that, different approaches and techniques are used to recover valuable minerals with a wide range of industrial applications. A series of laboratory, pilot, and industrial scale technologies have been employed for the treatment of AMD with value recovery, beneficiation, and material synthesis. Despite a wide range of available AMD treatment technologies, secondary sludge, which is complex, heterogeneous, and highly mineralized, require innovative approaches to harness value and advance the concept of circular economy. The need for sustainable and economic viable AMD treatment approaches has led to much focus on clean water reclamation, reuse, and recovery. Systematically, this chapter underscores (i) the composition of AMD, (ii) environmental and ecotoxicological impacts, (iii) developed remediation techniques, (iv) valorization routes, and (v) a case study in South Africa. Different configurations and approaches for the treatment of AMD with an option of acquiring valuable secondary minerals have been reported as well. Furthermore, the employment of membrane technology for AMD treatment has been emphasized. The advantages and disadvantages of individual approaches and their synergistic products were also discussed. Overall, this chapter gives insights to various routes for AMD valorization and how they can respond to circular economy. Future perspective and research avenues were also discussed.

Keywords: Acid mine drainage, valorization, beneficiation, treatment, circular economy, value recovery, reclamation, minerals synthesis

**Corresponding author*: masindivhahangwele@gmail.com [ORCID: 0000-0002-5440-8373]
[ORCID: Tekere, 0000-0002-0606-7826]

Elvis Fosso-Kankeu, Christian Wolkersdorfer and Jo Burgess (eds.) Recovery of Byproducts from Acid Mine Drainage Treatment, (189–218) © 2020 Scrivener Publishing LLC

7.1 Introduction

7.1.1 Problem Description

Intensive mining of mineral resources results in numerous environmental problems such as contamination of surface and subsurface water. Specifically, the generation of acid mine drainage (AMD), which is a product of the oxidation of di-sulfide bearing minerals. These minerals can produce different types of drainages ranging from acid to circumneutral and alkaline drainages [1]. However, the defined pH and drainage nomenclature depend on the residual pH and the influence of the surrounding geological setting [1, 2]. As a result of differing geological settings, host sulfide rocks can contain different minerals (Table 7.1) and their resultant drainages are consequently rich in chemical species resulting from these phases.

The chemical constituents present in mine drainages depend on the composition of the weathered host rock and associated geological strata. As such, this results in drainage, which constitutes of many elements. Moreover, the recovery, synthesis, and reclamation of minerals also depends on the minerals being weathered to produce AMD. Thus, AMD from the weathering of pyrite will be rich in Fe and S among other contaminants [9–17]. This makes this kind of AMD to be a viable source for Fe and S recovery. A wide array of documented studies has shown the chemistry of pyrite oxidation [2, 18–20].

Table 7.1 Sulfide-bearing minerals that contribute to the formation and the chemical composition of AMD [2–8].

Mineral phase	Formula
Pyrite	FeS_2
Marcasite	FeS_2
Pyrrhotite	Fe_xS_x
Chalcocite	Cu_2S
Covellite	CuS
Chalcopyrite	$CuFeS_2$
Pentlandite	$(Fe,Ni)_9S_8$
Molybdenite	MoS_2
Millerite	NiS
Galena	PbS
Sphalerite	ZnS
Arsenopyrite	FeAsS
Cinnabar	HgS

7.1.2 Physico-Chemical-Microbiological Properties of AMD

The formation of AMD from pyrite can be represented by three chemical reactions: (i) pyrite oxidation, (ii) ferrous oxidation, and (iii) iron hydrolysis (Table 7.2). Sulfur is oxidized to sulfate, and ferric iron is released during pyrite oxidation. The acid formed promotes leaching of (semi-)metals and toxic chemical species into the surrounding waterbody. Furthermore, the AMD formation process is governed by the oxidation of two chemical species, which are the ferrous iron and sulfur. Specifically, pyrite is exposed to water dissolved oxygen, thus promoting the leaching of ferrous iron to the hydrosphere [18, 19, 21–26]. Reaction (7.1) proceeds abiotically and by direct bacterial oxidation under the influence of microorganisms.

$$2\ FeS_2 + 7\ O_2 + 2\ H_2O \rightarrow 2\ Fe^{2+} + 4\ SO_4^{2-} + 4\ H^+ \tag{7.1}$$

$$4\ Fe^{2+} + O_2 + 4\ H^+ \rightarrow 4\ Fe^{3+} + 2\ H_2 \tag{7.2}$$

$$4\ Fe^{3+} + 12\ H_2O \rightarrow 4\ Fe\ (OH)_3 + 12\ H^+ \tag{7.3}$$

$$FeS_2 + 14\ Fe^{3+} + 8H_2O \rightarrow 15\ Fe^{2+} + 2SO_4^{2-} + 16\ H^+ \tag{7.4}$$

Table 7.2 A summary of chemical reactions and processes taking place in AMD formation.

Equations	Reactions
1	Exposure of pyrite (FeS_2) to oxygen and water culminate to the formation of ferric iron (Fe^{3+}), sulfate (SO_4^{2-}), and acid (H^+) as predominant chemical species. Ferrous water presents a darkish to clear water. Microbiologically, the reaction proceeds abiotically and by direct bacterial oxidation.
2	Aqueous ferrous (Fe^{2+}) ions react with oxygen and protons to form ferric (Fe^{3+}) ions and water. The conversion of ferrous to ferric iron consumes four moles of protons. Microbiologically, reaction (7.3) proceeds abiotically and slows down as pH increase. Thus, the chemistry depends on the pH which is above 4.5, high sulfate, low Fe, little or no acidity.
3	The generated ferric species hydrolyze in surrounding water to form ferric hydroxide. Three moles of acidity are generated. Ferric hydroxide formation (yellow boy) promotes the formation of acidity due to the consumption of OH^-. Due to a fractional precipitation study undertaken, solids form at pH ≥3.5, but at pH ≤3.5, minute solids will also precipitate [14].
4	Oxidation of additional pyrite by ferric iron can take place. Steps 1 and 2 generate ferric iron. This is the circulating and autonomous reaction, that continuous takes place until either ferric iron or pyrite and other metals that contribute to AMD formation are depleted. Iron (III) is the oxidizing agent, for this reaction not oxygen. Acidity from Fe^{2+}/Fe^{3+} in solution is the source of proton as attributed to the chemical reactions taking place and this will lead to precipitation of iron species, thus releasing protons *via* hydroxyl group binding to the surface hence they are known as acid forming metals.

Microorganisms catalyze the chemical reactions at different pH values, thus speeding up the reaction rates substantially. They comprise a relatively large number of various species (Table 7.3).

Depending on the local geological setting and mineral phases being mined, the chemistry of the resultant mine drainage will vary [22–27]. As such, there are different types of drainages and they range from AMD to circumneutral drainage and basic drainage (Figure 7.1) [1], which can be used to define them.

AMD is dominated by metals and sulfate while circumneutral drainage is rich in selected metals and anions such as sulfate and the basic drainage is rich in base metals and sulfate (Table 7.4 and Figure 7.1). The reduction in metals concentrations with an increase in pH is attributed to precipitation of metals with varying pH gradients [1].

According to literature, drainages from mining activities in South Africa are composed of Fe, Al, Mn, Ca, Na, and Mg in addition to traces of Cu, Co, Zn, Pb, and Ni (Table 7.4). These concentrations are far above the prescribed legal requirements. Furthermore, circumneutral drainage can be richer in sulfate than metals due to their precipitation at elevated pH gradients. The effluents of AMD from coal and gold mine are rich in sulfate, acid, and (semi)metals. Moreover, the Gold AMD is richer in potentially toxic (semi) metals than the Coal AMD, which is rich in Fe, Al, Mn, and sulfate, hence making it ideal for the recovery of valuable minerals. This makes the coal AMD a viable option for Fe and sulfate recovery. Notable concentrations of chemical species were observed, and they include Al (0.01–474 mg/L), Cu (0–7.8 mg/L), Fe (0.07–8158 mg/L), Mn (25–192 mg/L), Ni (0.21–16.6 mg/L), Zn (0.16–7.9 mg/L), among others. Sulfate ranged from 4,603 to 42,862 mg/L.

The color of AMD can be colorless or black to red, which mainly depends on the dissolved chemical species. The dominant metal present in AMD is iron, which may give water a typically reddish color, while in streams with a higher pH (pH 5–6.5), orange-yellow ferric iron-rich sediments ("yellow boy") are present. The white color of AMD is determined by

Table 7.3 Oxidizing bacteria and their growth conditions [21, 23, 27–29].

Bacteria species	pH, -	Temp, °C	Classification	Feeding mode
Acidithiobacillus thioparus	4.5–10	10–37	Aerobic	Autotrophic
Acidithiobacillus ferrooxidans	0.5–6.1	15–25	Aerobic	Autotrophic
Acidithiobacillus thiooxidans	0.5–6.0	10–37	Aerobic	Autotrophic
Thiobacillus neapolitanus	3.0–8.5	8–38	Aerobic	Autotrophic
Thiobacillus denitrificans	4.0–9.5	10–37	Aerobic	Autotrophic
Thiobacillus novellus	5.0–9.2	25–35	Aerobic	Autotrophic
Acidithiobacillus intermedius	1.9–7.0	25–35	Aerobic	Autotrophic
Acidithiobacillus perometabolis	2.8–6.8	25–35	Aerobic	Autotrophic
Sulfolobus acidocalderius	2.0–5.0	55–85	Aerobic	Autotrophic
Desulfovibrio desulfuricans	5.0–9.0	10–45	Anaerobic	Heterotrophic

Table 7.4 Parameters of selected mine waters. EC: electrical conductivity (μS/cm), all other elements in mg/L, pH: without units.

Parameters	Gold mine water*	Coal mine water**	Circumneutral mine water†
pH	2.3	2.5	6.5
EC	22,713	13,980	500
Na	248.4	70.5	20.1
K	21.6	34.2	29.1
Mg	2.3	398.9	861.8
Ca	710.8	598.7	537.5
Al	134.4	473.9	0.01
Fe	1243	8,158.2	0.07
Mn	91.5	88.2	25.0
Cu	7.8	–	–
Zn	7.9	8.36	0.16
Pb	6.3	–	–
Co	41.3	1.89	0.29
Ni	16.6	2.97	0.21
SO_4^{2-}	4,635	42,862	4,603

Note: *) Gold mining AMD [30], **), Coal mining AMD, and †), circumneutral drainage water [9, 10, 31–35].

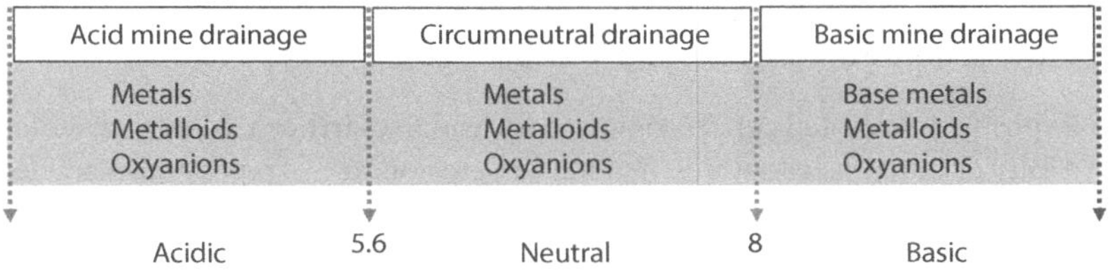

Figure 7.1 Mine drainage and their respective pH (modified from Nordstrom *et al.* [26]).

the presence of Al (Gibbsite). The black color arises by the presence of Mn [14, 18, 21, 36]. Their oxidation imparts color to the waterbodies.

7.2 Health Effects Associated with Contaminants in AMD

Different eco-toxicological and epidemiological studies have been pursued to point out health effects associated with metals present in AMD (Table 7.5) and a number of limits were proclaimed to be above the tolerance concentrations of living organisms [37–41].

Table 7.5 Selected elements in AMD and their eco-toxicological effects [42, 43].

Element	Ecotoxicological effects
As	May lead to chronic poisoning of the skin lesions, hyperpigmentation, and cancer, also, death due to upper respiratory, pulmonary, gastrointestinal, and cardiovascular failure. Furthermore, the nerve damage and sensory loss in the peripheral nervous system may also be linked to As exposure.
Al	May lead to chronic neurological disorders which include dialysis dementia and Alzheimer's disease.
Cr	May lead to nasal septum and skin ulcers, and lung cancer. Ingestion of Cr may pose taste effects and nausea at elevated concentrations. Documented epidemiological studies for carcinogenesis *via* the oral route are equivocal, and gastrointestinal cancer.
Fe	Can cause undesired metallic taste and coating of plumbing materials. Staining of clothes and water may also occur due to elevated concentrations of Fe in water.
Mn	May cause severe, aesthetically, and unacceptable staining. Some chance of manganese toxicity under unusual conditions.
Na	Elevated concentrations may cause hypertension and cardiovascular or renal diseases.
Cu	Exposure of living organisms may lead to gastrointestinal discomfort, nausea, and vomiting. May also cause taste and staining problems. To a certain extent, severe poisoning with possible fatalities may be experienced as well.
Mg	Exposure to excess concentrations can lead to the central nervous system and heart function suppression. Furthermore, notable toxicological effects may be linked to diarrhoea when magnesium sulfate is consumed. Other effects include scaling and high water hardness.
Zn	Exposure or consumption may lead to bitter taste; and milky appearance. Furthermore, gastrointestinal, irritation, nausea, and vomiting may be experienced. Acute toxicity with electrolyte disturbances and renal damage may be experienced.
Sulfate	Exposure to elevated concentrations may lead to diarrhoea. Moreover, sulfate imparts a salty or bitter taste to water. In addition, erosion rate of pipes is also accelerated.
TDS	May lead to bad taste, skin dryness, scum in water and high soap consumption, and kidney damages. Causes hypertension as well.

Exposure of living organisms to these potentially toxic and hazardous chemical species have been reported to pose numerous toxicological effects to living organisms and this has been shown in several dose-response tolerance studies.

7.3 Abatement of AMD

As an initial step toward managing the formation of AMD, different techniques are employed for its abatement. This is mainly attributed to high requirement of O_2, water, and microorganisms for the pyrite oxidation. It has grown easy to control the formation of AMD due to elimination of single or dual ingredients for AMD formation. These techniques were developed to prevent the formation of mine drainage and eliminate the problems of AMD. Commonly the used prevention techniques include: (i) top covers such as clay minerals that have low water permeability hence limiting the reaction to take place and for AMD to form, (ii) *in situ* neutralization which involves the addition of alkaline materials to the void or tailings that has high potential to produce AMD, (iii) passivation by using a coating material to prevent the reactive fractions to be exposed to oxidizing conditions, and (iv) lastly, bactericides to reduce the biological activities of bacterial community harbored in the tailings or the mining voids [2, 4, 5].

7.4 Techniques for AMD Treatment

7.4.1 Overview

As second step toward the management of AMD, different technologies with different contaminants removal mechanisms have been developed, deployed, and employed for AMD treatment (Figure 7.2). These technologies include:

- precipitation to remove metals at different pH gradients;
- adsorption of contaminants using an adsorbent which is a surface reaction that depends on the charge difference;
- filtration using microfiltration (MF), ultrafiltration (UF), nanofiltration (NF), reverse osmosis (RO), and membrane distillation (MD).
- Phytoremediation using plants with different affinities.

7.4.2 Chemical Precipitation

Chemical precipitation using alkaline material raise the pH of the water, thus promoting the formation of hydroxides, sulfides, and carbonates as precipitates [7, 8, 12, 13,

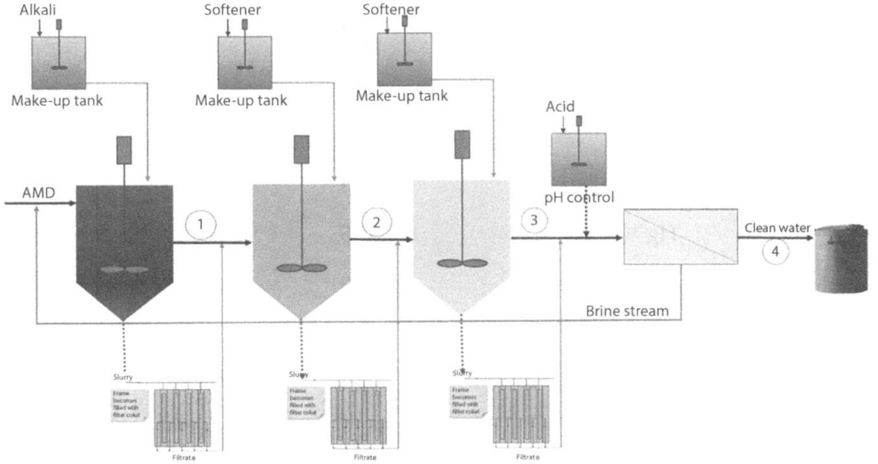

Figure 7.2 A typical multi-functional and process AMD treatment plant (modified from Masindi *et al.* [17]).

15, 44–50]. The dissolved chemical species will precipitate at varying pH values (Table 7.6) [51–53]. When metals recovery needs to be pursued, this mechanism is desirable. Chemicals used for precipitation are Ca-based (lime, limestone, hydrated lime, and dolomite), Mg-based (magnesite, brucite, periclase, magnesium bicarbonate, and calcined magnesite), and Na-based (soda ash, caustic soda, and sodium bicarbonate) and can be other materials rich in Ca, Mg, and Na. Variation in pH pushes the metals to an insoluble state, hence leading to metals precipitating as hydroxides, carbonates, and sulfides depending on the seeding material.

The chemical composition influences the type of reagents that need to be used for the recovery of chemical species in AMD [35, 54, 55]. Metals removal can be shown by the following equation:

$$M^{n+} + nOH^{-} \rightarrow M(OH)_n \downarrow \tag{7.5}$$

Silva *et al.* [56] recovered Fe(III) from AMD and explored its application for the production of yellow pigment using sodium hydroxide. Selective precipitation of iron was so precise in a way that the contamination of Al, Zn, and Mn was avoided. Akinwekomi *et al.* [49] reported the use of $NaCO_3$ and NaOH for the recovery of Fe(III), Fe(II), and Al(III) from AMD at varying pH values. Their study was effective in recovering quantifiable volumes of metals and employed them for the synthesis of goethite, hematite, and magnetite.

Table 7.6 The pH values at which metals precipitate in the hydrosphere [14].

Metal ion	pH
Al^{3+}	4.5
Fe^{3+}	3.5
Mn^{2+}	8.5
Cr^{3+}	5.5
Cd^{2+}	7.0
Fe^{2+}	8.5
Ni^{2+}	9.5
Hg^{2+}	7.5
Na^{+}	7.0
Pb^{2+}	6.0
Zn^{2+}	7.0
Cu^{2+}	5.5
Mg^{2+}	10.0

7.4.3 Adsorption

In aqueous media, substances show different behavior in terms of surface charges. Adsorption is a surface reaction, whereby an adsorbent adsorbs contaminants from aqueous solution using charge differences. A positively charged substance will adsorb a negatively charged compound, hence Al/Fe^{3+} substances would adsorb anionic substances such as fluoride, arsenic, and chromium from aqueous solution [33, 57]. Due to its simplicity, resource availability, and efficacy, adsorption is widely employed for contaminants recovery and removal. This process can be run on batch and column approach. Furthermore, this is a mechanism that is widely employed for the removal of pollutants. Activated carbon can be used for the removal of organic compounds and metals *via* different adsorption mechanisms [58–60]. It is a surface reaction whereby a compound that is negatively charged will attract positively charged surface [61], hence making it a charge driven process. Other mechanisms such as ion exchange, precipitation, and complexation may govern the removal of contaminants from the hydrosphere [62] (Figure 7.3). A number of charged materials such as clay minerals, biochar, powdered activated carbon, granular activated carbon, metals (e.g., Al, Fe, Mn, Mg, and Ca), and modified compounds are widely employed for the removal of contaminants from aqueous solution [63–67]. Minerals or element recovery from this mechanism may be governed by desorption [69]. Masindi *et al.* [33] assessed the use of Al^{3+} modified bentonite clay for the removal of fluoride from groundwater. The study proved effective at 60 min of mixing, 1 g:100 ml S/L ratios and room temperature. On a different study, Gitari *et al.* [57] successfully employed similar conditions for the removal of fluoride using Fe^{3+}modified bentonite clay. Furthermore, Masindi *et al.* [70–73] effectively evaluated the use of magnesite, pre-treated magnesite, and their bentonite clay composites for the removal of phosphate, arsenic and boron from aqueous solution. These materials were proven highly effective and efficient in the removal of contaminants using biochar adsorption (Figure 7.3) [74].

Masindi and Gitari [75] evaluated the use of magnesite-bentonite clay composite for the removal of (semi)metals from aqueous solutions. The authors highlighted adsorption, ion exchange, and precipitation governing the removal of contaminants from the hydrosphere. To model the mechanisms governing the removal of contaminants, scientists have developed a number of models to point out modes that governs the removal of pollutants

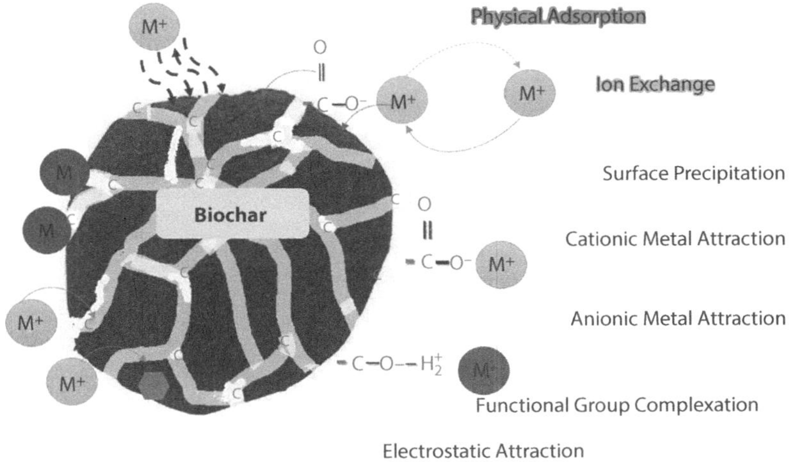

Figure 7.3 Mechanisms of pollutants removal from aqueous solution using biochar (modified from Tran *et al.* [74]).

[76–78]. These models include: adsorption kinetics, isotherms, and thermodynamics [62] (Figure 7.4). Furthermore, adsorption of contaminants from aqueous solutions can be influenced by a number of parameters such as contact time, adsorbent dosage, pH, concentration, and co-existing ions as shown in Table 7.7. These parameters play a crucial role in establishing optimum conditions that are suitable for contaminants attenuation.

7.4.4 Filtration

This section will discuss different technologies that are used for the removal of contaminants from aqueous solution using filtration technologies.

7.4.4.1 Introduction to Membrane Technologies

The filtration of chemicals using membrane technologies have been widely explored and reported in literature [63, 89–91]. Membrane filtration utilizes the combination of a driving

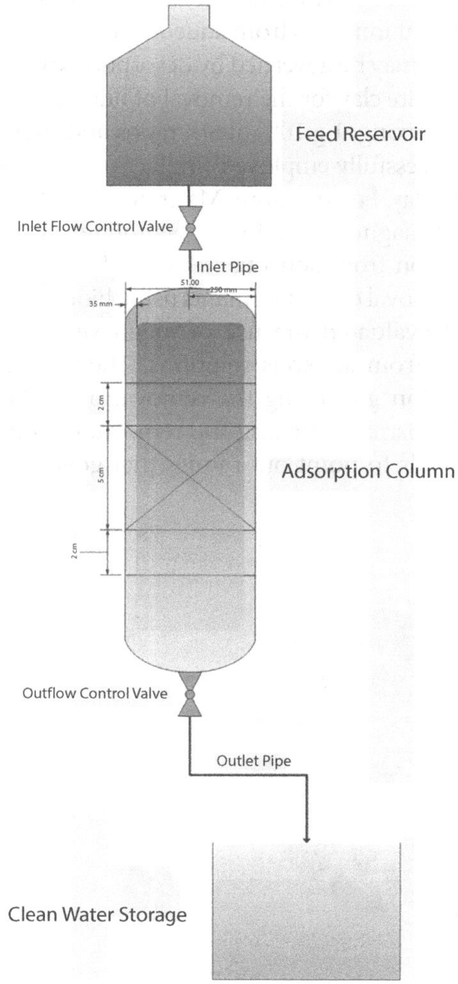

Figure 7.4 A column setup for the removal of contaminants from the hydrosphere.

Table 7.7 Factors that affect the adsorption technology [62, 79–88].

Parameter	Justification
Contact time	Adsorption process is time dependent. Some adsorbent takes time to remove contaminants, and some are very much quicker. Reaction kinetics and diffusion are dependent on the time at which the adsorption process is progressing.
Dosage	Available sites are dependent on the dosage. The higher the dosage, the higher the adsorption affinity. This is also related to surface area.
Concentration	Concentration also determines the amount of adsorbate that the material can adsorb. This is also used to determine the adsorption capacity of the material. Available adsorption site also determines the amount of contaminants that will be adsorbed.
pH	pH influences the type of chemical species to be available in the hydrosphere. Different contaminants have different charges at different pH ranges and that influences their adsorption. The point of Zero Charge (PZC) is also dependent on pH.
Temperature	The temperature influences the solubility of materials. Some minerals dissolve faster at high temperature and they behave in an opposite manner at low temperature. Thermodynamics and isotherms are dependent on temperature.

force, difference in concentration, and a membrane to achieve the separation of dissolved ions. This technique has been widely explored for water purification, minerals recovery, medical science, minerals synthesis, and minerals production but it depends on pore size and diameter (Figure 7.5). Commonly employed filtration techniques are UF, NF, RO, and MD. To a certain extent, they can be used in a combination to acquire a synergy. MF, UF, and NF are all pressure-driven processes where a pressure gradient is the driving force used to separate particulate, colloidal, or dissolved ions and volatile compounds through a porous to semi-porous membrane. The major difference between these three techniques is

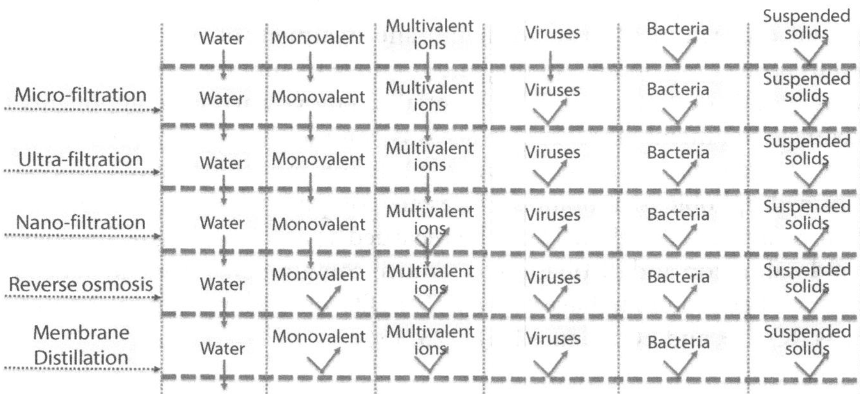

Figure 7.5 Membrane properties and their efficacy in terms of contaminants removal.

the pore size of the membrane used resulting in size-based separation efficacy. RO is similar to UF, MF, and NF since it is pressure driven but in a reverse fashion. It is one of the most commercially established and widely employed membrane filtration processes for the treatment of waste and raw water. Osmosis is the tendency of water to pass spontaneously from a high concentration potential to a low potential across a semipermeable membrane. If a greater pressure is applied to the concentrated solution, liquid water (and some monovalent ions) will flow in the opposite direction through the membrane to the dilute liquid. This pressure-driven osmosis process is known as RO. One of the differences between UF, MF, and NF is that RO has a stringent separation range of < 1 nm [92–105].

Different membranes have different efficacies in terms of contaminants removal (Figure 7.5). Suspended solids and bacteria are removed by different filtration techniques. All membranes except for MF can also remove viruses. Furthermore, MF, UF, and NF fail with multivalent. Carrying on, MF, UF, and NF fail with monovalent. It is for this reason that this technology can easily produce a concentrated dissolved salt stream and a clean water stream from wastewater. RO and, to some degree, NF are suitable and applicable technologies for concentrating (semi)metals which can be recycled or recovered by coupling with other methods such as electrolytic recovery or evaporation. Despite membrane technologies being a promising technology for effluent discharge reduction and water minimization through wastewater reclamation, the use of membrane technology for AMD treatment is not common due to the relatively high membrane costs and membrane fouling. These membrane systems are also susceptible to low pH found in AMD and (semi)metals poisoning. There are, however, applications where NF and RO processes have attracted the attention of researchers for AMD treatment [106–109]. AMD from gold treatment was evaluated using RO and NF and it was found that the NF had a higher permeate flux and pleasing solutes retention efficiencies when compared to RO [15, 106, 110, 111]. However, it should be noted that high membrane fouling was observed due a high permeate flux decrease after a maximum water recovery rate of 60% was achieved. The study was done at optimized conditions with a NF270 membrane. Another major drawback of membrane filtration technique is the generation of brines. Their treatment and disposal is often very expensive, because these brines are often high in total dissolved solids and they increase the environment footprint [112–119]. Therefore, along with combating the high capital costs and low fluxes (due to membrane fouling) for AMD treatment, a suitable brine solution treatment process will also need to be developed. To protect the poisoning of the membrane, upfront pre-treatment prior to membrane treatment of AMD should be strongly considered to protect membrane fouling and improve membrane performance [120–124]. Furthermore, good pre-treatment can potentially remove harmful components that may severely damage the membranes. MD process is commonly employed for drinking water reclamation and valuable minerals recovery [125]. This is a vapor-driven process whereby two sides of hydrophobic membrane surfaces are composed of a non-wetted microporous polymeric membrane. As such, MD can be regarded as an upcoming technology that uses a hybrid of desalination and distillation to remove contaminants from an aqueous system. Moreover, it has been found to have superior benefits when compared to conventional systems such as RO, UF, and FO (Figure 7.5) [125, 126]:

- Has the ability to reject close to 100% of non-volatile and inorganic compounds.
- Require low pressure and temperature as compared to traditional technologies.

- The membranes are made of polymers, hence polymeric, which is widely available.
- Has the ability to get close to 100% salts rejection, a superior quality to standards.
- This system can use waste-heat or rely on solar energy, hence reducing the eco-impacts.
- Has reduced scaling due to limited interaction of solute/ions with the membrane.

According to documented studies and emerging research, MD can be employed in a number of industrial applications such as mine water treatment and brine treatment [93, 127–129], among others. Ryu, Naidu, Hasan Johir, Choi, Jeong, and Vigneswaran [127] reported the use of integrated submerged direct contact MD (DCMD), a zeolite sorption system for the treatment of AMD. In their study, the authors proved that modified zeolite could achieve over 26–30% removal efficacy for contaminants as compared to the natural zeolite. The removal efficacy for modified zeolite showed the followed affinity Fe > Al > Zn > Cu > Ni. Furthermore, adjustment of pH from 2 to 4 further increased the removal of Fe and Al to about 100%. This was explained by a combination of adsorption and precipitation as the pH increase. In addition, above 50% water was recovered within 30 h of treatment. Jimenez and Ulbricht [130] reported the use of DCMD for the treatment of effluents emanating from copper mining process. The authors revealed that this approach was able to reclaim clean water, concentrated sulfuric acid and different metals.

7.4.5 Phyto Remediation

7.4.5.1 Theory of the MD Process

MD is regarded a thermally driven process where separate aqueous hot feed solutions at different temperatures and compositions using a hydrophobic microporous membrane [21]. Due to the difference in temperature and chemical composition which forms a vapor pressure difference across the membrane surface, thus enabling the transfer of vapor molecules from high vapor pressure side (hot side) through perforated polymeric membranes to the low vapor pressure side (cold side) [125]. Literature indicates the availability of four well-known MD configurations, and they include DCMD, air-gap MD, sweeping gas MD (SGMD), and vacuum MD (VMD).

More prevalently, DCMD configurations are popular and have been widely employed for wastewater treatment. In this configuration, a concentrated hot wastewater stream contacts one side of the membrane and a cold permeate stream contacts the opposite side of the membrane. The principles governing the attenuation of the contaminants occurs when the vapor molecules move from a hot side through the perforated polymer membranes and condenses at the permeate side. This is carried with the carrier water as the permeate. With AGMD, the air gap separates the permeate side as compared to the DCMD. However, the condensate gets removed by the cooling plate since this is a mass-heat transfer process. This can also lead to an increase in mass and heat transfer. The SGMD has a configuration similar to AGMD, but the difference is the use of inert gas to remove the vapor molecules. However, the condensate is removed using a condenser for vapor molecules. VMD is similar to SGMD, except the fact that the process relies on vacuum conditions. In light of the above, DCMD has demonstrated some superior advantages due to the fact that it is simple in its design, does not require external equipment (condensate), and permeate condensation occurs inside the membrane system [125, 126, 128, 130].

7.4.6 Phytoremediation

Phytoremediation is derived from the words phyto, which means plant, and remediation, meaning recovery [131, 132]. The term relates to the employment of plants to recover contaminants from different compartments of the environment, and it has been widely explored for (semi)metals removal from the edaphic- and hydrosphere [133–136]. The likes of floating plants for metals' removal and subsequent recovery are *via* different digestion processes and technologies to recover the absorbed nutrients. This method has been reported to be effective, since plants uptake (semi)metals from water. Bwapwa *et al.* [137] assessed the bioremediation of AMD using algae strains and highlighted the feasibility of recovering adsorbed metals due to the ability of algae to hyper-accumulate and hyper-absorb (semi) metals. However, the following challenges have been identified as limiting factors [133, 134, 138–140]:

- long residence time and limited capacity;
- sensitive to water quality conditions and too laborious;
- requires large space of land and a highly controlled environment;
- very difficult to recover minerals post absorption process.

7.5 Valorization of AMD

7.5.1 Aims of Valorization

AMD contains valuable compounds that can be recovered, reclaimed and synthesized using different approaches (Table 7.8). These compounds can make AMD a valuable resource and ensures that the environmental footprints of AMD are minimized, hence ensuring that the concept of circular economy and sustainable development are duly fostered and prioritized.

Table 7.8 Different techniques for minerals recovery, and their advantages and disadvantages.

Technique	Advantage	Disadvantage	References
Precipitation	Independent of concentration	Complex and heterogeneous sludge	[20, 143–145]
Adsorption	Cheap and rely on locally available materials	Quick saturation and charge dependency	[4, 63, 80, 146–148]
Filtration	Effective and efficient	Energy consuming and brine generation	[106, 144, 149–151]
Phytoremediation	Effective and selective	Limited removal efficacy	[138, 152–154]

As such, AMD will be perceived as a resource not a waste material. Various studies and reports have shown substantial advancement in the valorization of AMD [17, 56]. Its beneficiation ranges from drinking water reclamation [12, 13, 15, 17, 141], synthesis of valuable minerals, recovery of valuable minerals [17, 142], and its application for other uses. The main challenge of metals and valuable minerals recovery is cross-contamination during recovery due to the presence of sensitive chemical species that easily precipitate, co-precipitate, and co-adsorb, hence contaminating the resultant product.

7.5.2 Reclamation of Drinking Water

The dissolved solids in AMD cause a high electrical conductivity. As such, drinking water can be reclaimed by removing dissolved solids for their beneficiation. In a similar manner, drinking water, which is suitable for many applications, will also be reclaimed. A number of systems have been used to reclaim drinking water, but filtration has been identified as a sole technique that can reclaim drinking water [128, 155]. Other technologies such as adsorption and precipitation pose secondary pollution and are selective in their efficacies. Adsorption fails drastically but precipitation manages to reclaim water suitable for discharge or irrigation. Due to high concentrations of dissolved metals and hardness of the water, filtration technique experience challenges of scaling, fouling, and damage due to different chemicals and pH [13, 15, 20, 128]. As such, researchers, engineers, and the scientific community has focused on the use of different pre-treatment technologies and drinking water reclamation. Aguiar *et al.* [106] successfully explored the treatment of AMD by NF focusing on membrane fouling, chemical cleaning, and membrane ageing. However, drinking water was reclaimed. Andrade *et al.* [110] evaluated the use of UF-NF for the treatment of wastewater from a gold mine. The primary aim of using a hybrid was to prevent fouling control *via* a two-phase treatment process. Moreover, the integration inflated the cost of the treatment process, thus making this technology undesired. Ricci *et al.* [111] reported the feasibility of using a sequence of MF, NF, and RO to recover sulfuric acid, separate noble metals, and produce high quality reuse water from wastewater emanating from gold mining. The driving mechanisms were pressure-oxidation process for the highlighted wastewater stream.

7.5.3 Recovery of Valuable Minerals

Valuable substances to be recovered from AMD include drinking water, metals, and gypsum. Moreover, contamination of the recovered minerals from AMD has been an issue of prime concern and this affects the purity of the product minerals. As such, precise initiatives need to be considered when pursuing metals recovery [20, 128, 144, 145, 156]. Ricci *et al.* [111] reported the feasibility of integrating MF, NF and RO to produce sulfuric acid, separate noble metals, and reclaim clean water from gold mine drainage. The entire process was governed by pressure-oxidation modality [128]. The recovery of Fe(II) and Fe(III) species has been explored widely as well [49, 157–159]. Mohan and Chander [158] evaluated the use of lignite for AMD treatment. Sorption of Fe(II)/(III), Mn, Zn, and Ca from aqueous system was successfully studied, and the adsorbed metals were recovered as regenerants. This technology has been proven effective, but its weakness lies with quick saturation.

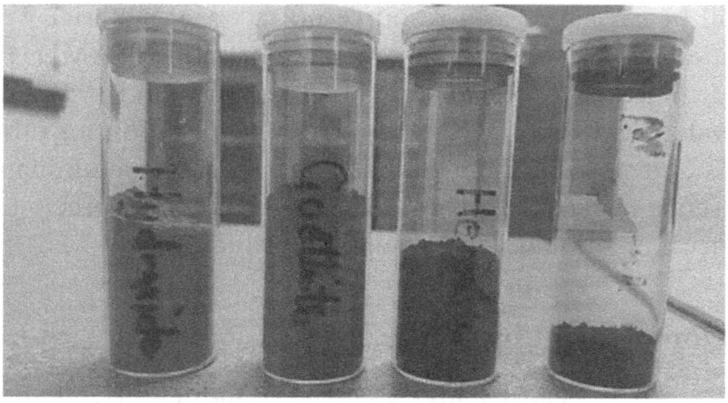

Figure 7.6 Fe-hydroxide, goethite, hematite, and magnetite synthesized from AMD.

7.5.4 Synthesis of Valuable Minerals

Fe, Mn, Zn, sulfate, and other chemical species in AMD make it a viable source of (semi) metals. Researchers have pursued the recovery of Fe^{3+}/Fe^{2+} from AMD for the synthesis of goethite, hematite, and magnetite (Figure 7.6) [48, 49]. Goethite can be used for pigments in the clays and painting industries, and it has been used as an adsorbent for (semi) metals [81, 160]. Hematite has been employed in pigments synthesis, removal of (semi) metals and oxyanions, and shielding of radiation, among other applications [161–163]. Similarly, magnetite can be used in ferro-fluid technology, storage of information, photodegradation for organic contaminants, photo-anode, catalyst, biomedicine, controlled drug delivery, removal of water contaminants, and magnetic nanoparticles formation [49, 68, 164–165].

7.6 Case Study

The study described here was done in response to the inter-ministerial committee assembled by the president to tackle the challenges of AMD in gold and coal reserves within South Africa. The Council for Scientific and Industrial Research (CSIR) has invested resources in the treatment of AMD *via* a multi-stage approach in a multi-functional and state-of-art pilot plant for the treatment of AMD and recovery of valuable minerals (Figure 7.7). This is a five staged technological approach: (1) neutralization with calcined cryptocrystalline magnesite to increase the pH and selectively recover the metals at varying pH gradients, (2) softening using lime to remove residual sulfate and magnesium, (3) addition of soda ash to remove residual calcium and further soften the water, (4) pH balance and control system to protect the membrane, and (5) RO system for the reclamation of drinking water that meet SANS 241 and WHO drinking water specifications and standards. Valuable minerals recovered and synthesized are (1) Fe-based minerals such as goethite, hematite, and magnetite; (2) Mg rich gypsum from the residual sulfate; (3) limestone from the soda as softening stage; and lastly, (4) the drinking water that meet the prescribed limits as outlined in water quality guidelines (WQGs), specifications, limits, and standards. This plant can treat 3.5 m^3/h of mine water to drinking water. Valuable minerals recovered include magnetite, hematite, goethite, gypsum,

Figure 7.7 A 3.5 m^3/h pilot plant at the CSIR premises, South Africa (image Vhahangwele Masindi, system inventor).

and limestone. These minerals can be used to reduce the operating costs of the technology, thus rendering this technology self-sustainable, economically viable, and more efficient.

The type of drainage treated is acid (Table 7.9) and is predominated by Fe, Al, Mn, and sulfate among traces of other metals. These results corroborate coal AMD results (Table 7.4) as highlighted by numerous authors [10, 12, 141]. After contacting the calcined magnesite, (semi)metals were completely and sulfate partially removed and increased concentrations of base metals (Mg and Ca). The addition of hydrated lime further removed sulfate and Mg from the mine water, hence reducing the total dissolved solids. However, the concentration of Ca increased resulting from the addition of a Ca-based reagent to facilitate the chemical reaction. Softening was achieved using Na-based material and all the bivalents were removed. Lastly, RO removed residual impurities. The brine was concentrated; hence, MD has been piloted on the same brine. The study was proven effective and efficient. The primary aim was to minimize water loss and improve water recovery or reclamation. Furthermore, the

Table 7.9 Chemical composition of AMD before and after treatment [17], EC: electrical conductivity, TDS: total dissolved solids.

Parameters	Units	A	B (AMD)	C	D	E	F	G
pH	–	5–9.5	2.38 ± 0.2	10.00 ± 0.001	12.0 ± 0.001	8.5 ± 0.5	7.5 ± 0.01	8 ± 0.01
EC	mS/m	170	883 ± 0.03	904 ± 0.3	508 ± 0.006	735 ± 0.05	806 ± 0.5	1,391 ± 0.51
TDS	mg/L	< 1,200	5,810 ± 0.8	5,857 ± 0.8	2,927 ± 0.8	4,332 ± 0.1	3,940 ± 0.5	7,653 ± 0.9
Turbidity	NTU	< 5	43 ± 2	66 ± 0.001	110 ± 0.04	75 ± 0.01	0.62 ± 0.02	0.88 ± 0.09
Color	mg/lPt	15	600 ± 0.03	5 ± 0.01	7 ± 0.001	< 5.0 ± 0.001	6 ± 0.01	6 ± 0.01
Aluminium	mg/L	0.3	70 ± 0.01	0.086 ± 0.05	0.038 ± 0.001	0.001 ± 0.01	0.001 ± 0.001	0.001 ± 0.001
Ammonia	mg/L	1.5	7.7 ± 0.05	8.2 ± 0.05	6.3 ± 0.09	6.5 ± 0.9	6.8 ± 7	12 ± 1
Cadmium	mg/L	0.003	0.004 ± 0.03	0.001 ± 0.02	0.001 ± 0.001	0.001 ± 0.001	0.001 ± 0.001	0.001 ± 0.001
Calcium	mg/L	–	250 ± 0.09	600 ± 0.09	800 ± 0.001	50 ± 0.006	5 ± 0.05	100 ± 0.67
Chloride	mg/L	300	3.8 ± 0.01	2.1 ± 0.05	6.3 ± 0.005	6.3 ± 0.006	32 ± 0.7	28 ± 1.5
Chlorine (free)	mg/L	5	1.1 ± 0.05	0.25 ± 0.03	0.05 ± 0.003	0.05 ± 0.006	0.05 ± 0.045	0.04 ± 0.003
Chromium	mg/L	0.005	0.045 ± 0.02	0.001 ± 0.01	0.001 ± 0.001	0.001 ± 0.001	0.001 ± 0.001	0.001 ± 0.001

(*Continued*)

Table 7.9 Chemical composition of AMD before and after treatment [17], EC: electrical conductivity, TDS: total dissolved solids. (*Continued*)

Parameters	Units	A	B (AMD)	C	D	E	F	G
Copper	mg/L	2	0.6 ± 0.02	0.004 ± 0.01	0.001 ± 0.07	0.001 ± 0.003	0.001 ± 0.01	0.001 ± 0.01
Iron	mg/L	2	1000 ± 0.9	1 ± 0.01	0.002 ± 0.005	0.001 ± 0.001	0.001 ± 0.01	0.001 ± 0.001
Lead	mg/L	0.01	0.03 ± 0.05	0.001 ± 0.01	0.001 ± 0.006	0.001 ± 0.001	0.001 ± 0.01	0.005 ± 0.01
Magnesium	mg/L	–	800 ± 0.5	2000 ± 2	2 ± 3	0.5 ± 1	0.5 ± 4	50 ± 0.98
Manganese	mg/L	0.04	50 ± 1	0.5 ± 0.05	0.007 ± 0.01	0.1 ± 0.01	0.001 ± 0.09	0.002 ± 0.5
Mercury	mg/L	0.006	0.002 ± 0.08	0.002 ± 0.05	0.002 ± 0.01	0.002 ± 0.001	0.002 ± 1	0.008 ± 0.6
Nickel	mg/L	0.07	1 ± 0.001	0.001 ± 0.05	0.001 ± 0.01	0.001 ± 0.001	0.001 ± 0.2	0.002 ± 0.3
Selenium	mg/L	0.04	0.022 ± 0.1	0.001 ± 0.01	0.001 ± 0.03	0.001 ± 0.12	0.001 ± 0.5	0.005 ± 0.6
Sodium	mg/L	< 200	91 ± 0.1	110 ± 0.7	103 ± 0.006	1624 ± 5	1519 ± 2	2803 ± 4
Sulfate	mg/L	< 500	11681 ± 2	11000 ± 6	621 ± 6	71 ± 9	8 ± 0.5	15319 ± 0.4

A, SANS 241 specifications; B, AMD; C, Pre-treated magnesite; D, Hydrated lime; E, soda ash; F, permeate; and G, brine.

direct field cost of the technology was calculated to be R 65.60/m^3. Fourteen percent of this cost can be recuperated from the sale of the valorization products, thus reducing the AMD DFC treatment cost to R 56.60/m^3 [17]. The cost is a bit high for this technology, but further optimization of the process can reduce the DFC of the technology, hence making it more attractive. On another note, the drying of products minerals has been a challenging issue.

7.7 Challenges Relating to Valorization

Valorization and beneficiation of AMD is a promising possibility toward circular economy. This will ensure that the process of sustainable development is achieved, and the environmental footprints of AMD are minimized. Moreover, people will start to perceive AMD as a resource not a waste material. However, these are few challenges which are related to AMD valorization:

- generation of complex, highly mineralized, and heterogeneous sludge on precipitation technology and high demand for non-selective contaminants removal technology;
- saturation of the adsorbent and frequent regeneration due to quick saturation;
- generation of brine that is hard and difficult to dispose onto the environment;
- high costs associated with value extraction from AMD and integrated processes;
- insignificant quantities of the recovered minerals that limit products marketability;
- poor purity of the recovered and synthesized materials, which limit market competition;
- technical requirements for operation;
- high-energy demand and environmental footprint, which inflate the cost.

7.8 Conclusions and Future Perspectives

As could be seen, AMD can be a viable resource. Valuable minerals such as goethite, hematite, and magnetite can be synthesized and drinking water can be reclaimed. Gypsum, limestone, and dolomite can be synthesized from the treatment process. This can make this system self-sustainable, since the revenue generated from the reclaimed, recovered, and synthesized products can off-set the running costs of the treatment process. Among the value recovery technology, precipitation has been the main technique for minerals recovery but for drinking water reclamation, filtration technologies are the best. Challenges associated with minerals recovery and AMD valorization are the generation of heterogeneous sludge, the costs associated with minerals recovery and feasibility considering the complexity of AMD. Adsorption has been identified as unviable technology for minerals recovery and beneficiation due to quick saturation of the sorbents and regeneration requirement, which is unselective. As a future perspective, studies should focus on sequential and fractional recovery of valuable minerals from mine drainages using different techniques. Moreover, an integrated approach would be the future in terms of minerals recovery from AMD.

References

1. Nordstrom, D.K., Blowes, D.W., Ptacek, C.J., Hydrogeochemistry and microbiology of mine drainage: An update. *Appl. Geochem.*, 57, 3–16, 2015. https://doi.org/10.1016/j.apgeochem.2015.02.008.
2. Simate, G.S. and Ndlovu, S., Acid mine drainage: Challenges and opportunities. *J. Environ. Chem. Eng.*, 2, 1785–1803, 2014. https://doi.org/10.1016/j.jece.2014.07.021.
3. Wang, X., Jiang, H., Fang, D., Liang, J., Zhou, L., A novel approach to rapidly purify acid mine drainage through chemically forming schwertmannite followed by lime neutralization. *Water Res.*, 151, 515–522, 2019. https://doi.org/10.1016/j.watres.2018.12.052.
4. Park, I., Tabelin, C.B., Jeon, S., Li, X., Seno, K., Ito, M., Hiroyoshi, N., A review of recent strategies for acid mine drainage prevention and mine tailings recycling. *Chemosphere*, 219, 588–606, 2019. https://doi.org/10.1016/j.chemosphere.2018.11.053.
5. Kefeni, K.K., Msagati, T.A.M., Mamba, B.B., Acid mine drainage: Prevention, treatment options, and resource recovery: A review. *J. Clean. Prod.*, 151, 475–493, 2017. https://doi.org/10.1016/j.jclepro.2017.03.082.
6. García, V., Häyrynen, P., Landaburu-Aguirre, J., Pirilä, M., Keiski, R.L., Urtiaga, A., Purification techniques for the recovery of valuable compounds from acid mine drainage and cyanide tailings: application of green engineering principles. *J. Chem. Technol. Biotechnol.*, 89, 803–813, 2014. https://doi.org/10.1002/jctb.4328.
7. Younger, P.L., Banwart, S.A., Hedin, R.S., *Mine Water – Hydrology, Pollution, Remediation*, p. 464, Kluwer, Dordrecht, 2002, https://doi.org/10.1007/978-94-010-0610-1.
8. Wolkersdorfer, C., *Water Management at Abandoned Flooded Underground Mines – Fundamentals, Tracer Tests, Modelling, Water Treatment*, p. 466, Springer, Heidelberg, 2008, https://doi.org/10.1007/978-3-540-77331-3.
9. Madzivire, G., Gitari, W.M., Vadapalli, V.R.K., Ojumu, T.V., Petrik, L.F., Fate of sulphate removed during the treatment of circumneutral mine water and acid mine drainage with coal fly ash: Modelling and experimental approach. *Miner. Eng.*, 24, 1467–1477, 2011. https://doi.org/10.1016/j.mineng.2011.07.009.
10. Madzivire, G., Maleka, P.P., Vadapalli, V.R.K., Gitari, W.M., Lindsay, R., Petrik, L.F., Fate of the naturally occurring radioactive materials during treatment of acid mine drainage with coal fly ash and aluminium hydroxide. *J. Environ. Manage.*, 133, 12–17, 2014. https://doi.org/10.1016/j.jenvman.2013.11.041.
11. Masindi, V., A novel technology for neutralizing acidity and attenuating toxic chemical species from acid mine drainage using cryptocrystalline magnesite tailings. *J. Water Process Eng.*, 10, 67–77, 2016. https://doi.org/10.1016/j.jwpe.2016.02.002.
12. Masindi, V., Chatzisymeon, E., Kortidis, I., Foteinis, S., Assessing the sustainability of acid mine drainage (AMD) treatment in South Africa. *Sci. Total Environ.*, 635, 793–802, 2018. https://doi.org/10.1016/j.scitotenv.2018.04.108.
13. Masindi, V., Madzivire, G., Tekere, M., Reclamation of water and the synthesis of gypsum and limestone from acid mine drainage treatment process using a combination of pre-treated magnesite nanosheets, lime, and CO_2 bubbling. *Water Resour. Ind.*, 20, 1–14, 2018. https://doi.org/10.1016/j.wri.2018.07.001.
14. Masindi, V., Ndiritu, J.G., Maree, J.P., Fractional and step-wise recovery of chemical species from acid mine drainage using calcined cryptocrystalline magnesite nano-sheets: An experimental and geochemical modelling approach. *J. Environ. Chem. Eng.*, 6, 1634–1650, 2018. https://doi.org/10.1016/j.jece.2018.02.005.
15. Masindi, V., Osman, M.S., Abu-Mahfouz, A.M., Integrated treatment of acid mine drainage using BOF slag, lime/soda ash and reverse osmosis (RO): Implication for the production of drinking water. *Desalination*, 424, 45–52, 2017. https://doi.org/10.1016/j.desal.2017.10.002.

16. Masindi, V., Osman, M.S., Mbhele, R.N., Rikhotso, R., Fate of pollutants post treatment of acid mine drainage with basic oxygen furnace slag: Validation of experimental results with a geochemical model. *J. Clean. Prod.*, 172, 2899–2909, 2016. https://doi.org/10.1016/j.jclepro.2017.11.124.
17. Masindi, V., Osman, M.S., Shingwenyana, R., Valorization of acid mine drainage (AMD): A simplified approach to reclaim drinking water and synthesize valuable minerals – Pilot study. *J. Environ. Chem. Eng.*, 7, 103082, 2019. https://doi.org/10.1016/j.jece.2019.103082.
18. Amos, R.T., Blowes, D.W., Bailey, B.L., Sego, D.C., Smith, L., Ritchie, A.I.M., Waste-rock hydrogeology and geochemistry. *Appl. Geochem.*, 57, 140–156, 2015. https://doi.org/10.1016/j.apgeochem.2014.06.020.
19. Johnson, D.B. and Hallberg, K.B., Acid mine drainage remediation options: a review. *Sci. Total Environ.*, 338, 3–14, 2005. https://doi.org/10.1016/j.scitotenv.2004.09.002.
20. Nleya, Y., Simate, G.S., Ndlovu, S., Sustainability assessment of the recovery and utilisation of acid from acid mine drainage. *J. Clean. Prod.*, 113, 17–27, 2016. https://doi.org/10.1016/j.jclepro.2015.11.005.
21. Zhao, H., Xia, B., Qin, J., Zhang, J., Hydrogeochemical and mineralogical characteristics related to heavy metal attenuation in a stream polluted by acid mine drainage: A case study in Dabaoshan Mine, China. *J. Environ. Sci.*, 24, 979–989, 2012. https://doi.org/10.1016/S1001-0742(11)60868-1.
22. Sracek, O., Choquette, M., Gélinas, P., Lefebvre, R., Nicholson, R.V., Geochemical characterization of acid mine drainage from a waste rock pile, Mine Doyon, Québec, Canada. *J. Contam. Hydrol.*, 69, 45–71, 2004. https://doi.org/10.1016/S0169-7722(03)00150-5.
23. Sheoran, V., Sheoran, A.S., Choudhary, R.P., Biogeochemistry of acid mine drainage formation: A review, in: *Mine Drainage and Related Problems*, B.C. Robinson (Ed.), pp. 119–154, Nova Science Publishers Inc, Hauppauge, New York, 2011, https://doi.org/10.3923/jest.2012.119.127.
24. Sheoran, A.S. and Sheoran, V., Heavy metal removal mechanism of acid mine drainage in wetlands: A critical review. *Miner. Eng.*, 19, 105–116, 2006. https://doi.org/10.1016/j.mineng.2005.08.006.
25. Peretyazhko, T., Zachara, J.M., Boily, J.F., Xia, Y., Gassman, P.L., Arey, B.W., Burgos, W.D., Mineralogical transformations controlling acid mine drainage chemistry. *Chem. Geol.*, 262, 169–178, 2009. https://doi.org/10.1016/j.chemgeo.2009.01.017.
26. Nordstrom, D.K., Blowes, D.W., Ptacek, C.J., Hydrogeochemistry and microbiology of mine drainage: An update. *Appl. Geochem.*, 57, 3–16, 2015. https://doi.org/10.1016/j.apgeochem.2015.02.008.
27. Natarajan, K.A., Microbial aspects of acid mine drainage and its bioremediation. *Trans. Nonferrous Met. Soc. China*, 18, 1352–1360, 2008. https://doi.org/10.1016/S1003-6326(09)60008-X.
28. Fang, J., Hasiotis, S.T., Gupta, S.D., Brake, S.S., Bazylinski, D.A., Microbial biomass and community structure of a stromatolite from an acid mine drainage system as determined by lipid analysis. *Chem. Geol.*, 243, 191–204, 2007. https://doi.org/10.1016/j.chemgeo.2007.06.001.
29. Lacelle, D. and Léveillé, R., Acid drainage generation and associated Ca-Fe-SO_4 minerals in a periglacial environment, Eagle Plains, Northern Yukon, Canada: A potential analogue for low-temperature sulfate formation on Mars. *Planet. Space Sci.*, 58, 509–521, 2010. https://doi.org/10.1016/j.pss.2009.06.009.
30. Tutu, H., McCarthy, T.S., Cukrowska, E., The chemical characteristics of acid mine drainage with particular reference to sources, distribution and remediation: The Witwatersrand Basin, South Africa as a case study. *Appl. Geochem.*, 23, 3666–3684, 2008. https://doi.org/10.1016/j.apgeochem.2008.09.002.
31. Madzivire, G., Petrik, L.F., Gitari, W.M., Ojumu, T.V., Balfour, G., Application of coal fly ash to circumneutral mine waters for the removal of sulphates as gypsum and ettringite. *Miner. Eng.*, 23, 252–257, 2010. https://doi.org/10.1016/j.mineng.2009.12.004.

32. Madzivire, G., Maleka, P.P., Vadapalli, V.R.K., Gitari, W.M., Lindsay, R., Petrik, L.F., Fate of the naturally occurring radioactive materials during treatment of acid mine drainage with coal fly ash and aluminium hydroxide. *J. Environ. Manage.*, 133, 12–17, 2014. https://doi.org/10.1016/j.jenvman.2013.11.041.
33. Masindi, V., Gitari, W.M., Ngulube, T., Defluoridation of drinking water using Al^{3+}-modified bentonite clay: optimization of fluoride adsorption conditions. *Toxicol. Environ. Chem.*, 96, 9, 1–16, 2014. https://doi.org/10.1080/19443994.2013.855669.
34. Gitari, W.M., Petrik, L.F., Etchebers, O., Key, D.L., Iwuoha, E., Okujeni, C., Passive neutralisation of acid mine drainage by fly ash and its derivatives: A column leaching study. *Fuel*, 87, 1637–1650, 2008. https://doi.org/10.1016/j.fuel.2007.08.025.
35. Gitari, M., Petrik, L., Etchebers, O., Key, D., Iwuoha, E., Okujeni, C., Treatment of acid mine drainage with fly ash: Removal of major contaminants and trace elements. *J. Environ. Sci. Health - Part A.*, 41, 1729–1747, 2006. https://doi.org/10.1080/10934520600754425.
36. Cheng, H., Hu, Y., Luo, J., Xu, B., Zhao, J., Geochemical processes controlling fate and transport of arsenic in acid mine drainage (AMD) and natural systems. *J. Hazard. Mat.*, 165, 13–26, 2009. https://doi.org/10.1016/j.jhazmat.2008.10.070.
37. Dabrowski, J.M., Dabrowski, J., Hill, L., Macmillan, P., Oberholster, P.J., Fate, transport and effects of pollutants originating from acid mine drainage in the olifants river, South Africa. *River Res. Appl.*, 31, 10, 2014. https://doi.org/10.1002/rra.2833.
38. Dabrowski, J., Oberholster, P.J., Dabrowski, J.M., Le Brasseur, J., Gieskes, J., Chemical characteristics and limnology of Loskop Dam on the Olifants River (South Africa), in light of recent fish and crocodile mortalities. *Water SA*, 39, 675–686, 2013. https://doi.org/10.4314/wsa.v39i5.12.
39. Dabrowski, J.M. and de Klerk, L.P., An assessment of the impact of different land use activities on water quality in the upper Olifants River catchment. *Water SA*, 39, 231–244, 2013. https://doi.org/10.4314/wsa.v39i2.6.
40. de Villiers, S. and Mkwelo, S.T., Has monitoring failed the Olifants River, Mpumalanga? *Water SA*, 35, 671–676, 2009. https://doi.org/10.4314/wsa.v35i5.49193.
41. Van Zyl, H.C., Maree, J.P., Van Niekerk, A.M., Van Tonder, G.J., Naidoo, C., Collection, treatment and re-use of mine water in the Olifants River Catchment. *J. S. Afr. Inst. Min. Metall.*, 101, 41–46, 2001.
42. Department of Water Affairs and Forestry, South African Water Quality Guidelines (first edition), Field Guide, *Department of Water Affairs and Forestry*, vol. 8, South Africa, 1996.
43. World Health Organization, *Guidelines for drinking-water quality: fourth edition incorporating first addendum*, 4th ed + 1st add ed., World Health Organization, Geneva, 2017.
44. Aklil, A., Mouflih, M., Sebti, S., Removal of heavy metal ions from water by using calcined phosphate as a new adsorbent. *J. Hazard. Mat.*, 112, 183–190, 2004. https://doi.org/10.1016/j.jhazmat.2004.05.018.
45. Alcolea, A., Vázquez, M., Caparrós, A., Ibarra, I., García, C., Linares, R., Rodríguez, R., Heavy metal removal of intermittent acid mine drainage with an open limestone channel. *Miner. Eng.*, 26, 86–98, 2012. https://doi.org/10.1016/j.mineng.2011.11.006.
46. Blais, J.F., Djedidi, Z., Cheikh, R.B., Tyagi, R.D., Mercier, G., Metals precipitation from effluents: Review. *Pract. Period. Hazard. Toxic Radioact. Waste Manage.*, 12, 135–149, 2008. https://doi.org/10.1061/(ASCE)1090-025X(2008)12:3(135).
47. Charerntanyarak, L., Heavy metals removal by chemical coagulation and precipitation. *Water Sci. Technol.*, Vol. 39, No. 10–11, pp. 135–138, 1999. https://doi.org/10.2166/wst.1999.0642.
48. Akinwekomi, V., Maree, J.P., Masindi, V., Zvinowanda, C., Osman, M.S., Foteinis, S., Mpenyana-Monyatsi, L., Chatzisymeon, E., Beneficiation of acid mine drainage (AMD): A viable option for the synthesis of goethite, hematite, magnetite, and gypsum – Gearing towards a circular economy concept. *Miner. Eng.*, 148, 106204, 2020. https://doi.org/10.1016/j.mineng.2020.106204.

49. Akinwekomi, V., Maree, J.P., Zvinowanda, C., Masindi, V., Synthesis of magnetite from iron-rich mine water using sodium carbonate. *J. Environ. Chem. Eng.*, 5, 2699–2707, 2017. https://doi.org/10.1016/j.jece.2017.05.025.
50. Masindi, V., Gitari, M.W., Tutu, H., De Beer, M., Neutralization and attenuation of metal species in acid mine drainage and mine leachates using magnesite: a batch experimental approach, in: *An Interdisciplinary Response to Mine Water Challenges*, Sui, Sun, Wang (Eds.), pp. 640–644, China University of Mining and Technology Press, Xuzhou, 2014.
51. Yang, J.E., Skousen, J.G., Ok, Y.S., Yoo, K.Y., Kim, H.J., Reclamation of abandoned coal mine waste in Korea using lime cake by-products. *Mine Water Environ.*, 25, 227–232, 2006. https://doi.org/10.1007/s10230-006-0137-z.
52. Skousen, J., McDonald, L., MacK, B., Demchak, J., Water quality from above-drainage underground mines over a 35-year period, in: *7th International Conference on Acid Rock Drainage 2006, ICARD – Also Serves as the 23rd Annual Meetings of the American Society of Mining and Reclamation*, pp. 2044–2054, 2006, https://doi.org/10.21000/JASMR06022044.
53. Demchak, J., Morrow, T., Skousen, J., Treatment of acid mine drainage by four vertical flow wetlands in Pennsylvania. *Geochem. Explora. Environ. Anal.*, 1, 71–80, 2001. https://doi.org/10.1144/geochem.1.1.71.
54. Maree, J.P., Van Tonder, G.J., Adlem, C., Millard, P., De Beer, M., Strydom, W.F., Pilot Plant Studies on Limestone Neutralization and Gypsum Crystallization of Acidic Colliery Effluent, *Confidential CSIR Report*, 1997.
55. Maree, J.P. and Du Plessis, P., Neutralization of acid mine water calcium carbonate. *Proc. IAWQ Conference on Pre-treatment of Industrial Wastewaters, Athens, Water Sci. Technol.*, vol. 29, 7, pp. 285–296, 1993.
56. Silva, R.d.A., Secco, M.P., Lermen, R.T., Schneider, I.A.H., Hidalgo, G.E.N., Sampaio, C.H., Optimizing the selective precipitation of iron to produce yellow pigment from acid mine drainage. *Miner. Eng.*, 135, 111–117, 2019. https://doi.org/10.1016/j.mineng.2019.02.040.
57. Gitari, W.M., Ngulube, T., Masindi, V., Gumbo, J.R., Defluoridation of groundwater using Fe^{3+}-modified bentonite clay: optimization of adsorption conditions. *Desalin. Water Treat.*, 53, 1578–1590, 2015. https://doi.org/10.1080/19443994.2013.855669.
58. Alagumuthu, G. and Sumathi, T., Adsorption of arsenic onto activated carbon of clerodendrum inerme. *Ecol. Environ. Conserv.*, 20, 595–603, 2014.
59. Dias, J.M., Alvim-Ferraz, M.C.M., Almeida, M.F., Rivera-Utrilla, J., Sánchez-Polo, M., Waste materials for activated carbon preparation and its use in aqueous-phase treatment: A review. *J. Environ. Manage.*, 85, 833–846, 2007. https://doi.org/10.1016/j.jenvman.2007.07.031.
60. Dong, L., Liu, W., Jiang, R., Wang, Z., Study on the adsorption mechanism of activated carbon removing low concentrations of heavy metals. *Desalin. Water Treat.*, 57, 7812–7822, 2016. https://doi.org/10.1080/19443994.2015.1100140.
61. Langmuir, D., *Aqueous Environmental Geochemistry*, Prentice Hall, New Jersey, 1997.
62. Tran, H.N., You, S.-J., Hosseini-Bandegharaei, A., Chao, H.-P., Mistakes and inconsistencies regarding adsorption of contaminants from aqueous solutions: A critical review. *Water Res.*, 120, 88–116, 2017. https://doi.org/10.1016/j.watres.2017.04.014.
63. Fu, F. and Wang, Q., Removal of heavy metal ions from wastewaters: A review. *J. Environ. Manage.*, 92, 407–418, 2011. https://doi.org/10.1016/j.jenvman.2010.11.011.
64. Huang, W., Zhang, Y., Li, D., Adsorptive removal of phosphate from water using mesoporous materials: A review. *J. Environ. Manage.*, 193, 470–482, 2017. https://doi.org/10.1016/j.jenvman.2017.02.030.
65. Kobielska, P.A., Howarth, A.J., Farha, O.K., Nayak, S., Metal–organic frameworks for heavy metal removal from water. *Coordin. Chem. Reviews*, 358, 92–107, 2018. https://doi.org/10.1016/j.ccr.2017.12.010.

66. Reddy, D.H.K. and Yun, Y.-S., Spinel ferrite magnetic adsorbents: Alternative future materials for water purification? *Coordin. Chem. Reviews*, 315, 90–111, 2016. https://doi.org/10.1016/j.ccr.2016.01.012.
67. Siyal, A.A., Shamsuddin, M.R., Khan, M.I., Rabat, N.E., Zulfiqar, M., Man, Z., Siame, J., Azizli, K.A., A review on geopolymers as emerging materials for the adsorption of heavy metals and dyes. *J. Environ. Manage.*, 224, 327–339, 2018. https://doi.org/10.1016/j.jenvman.2018.07.046.
68. Wei, X. and Viadero, R.C., Jr., Synthesis of magnetite nanoparticles with ferric iron recovered from acid mine drainage: Implications for environmental engineering. *Colloids Surf. A: Physicochem. Eng. Aspects*, 294, 280–286, 2007. https://doi.org/10.1016/j.colsurfa.2006.07.060.
69. Sparks, D.L., *Environmental Soil Chemistry*, Academic Press, California, 2003, https://doi.org/10.1016/B978-012656446-4/50001-3.
70. Masindi, V., Gitari, W.M., Pindihama, K.G., Adsorption of phosphate from municipal effluents using cryptocrystalline magnesite: complementing laboratory results with geochemical modelling. *Desalin. Water Treat.*, 57, 20957–20969, 2016. https://doi.org/10.1080/19443994.2015.1110720.
71. Masindi, V., Gitari, M.W., Tutu, H., De Beer, M., Application of magnesite–bentonite clay composite as an alternative technology for removal of arsenic from industrial effluents. *Toxicol. Environ. Chem.*, 96, 1435–1451, 2014. https://doi.org/10.1080/02772248.2014.966714.
72. Masindi, V., Gitari, M.W., Tutu, H., Debeer, M., Removal of boron from aqueous solution using magnesite and bentonite clay composite. *Desalin. Water Treat.*, 57, 8754–8764, 2016. https://doi.org/10.1080/19443994.2015.1025849.
73. Masindi, V., Gitari, W.M., Pindihama, K.G., Synthesis of cryptocrystalline magnesite/bentonite clay composite and its application for removal of phosphate from municipal wastewaters. *Environ. Technol.*, 37, 603–612, 2016. https://doi.org/10.1080/09593330.2015.1075598.
74. Anawar, H.M., Akter, F., Solaiman, Z.M., Strezov, V., Biochar: An Emerging Panacea for Remediation of Soil Contaminants from Mining, Industry and Sewage Wastes. *Pedosphere*, 25, 654–665, 2015. https://doi.org/10.1016/S1002-0160(15)30046-1.
75. Masindi, V. and Gitari, W.M., Simultaneous removal of metal species from acidic aqueous solutions using cryptocrystalline magnesite/bentonite clay composite: An experimental and modelling approach. *J. Clean. Prod.*, 112, 1077–1085, 2016. https://doi.org/10.1016/j.jclepro.2015.07.128.
76. Godlewska, P., Schmidt, H.P., Ok, Y.S., Oleszczuk, P., Biochar for composting improvement and contaminants reduction. A review. *Bioresour. Technol.*, 246, 193–202, 2017. https://doi.org/10.1016/j.biortech.2017.07.095.
77. Oliveira, F.R., Patel, A.K., Jaisi, D.P., Adhikari, S., Lu, H., Khanal, S.K., Environmental application of biochar: Current status and perspectives. *Bioresour. Technol.*, 246, 110–122, 2017. https://doi.org/10.1016/j.biortech.2017.08.122.
78. Sophia, A.,.C. and Lima, E.C., Removal of emerging contaminants from the environment by adsorption. *Ecotoxicol. Environ. Safety*, 150, 1–17, 2018. https://doi.org/10.1016/j.ecoenv.2017.12.026.
79. Ngulube, T., Gumbo, J.R., Masindi, V., Maity, A., An update on synthetic dyes adsorption onto clay based minerals: A state-of-art review. *J. Environ. Manage.*, 191, 35–57, 2017. https://doi.org/10.1016/j.jenvman.2016.12.031.
80. Demirbas, A., Heavy metal adsorption onto agro-based waste materials: A review. *J. Hazard. Mat.*, 157, 220–229, 2008. https://doi.org/10.1016/j.jhazmat.2008.01.024.
81. Mohan, D. and Pittman, C.U., Jr., Arsenic removal from water/wastewater using adsorbents—A critical review. *J. Hazard. Mat.*, 142, 1–53, 2007. https://doi.org/10.1016/j.jhazmat.2007.01.006.
82. Mu, B. and Wang, A., Adsorption of dyes onto palygorskite and its composites: A review. *J. Environ. Chem. Eng.*, 4, 1274–1294, 2016. https://doi.org/10.1016/j.jece.2016.01.036.

83. Sen Gupta, S. and Bhattacharyya, K.G., Adsorption of heavy metals on kaolinite and montmorillonite: A review. *Phys. Chem. Chem. Phys.*, 14, 6698–6723, 2012. https://doi.org/10.1039/c2cp40093f.
84. Theiss, F.L., Couperthwaite, S.J., Ayoko, G.A., Frost, R.L., A review of the removal of anions and oxyanions of the halogen elements from aqueous solution by layered double hydroxides. *J. Colloid Interface Sci.*, 417, 356–368, 2014. https://doi.org/10.1016/j.jcis.2013.11.040.
85. Vakili, M., Deng, S., Cagnetta, G., Wang, W., Meng, P., Liu, D., Yu, G., Regeneration of chitosan-based adsorbents used in heavy metal adsorption: A review. *Sep. Purif. Technol.*, 224, 373–387, 2019. https://doi.org/10.1016/j.seppur.2019.05.040.
86. Vakili, M., Rafatullah, M., Salamatinia, B., Abdullah, A.Z., Ibrahim, M.H., Tan, K.B., Gholami, Z., Amouzgar, P., Application of chitosan and its derivatives as adsorbents for dye removal from water and wastewater: A review. *Carbohydr. Polym.*, 113, 115–130, 2014. https://doi.org/10.1016/j.carbpol.2014.07.007.
87. Wang, B., Guo, X., Bai, P., Removal technology of boron dissolved in aqueous solutions – A review. *Colloids Surf. A: Physicochem. Eng. Aspects*, 444, 338–344, 2014. https://doi.org/10.1016/j.colsurfa.2013.12.049.
88. Yagub, M.T., Sen, T.K., Afroze, S., Ang, H.M., Dye and its removal from aqueous solution by adsorption: A review. *Adv. Colloid Interface Sci.*, 209, 172–184, 2014. https://doi.org/10.1016/j.cis.2014.04.002.
89. Barakat, M.A., New trends in removing heavy metals from industrial wastewater. *Arabian J. Chem.*, 4, 361–377, 2011. https://doi.org/10.1016/j.arabjc.2010.07.019.
90. Joseph, L., Jun, B.-M., Flora, J.R.V., Park, C.M., Yoon, Y., Removal of heavy metals from water sources in the developing world using low-cost materials: A review. *Chemosphere*, 229, 142–159, 2019. https://doi.org/10.1016/j.chemosphere.2019.04.198.
91. Patil, D.S., Chavan, S.M., Oubagaranadin, J.U.K., A review of technologies for manganese removal from wastewaters. *J. Environ. Chem. Eng.*, 4, 468–487, 2016. https://doi.org/10.1016/j.jece.2015.11.028.
92. Adam, M.R., Othman, M.H.D., Abu Samah, R., Puteh, M.H., Ismail, A.F., Mustafa, A., A. Rahman, M., Jaafar, J., Current trends and future prospects of ammonia removal in wastewater: A comprehensive review on adsorptive membrane development. *Sep. Purif. Tech.*, 213, 114–132, 2019. https://doi.org/10.1016/j.seppur.2018.12.030.
93. Agboola, O., Maree, J., Mbaya, R., Characterization and performance of nanofiltration membranes. *Environ. Chem. Letters*, 12, 241–255, 2014. https://doi.org/10.1007/s10311-014-0457-3.
94. Cornelissen, E.R., Chasseriaud, D., Siegers, W.G., Beerendonk, E.F., van der Kooij, D., Effect of anionic fluidized ion exchange (FIX) pre-treatment on nanofiltration (NF) membrane fouling. *Water Res.*, 44, 3283–3293, 2010. https://doi.org/10.1016/j.watres.2010.03.007.
95. Cui, Y., Ge, Q., Liu, X.-Y., Chung, T.-S., Novel forward osmosis process to effectively remove heavy metal ions. *J. Membr. Sci.*, 467, 188–194, 2014. https://doi.org/10.1016/j.memsci.2014.05.034.
96. Hu, K. and Dickson, J.M., Nanofiltration membrane performance on fluoride removal from water. *J. Membr. Sci.*, 279, 529–538, 2006. https://doi.org/10.1016/j.memsci.2005.12.047.
97. Kabay, N., Köseoğlu, P., Yavuz, E., Yüksel, Ü., Yüksel, M., An innovative integrated system for boron removal from geothermal water using RO process and ion exchange-ultrafiltration hybrid method. *Desalination*, 316, 1–7, 2013. https://doi.org/10.1016/j.desal.2013.01.020.
98. Katsou, E., Malamis, S., Haralambous, K., Examination of zinc uptake in a combined system using sludge, minerals and ultrafiltration membranes. *J. Hazard. Mat.*, 182, 27–38, 2010. https://doi.org/10.1016/j.jhazmat.2010.05.101.
99. Phattaranawik, J., Jiraratananon, R., Fane, A.G., Effects of net-type spacers on heat and mass transfer in direct contact membrane distillation and comparison with ultrafiltration studies. *J. Membr. Sci.*, 217, 193–206, 2003. https://doi.org/10.1016/S0376-7388(03)00130-3.

100. Sarkar, S. and SenGupta, A.K., A new hybrid ion exchange-nanofiltration (HIX-NF) separation process for energy-efficient desalination: Process concept and laboratory evaluation. *J. Membr. Sci.*, 324, 76–84, 2008. https://doi.org/10.1016/j.memsci.2008.06.058.
101. Shen, J. and Schäfer, A., Removal of fluoride and uranium by nanofiltration and reverse osmosis: A review. *Chemosphere*, 117, 679–691, 2014. https://doi.org/10.1016/j.chemosphere.2014.09.090.
102. van Dijk, J.C., Munneke, B.R., Kramer, B., Wouters, J.W., Membrane filtration: A realistic option in the field of water supply? *Desalination*, 81, 229–247, 1991. https://doi.org/10.1016/0011-9164(91)85056-Z.
103. Wu, L., Shang, F., Gao, C., Hu, Y., The research of cleaning method of membrane modules in ultrafiltration pretreatment system. *Appl. Mech. Mat.*, 316–317, 927–932, 2013. https://doi.org/10.4028/www.scientific.net/AMM.316-317.927.
104. Xu, P., Bellona, C., Drewes, J.E., Fouling of nanofiltration and reverse osmosis membranes during municipal wastewater reclamation: Membrane autopsy results from pilot-scale investigations. *J. Membr. Sci.*, 353, 111–121, 2010. https://doi.org/10.1016/j.memsci.2010.02.037.
105. Yaroshchuk, A.E., Recent progress in the transport characterisation of nanofiltration membranes. *Desalination*, 149, 423–428, 2002. https://doi.org/10.1016/S0011-9164(02)00768-3.
106. Aguiar, A., Andrade, L., Grossi, L., Pires, W., Amaral, M., Acid mine drainage treatment by nanofiltration: A study of membrane fouling, chemical cleaning, and membrane ageing. *Sep. Purif. Technol.*, 192, 185–195, 2018. https://doi.org/10.1016/j.seppur.2017.09.043.
107. Lopez, J., Reig, M., Gibert, O., Valderrama, C., Cortina, J.L., Evaluation of NF membranes as treatment technology of acid mine drainage: metals and sulfate removal. *Desalination*, 440, 122–134, 2018. https://doi.org/10.1016/j.desal.2018.03.030.
108. Mhamdi, M., Elaloui, E., Trabelsi-Ayadi, M., Adsorption of zinc by a Tunisian Smectite through a filtration membrane. *Industr. Crops Products*, 47, 204–211, 2013. https://doi.org/10.1016/j.indcrop.2013.03.003.
109. Oyewo, O.A., Agboola, O., Onyango, M.S., Popoola, P., Bobape, M.F., Current Methods for the Remediation of Acid Mine Drainage Including Continuous Removal of Metals From Wastewater and Mine Dump, in: *Bio-Geotechnologies for Mine Site Rehabilitation*, M.N.V. Prasad, P.J.d.C. Favas, S.K. Maiti (Eds.), pp. 103–114, Elsevier, Amsterdam, 2018, https://doi.org/10.1016/B978-0-12-812986-9.00006-3.
110. Andrade, L.H., Aguiar, A.O., Pires, W.L., Grossi, L.B., Amaral, M.C.S., Comprehensive bench- and pilot-scale investigation of NF for gold mining effluent treatment: Membrane performance and fouling control strategies. *Sep. Purif. Technol.*, 174, 44–56, 2017. https://doi.org/10.1016/j.seppur.2016.09.048.
111. Ricci, B.C., Ferreira, C.D., Aguiar, A.O., Amaral, M.C.S., Integration of nanofiltration and reverse osmosis for metal separation and sulfuric acid recovery from gold mining effluent. *Sep. Purif. Technol.*, 154, 11–21, 2015. https://doi.org/10.1016/j.seppur.2015.08.040.
112. Adham, S., Hussain, A., Matar, J.M., Dores, R., Janson, A., Application of Membrane Distillation for desalting brines from thermal desalination plants. *Desalination*, 314, 101–108, 2013. https://doi.org/10.1016/j.desal.2013.01.003.
113. Afrasiabi, N. and Shahbazali, E., RO brine treatment and disposal methods. *Desalin. Water Treat.*, 35, 39–53, 2011. https://doi.org/10.5004/dwt.2011.3128.
114. Ahmad, N. and Baddour, R.E., A review of sources, effects, disposal methods, and regulations of brine into marine environments. *Ocean Coastal Manage.*, 87, 1–7, 2014. https://doi.org/10.1016/j.ocecoaman.2013.10.020.
115. Ahmed, M., Arakel, A., Hoey, D., Thumarukudy, M.R., Goosen, M.F.A., Al-Haddabi, M., Al-Belushi, A., Feasibility of salt production from inland RO desalination plant reject brine: A case study. *Desalination*, 158, 109–117, 2003. https://doi.org/10.1016/S0011-9164(03)00441-7.

116. Ahmed, M., Shayya, W.H., Hoey, D., Al-Handaly, J., Brine disposal from reverse osmosis desalination plants in Oman and the United Arab Emirates. *Desalination*, 133, 135–147, 2001. https://doi.org/10.1016/S0011-9164(01)80004-7.
117. Azadi Aghdam, M., Zraick, F., Simon, J., Farrell, J., Snyder, S.A., A novel brine precipitation process for higher water recovery. *Desalination*, 385, 69–74, 2016. https://doi.org/10.1016/j.desal.2016.02.007.
118. Giwa, A., Dufour, V., Al Marzooqi, F., Al Kaabi, M., Hasan, S.W., Brine management methods: Recent innovations and current status. *Desalination*, 407, 1–23, 2017. https://doi.org/10.1016/j.desal.2016.12.008.
119. Pramanik, B.K., Shu, L., Jegatheesan, V., A review of the management and treatment of brine solutions. *Environ. Sci.: Water Res. Technol.*, 3, 625–658, 2017. https://doi.org/10.1039/C6EW00339G.
120. Al-Amoudi, A.S., Factors affecting natural organic matter (NOM) and scaling fouling in NF membranes: A review. *Desalination*, 259, 1–10, 2010. https://doi.org/10.1016/j.desal.2010.04.003.
121. Drioli, E., Ali, A., Macedonio, F., Membrane distillation: Recent developments and perspectives. *Desalination*, 356, 56–84, 2015. https://doi.org/10.1016/j.desal.2014.10.028.
122. Naidu, G., Jeong, S., Vigneswaran, S., Hwang, T.M., Choi, Y.J., Kim, S.H., A review on fouling of membrane distillation. *Desalin. Water Treat.*, 57, 10052–10076, 2016. https://doi.org/10.1080/19443994.2015.1040271.
123. Pandey, S.R., Jegatheesan, V., Baskaran, K., Shu, L., Fouling in reverse osmosis (RO) membrane in water recovery from secondary effluent: A review. *Reviews Environ. Sci. Biotechnol.*, 11, 125–145, 2012. https://doi.org/10.1007/s11157-012-9272-0.
124. Tijing, L.D., Woo, Y.C., Choi, J.S., Lee, S., Kim, S.H., Shon, H.K., Fouling and its control in membrane distillation – A review. *J. Membr. Sci.*, 475, 215–244, 2015. https://doi.org/10.1016/j.memsci.2014.09.042.
125. Foureaux, A.F.S., Moreira, V.R., Lebron, Y.A.R., Santos, L.V.S., Amaral, M.C.S., Direct contact membrane distillation as an alternative to the conventional methods for value-added compounds recovery from acidic effluents: A review. *Sep. Purif. Technol.*, 236, 116251, 2020. https://doi.org/10.1016/j.seppur.2019.116251.
126. Osman, M.S., Masindi, V., Abu-Mahfouz, A.M., Computational and experimental study for the desalination of petrochemical industrial effluents using direct contact membrane distillation. *Appl. Water Sci.*, 9, 29, 2019. https://doi.org/10.1007/s13201-019-0910-3.
127. Ryu, S., Naidu, G., Hasan Johir, M.A., Choi, Y., Jeong, S., Vigneswaran, S., Acid mine drainage treatment by integrated submerged membrane distillation–sorption system. *Chemosphere*, 218, 955–965, 2019. https://doi.org/10.1016/j.chemosphere.2018.11.153.
128. Agboola, O., The role of membrane technology in acid mine water treatment: a review. *Korean J. Chem. Eng.*, 36, 1389–1400, 2019. https://doi.org/10.1007/s11814-019-0302-2.
129. Agboola, O., Maree, J., Kolesnikov, A., Mbaya, R., Sadiku, R., Theoretical performance of nanofiltration membranes for wastewater treatment. *Environ. Chem. Letters*, 13, 37–47, 2014. https://doi.org/10.1007/s10311-014-0486-y.
130. Jimenez, Y.P. and Ulbricht, M., Recovery of Water from Concentration of Copper Mining Effluents Using Direct Contact Membrane Distillation. *Industr. Eng. Chem. Res.*, 58, 19599–19610, 2019. https://doi.org/10.1021/acs.iecr.9b02499.
131. Cherian, S., Weyens, N., Lindberg, S., Vangronsveld, J., Phytoremediation of trace element-contaminated environments and the potential of endophytic bacteria for improving this process. *Crit. Reviews Environ. Sci. Technol.*, 42, 2215–2260, 2012. https://doi.org/10.1080/10643389.2011.574106.
132. Gardea-Torresdey, J.L., Peralta-Videa, J.R., De La Rosa, G., Parsons, J.G., Phytoremediation of heavy metals and study of the metal coordination by X-ray absorption spectroscopy. *Coordinat. Chem. Reviews*, 249, 1797–1810, 2005. https://doi.org/10.1016/j.ccr.2005.01.001.

133. Ali, H., Khan, E., Sajad, M.A., Phytoremediation of heavy metals—Concepts and applications. *Chemosphere*, 91, 869–881, 2013. https://doi.org/10.1016/j.chemosphere.2013.01.075.
134. Asad, S.A., Farooq, M., Afzal, A., West, H., Integrated phytobial heavy metal remediation strategies for a sustainable clean environment: A review. *Chemosphere*, 217, 925–941, 2019. https://doi.org/10.1016/j.chemosphere.2018.11.021.
135. Ashraf, S., Ali, Q., Zahir, Z.A., Ashraf, S., Asghar, H.N., Phytoremediation: Environmentally sustainable way for reclamation of heavy metal polluted soils. *Ecotoxicol. Environ. Safety*, 174, 714–727, 2019. https://doi.org/10.1016/j.ecoenv.2019.02.068.
136. Tangahu, B.V., Sheikh Abdullah, S.R., Basri, H., Idris, M., Anuar, N., Mukhlisin, M., A review on heavy metals (As, Pb, and Hg) uptake by plants through phytoremediation. *Int. J. Chem. Eng.*, 2011, 31, 2011. https://doi.org/10.1155/2011/939161.
137. Bwapwa, J.K., Jaiyeola, A.T., Chetty, R., Bioremediation of acid mine drainage using algae strains: A review. *S. Afr. J. Chem. Eng.*, 24, 62–70, 2017. https://doi.org/10.1016/j.sajce.2017.06.005.
138. Gu, J.-D., Mining, pollution and site remediation. *Int. Biodeteriorat. Biodegrad.*, 2018, 1–2, 2018. https://doi.org/10.1016/j.ibiod.2017.11.006.
139. Syranidou, E., Christofilopoulos, S., Kalogerakis, N., *Juncus* spp. – The helophyte for all (phyto) remediation purposes? *New Biotech.*, 38, 43–55, 2017. https://doi.org/10.1016/j.nbt.2016.12.005.
140. Wang, L., Ji, B., Hu, Y., Liu, R., Sun, W., A review on *in situ* phytoremediation of mine tailings. *Chemosphere*, 184, 594–600, 2017. https://doi.org/10.1016/j.chemosphere.2017.06.025.
141. Masindi, V., Recovery of drinking water and valuable minerals from acid mine drainage using an integration of magnesite, lime, soda ash, CO_2 and reverse osmosis treatment processes. *J. Environ. Chem. Eng.*, 5, 3136–3142, 2017. https://doi.org/10.1016/j.jece.2017.06.025.
142. Silva, R.d.A., Castro, C.D., Vigânico, E.M., Petter, C.O., Schneider, I.A.H., Selective precipitation/UV production of magnetite particles obtained from the iron recovered from acid mine drainage. *Miner. Eng.*, 29, 22–27, 2012. https://doi.org/10.1016/j.mineng.2011.12.013.
143. Cornelis, G., Johnson, C.A., Gerven, T.V., Vandecasteele, C., Leaching mechanisms of oxyanionic metalloid and metal species in alkaline solid wastes: A review. *Appl. Geochem.*, 23, 955–976, 2008. https://doi.org/10.1016/j.apgeochem.2008.02.001.
144. Fernando, W.A.M., Ilankoon, I.M.S.K., Syed, T.H., Yellishetty, M., Challenges and opportunities in the removal of sulphate ions in contaminated mine water: A review. *Miner. Eng.*, 117, 74–90, 2018. https://doi.org/10.1016/j.mineng.2017.12.004.
145. Lewis, A.E., Review of metal sulphide precipitation. *Hydrometallurgy*, 104, 222–234, 2010. https://doi.org/10.1016/j.hydromet.2010.06.010.
146. Anastopoulos, I., Bhatnagar, A., Lima, E.C., Adsorption of rare earth metals: A review of recent literature. *J. Molecul. Liq.*, 221, 954–962, 2016. https://doi.org/10.1016/j.molliq.2016.06.076.
147. Bhattacharyya, K.G. and Gupta, S.S., Adsorption of a few heavy metals on natural and modified kaolinite and montmorillonite: A review. *Adv. Colloid Interface Sci.*, 140, 114–131, 2008. https://doi.org/10.1016/j.cis.2007.12.008.
148. Masindi, V., Gitari, M.W., Tutu, H., DeBeer, M., Efficiency of ball milled South African bentonite clay for remediation of acid mine drainage. *J. Water Process Eng.*, 8, 227–240, 2015. https://doi.org/10.1016/j.jwpe.2015.11.001.
149. Abdullah, N., Yusof, N., Lau, W.J., Jaafar, J., Ismail, A.F., Recent trends of heavy metal removal from water/wastewater by membrane technologies. *J. Industr. Eng. Chem.*, 76, 17–38, 2019. https://doi.org/10.1016/j.jiec.2019.03.029.
150. Al-Zoubi, H., Rieger, A., Steinberger, P., Pelz, W., Haseneder, R., Härtel, G., Nanofiltration of acid mine drainage. *Desalin. Water Treat.*, 21, 148–161, 2010. https://doi.org/10.5004/dwt.2010.1316.
151. Sierra, C., Álvarez Saiz, J.R., Gallego, J.L.R., Nanofiltration of acid mine drainage in an abandoned mercury mining area. *Water Air Soil Pollut.*, 224, 1734, 2013. https://doi.org/10.1007/s11270-013-1734-7.

152. Guo, L., Ott, D.W., Cutright, T.J., Accumulation and histological location of heavy metals in Phragmites australis grown in acid mine drainage contaminated soil with or without citric acid. *Environ. Experiment. Bot.*, 105, 46–54, 2014. https://doi.org/10.1016/j.envexpbot.2014.04.010.
153. Kiiskila, J.D., Li, K., Sarkar, D., Datta, R., Metabolic response of vetiver grass (*Chrysopogon zizanioides*) to acid mine drainage. *Chemosphere*, 240, 124961, 2020. https://doi.org/10.1016/j.chemosphere.2019.124961.
154. Rojas, C., Martínez, C.E., Bruns, M.A., Fe biogeochemistry in reclaimed acid mine drainage precipitates—Implications for phytoremediation. *Environ. Pollut.*, 184, 231–237, 2014. https://doi.org/10.1016/j.envpol.2013.08.033.
155. Qu, D., Wang, J., Wang, L., Hou, D., Luan, Z., Wang, B., Integration of accelerated precipitation softening with membrane distillation for high-recovery desalination of primary reverse osmosis concentrate. *Sep. Purif. Technol.*, 67, 21–25, 2009. https://doi.org/10.1016/j.seppur.2009.02.021.
156. Ben Ali, H.E., Neculita, C.M., Molson, J.W., Maqsoud, A., Zagury, G.J., Performance of passive systems for mine drainage treatment at low temperature and high salinity: A review. *Miner. Eng.*, 134, 325–344, 2019. https://doi.org/10.1016/j.mineng.2019.02.010.
157. Chen, T., Yan, B., Lei, C., Xiao, X., Pollution control and metal resource recovery for acid mine drainage. *Hydrometallurgy*, 147–148, 112–119, 2014. https://doi.org/10.1016/j.hydromet.2014.04.024.
158. Mohan, D. and Chander, S., Removal and recovery of metal ions from acid mine drainage using lignite—A low cost sorbent. *J. Hazard. Mat.*, 137, 1545–1553, 2006. https://doi.org/10.1016/j.jhazmat.2006.04.053.
159. Yan, B., Mai, G., Chen, T., Lei, C., Xiao, X., Pilot test of pollution control and metal resource recovery for acid mine drainage. *Water Sci. Technol.*, 72, 2308–2317, 2015. https://doi.org/10.2166/wst.2015.429.
160. Basu, H., Singhal, R.K., Pimple, M.V., Reddy, A.V.R., Arsenic removal from groundwater by goethite impregnated calcium alginate beads. *Water Air Soil Poll.*, 226, 22, 2015. https://doi.org/10.1007/s11270-014-2251-z.
161. Bhateria, R. and Singh, R., A review on nanotechnological application of magnetic iron oxides for heavy metal removal. *J. Water Process Eng.*, 31, 100845, 2019. https://doi.org/10.1016/j.jwpe.2019.100845.
162. Duuring, P., Hagemann, S.G., Banks, D.A., Schindler, C., A synvolcanic origin for magnetite-rich orebodies hosted by BIF in the Weld Range District, Western Australia. *Ore Geol. Reviews*, 93, 211–254, 2018. https://doi.org/10.1016/j.oregeorev.2017.12.007.
163. Phuan, Y.W., Ong, W.-J., Chong, M.N., Ocon, J.D., Prospects of electrochemically synthesized hematite photoanodes for photoelectrochemical water splitting: A review. *J. Photochem. Photobiol., C*, 33, 54–82, 2017. https://doi.org/10.1016/j.jphotochemrev.2017.10.001.
164. Bui, T.Q., Ngo, H.T.M., Tran, H.T., Surface-protective assistance of ultrasound in synthesis of superparamagnetic magnetite nanoparticles and in preparation of mono-core magnetite-silica nanocomposites. *J. Sci. Adv. Mat. Devices*, 3, 323–330, 2018. https://doi.org/10.1016/j.jsamd.2018.07.002.
165. Kefeni, K.K., Mamba, B.B., Msagati, T.A.M., Magnetite and cobalt ferrite nanoparticles used as seeds for acid mine drainage treatment. *J. Hazard. Mat.*, 333, 308–318, 2017. https://doi.org/10.1016/j.jhazmat.2017.03.054.

8

Recovery of Critical Raw Materials from Acid Mine Drainage (AMD): The EIT-Funded MORECOVERY Project

Carlos Ruiz Cánovas[1]*, Jose Miguel Nieto[1], Francisco Macías[1,2], Maria Dolores Basallote[1], Manuel Olías[1], Rafael Pérez-López[1] and Carlos Ayora[2]

[1]Department of Earth Sciences and Research Center on Natural Resources, Health and the Environment, University of Huelva, Campus 'El Carmen', Fuerzas Armadas s/n, Huelva, Spain
[2]Institute of Environmental Assessment and Water Research, Spanish Council of Scientific Research (CSIC), Barcelona, Spain

Abstract
Acid mine drainage (AMD) is a worldwide problem generated by oxidation of metal sulfides, which causes the generation of acidic and metal-rich solutions. The strong water-rock interaction processes observed during sulfide oxidation lead to the release of high concentrations of sulfate (up to hundreds of g/L) and pollutants such as Fe, Al, and As (from hundreds of g/L to few mg/L). However, these waters may also contain other elements of economic interest such as Cu, Zn, or rare earth elements (REEs), which could be recovered for their economic interest. Unless the AMD water characterization and treatment, which is a relevant topic since the 1980's, the recovery of elements of economic interest from AMD is of more recent interest. The search of secondary sources of raw materials encouraged mainly by industry and governmental agents has fostered the research on this subject in the last years. This work describes the main routes of raw materials extraction from AMD in the last years, discussing the main advantages and drawbacks of each technology and placing emphasis on the European Institute of Technology and Innovation (EIT)-funded project MORECOVERY, which aims to improve the eco-efficient and sustainable use of natural resources by creating a modular recovery process service package for hydrometallurgy and water treatment.

Keywords: Technology critical elements (TCEs), mine water, metal recovery, passive treatment, sustainable management, selective extraction, disperse alkaline substrate (DAS) technology, Iberian Pyrite Belt (IPB)

8.1 Introduction

Acid mine drainage (AMD) is a worldwide problem generated by oxidation of metal sulfides leading to the generation of acidic and metal-rich solutions that may deteriorate the

**Corresponding author*: carlos.ruiz@dgeo.uhu.es [ORCID:0000-0002-2860-5154]
[ORCID: Cánovas, 0000-0002-2860-5154; Nieto, 0000-0002-0086-252X; Macías, 0000-0003-4083-6040; Basallote, 0000-0003-2011-3806; Olías, 0000-0001-5394-3449; Pérez-López, 0000-0001-5394-3449; Ayora, 0000-0003-0238-7723]

Elvis Fosso-Kankeu, Christian Wolkersdorfer and Jo Burgess (eds.) Recovery of Byproducts from Acid Mine Drainage Treatment, (219–234) © 2020 Scrivener Publishing LLC

quality of the receiving water bodies. AMD contamination is recognized by its long life due to the generation of pollutants that may continue for extended periods (i.e., hundreds to thousands of years) after the cease of mining activities [1]. Johnson and Hallberg (2005) [2] estimated that mining effluents severely affected in 1989 around 19,300 km of rivers and 720 km^2 of lakes and reservoirs worldwide. The blooming rate of mining activities observed in the last years due to the rising need of resources supply worldwide may have increased these figures, especially in developing countries by artisanal and small-scale mining, where the lack of environmental controls leads to water quality deterioration of receiving water bodies [3]. Younger and Wolkersdorfer (2004) [4] propose that forthcoming plans and measures adopted to limit the mine water pollution in the long-term be managed as a fundamental pillar within the catchment scale strategy management. The special characteristics of AMD, i.e., high acidity, sulfate, and metal concentrations, may pose an environmental concern due to its great pollutant potential. The low pH values observed in some AMD cases worldwide lead to the occurrence of extreme metal-rich waters. For example, up to 760 g/L of sulfate, 111 g/L of Fe, 23.5 g/L of Zn, and 340 mg/L of As were reported in AMD with negative pH values from the Richmond mine (California, USA) [5]. More recently, Moreno-González *et al.* (2020) [6] have reported similar concentrations (up to 408 g/L of sulfate, 194 g/L of Fe, 11 g/L of Zn, 2.2 g/L of As) in negative pH waters from the Iberian Pyrite Belt (IPB). Such great pollutant capacity and the longevity of AMD has led to intense research activity on AMD water treatment in the last years. A single search on the Web of Science web site, introducing the keyword "AMD water treatment", provides us with 616 references dealing with this subject. As can be seen in Figure 8.1, the number of works has progressively increased from the 1990s, when only two or three papers were published, to the last 5 years, when more than 50 papers were published every year on AMD treatment. This recent interest in the AMD treatment is also reflected in the number of patents. According to the search engine of the European Office of Patents, 149 international patents were recorded since the last 1990s, with a progressive increase in the number of patents, especially in the last 5 years, when up to 30 patents were registered each year (Figure 8.1).

The oxidation of different sulfide minerals (e.g., chalcopyrite, sphalerite, chalcocite, pentlandite, covellite, cobaltite, stibnite, and millerite) may release other main and trace elements, including several (semi-)metals. In addition, the extreme conditions and low pH values achieved during sulfide oxidation result in the release of other elements enclosed in host rocks such as Al, Ca, Mg, or Na [7]. Thus, AMD can also contain high concentrations of elements which recovery could be profitable. Some of these metals, such as Al, Ni, Cu, and Zn have a great economic importance (Figure 8.2) according to the report on criticality of raw materials supply for EU countries published by the EU. The concentration of these elements in AMD can range from few μg/L to thousands of mg/L. For example, concentrations of up to 16 g/L of Zn, 12 g/L of Al, 4.4 g/L of Cu, and 13 mg/L of Ni has been reported by Moreno-González *et al.* (2020) [6] in extreme AMD outflows in the Tharsis mines (IPB, SW Spain).

The risk of supply for some of these elements enhances their importance in the market. This is the case for those countries which do not possess primary deposits to obtain these raw materials and their supply depends strongly on external suppliers. Some examples of elements which have both an economic importance and show a high risk of supply for the EU are Mg, Co, Sb, and V (Figure 8.2). These elements are also commonly found in AMD; Moreno-González *et al.* (2020) [6] reported maximum concentrations of 11 g/L of

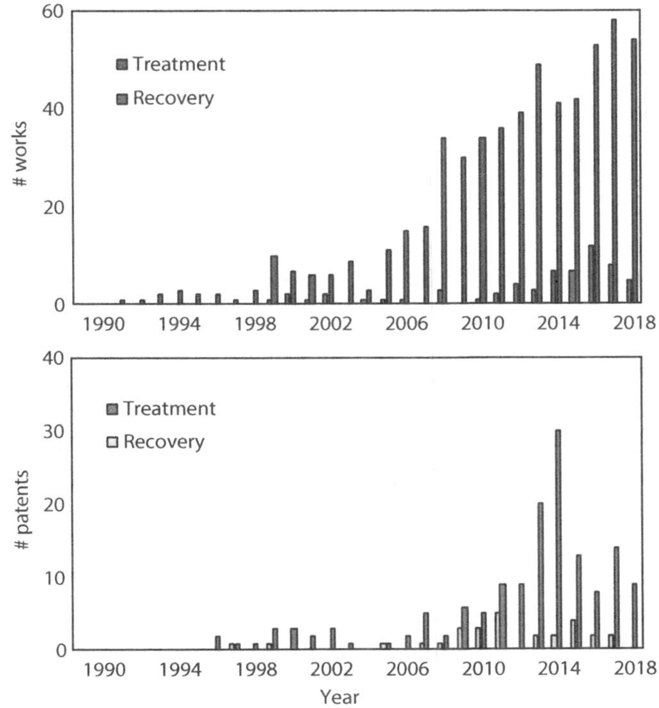

Figure 8.1 Evolution of the number of scientific papers (above) and patents (below) dealing with AMD treatment and metal recovery since 1990 (Web of Science, European Patent Office; keywords: AMD water treatment, AMD metal recovery).

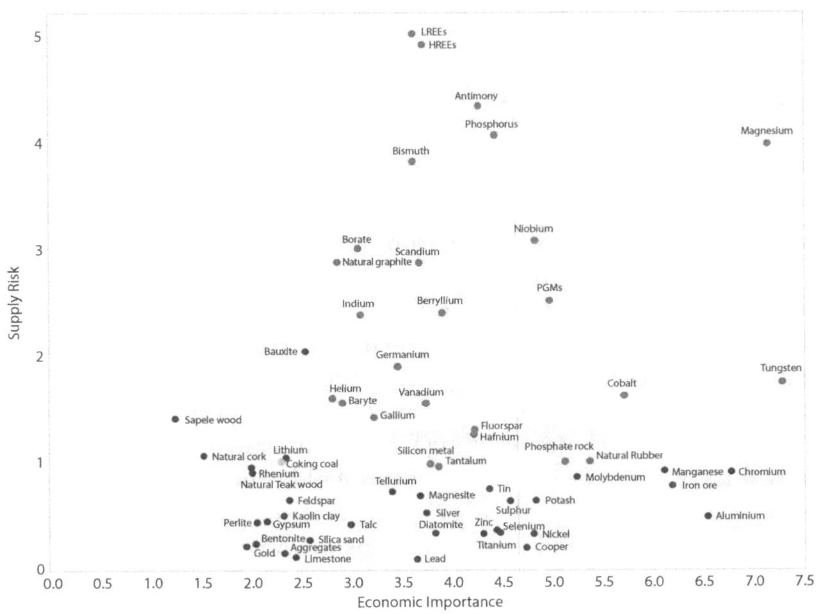

Figure 8.2 List of critical raw materials (CRMs) according to their supply risk and economic importance [11].

Mg, 53 mg/L of Co, 52 mg/L of V, and 12 mg/L of Sb. Even more interest has received the so-called "technology critical elements (TCEs)" whose production has bloomed in recent years for high-tech applications, i.e., rechargeable batteries, catalytic converters, magnets, semiconductors, electronic displays, and optical or photonics-related technologies. TCEs commonly include Ga, Ge, In, Te, Nb, Ta, Tl, the platinum group elements (PGEs), and the rare earth elements (REEs) along with Sc and Y [8]. These elements are at supply risk in many countries worldwide due to the absence of primary deposits. For this reason, governmental organizations encourage industry and researchers the search of secondary sources of these critical raw materials (CRMs). A promising secondary source of these elements could be AMD. Recently, Cánovas *et al.* (2020) [9] reported high concentrations of REEs (average concentration of 8.2 mg/L), Y (1.5 mg/L), Ga (80 μg/L), and Sc (53 μg/L), in AMD outflows from the IPB. These authors attribute such high concentrations, several orders of magnitude higher in some cases than those observed in natural waters, to the intensity of chemical weathering inside the mined zone. Other elements seem to show a lower mobility during sulfide oxidation processes. For example, Wierzbicka-Wieczorek *et al.* (2019) [10] reported that In is substantially accumulated in sulfide-rich mine wastes of the IPB and its mobility is to a certain degree restricted by oxygenation of AMD and resulting iron precipitation.

The occurrence of such elements of economic interest in AMD has attracted the interest of both industry and research organizations in the last years. Since 2000, 61 works have been published dealing with metal recovery from AMD, being especially intense in the last five years, with a maximum number of 12 scientific papers per year (Figure 8.1). This growing interest has been also reflected in industry; 28 patents about metal recovery from AMD have been registered from the same period (2000–2018; Figure 8.1). This work describes the main approaches followed in recovering CRMs from AMD, describing the advantages and drawbacks of such approaches and providing some insights into a sustainable way to recover these elements from AMD.

8.2 Recovery of CRMs from AMD

The environmental concern associated to AMD along with the need of searching alternative sources of CRMs has pushed the research on recovery of CRMs from AMD. Some initiatives focus on the recovery of CRMs from AMD treatment sludge rather than from the AMD effluent itself. The main reason behind such approach is the need of reducing environmental impacts as well as storage and management costs. Sludge management costs may account for around 5–20% of the operational water treatment cost [12] and the release of some hazardous compounds may take place if additional stabilization measures are not performed on the sludge. For example, Macías *et al.* (2017) [13] reported high concentrations of Cd, Zn, and Al in leachates resulting from the interaction of metal-rich sludge with rainfall and organics, which exceeded the risk threshold concentrations for aquatic life.

In this sense, Rakotonimaro *et al.* (2017) [14] reviewed several options to recover and reuse sludge from both active and passive systems for AMD treatment. For example, the extraction of Fe as pigment has been traditionally explored. In this context, Marcello *et al.* (2008) [15] studied the use of AMD sludge to obtain pigments that could be used for ceramic manufacture. This is probably the most explored route to recover CRMs from

AMD, especially in the USA, where the recycling of this Fe-rich sludge has been traditionally performed since 2001 for economic purposes [16]. However, it faces some difficulties such as transport from the mine site, the equipment and material supply or handling and the loading capacities [16]. Other limitation observed in this valorization route is the variability in color, variety, and morphology of the pigments, depending on the presence of pollutants. Silva *et al.* (2019) [17] evaluated several processes for cleaning iron sludge obtained from AMD active treatment to obtain yellow pigment of high quality. To accomplish this issue, these authors performed several experiments by precipitating with alkaline solutions and eliminating other impurities from the sludge after washing and filtration. They concluded that high-quality goethite-based pigment can be produced from AMD sludge, however also highlighted that the economic feasibility regarding the costs of the procedure must be considered. Similar conclusions were obtained by Ryan *et al.* (2017) [18], who studied the selective precipitation to recover purified iron oxide pigments from synthetic AMD. An increasing redness in pigment during drying due to Fe mineral transformation into more crystalline hematite was reported by these authors. The purest pigments were achieved when Fe precipitated at pH 3, with a distinct yellow color typical of jarosite, while darker pigments were obtained when trace metal impurities increased. The authors finally concluded that the choice to obtain higher quality Fe pigments at lower pH comes with the compromise of reduced Fe recovery efficiency. Sludge from AMD can be also used for other applications. Liu *et al.* (2019) [19] tested AMD treatment sludge as alternative sources of Fe and Ca to manufacture hexaferrite, which is widely used in magnets, microwaves devices, and magnetic recording media. However, they reported the importance of obtaining a clean sludge, as impurities in sludge affect phase composition and microstructure.

Metal and acid recovery from AMD has been traditionally less favored than other recycling approaches owing to expensive operational costs, being mainly adopted as alternative option, when disposal costs are sub-economic compared to metal recovery routes [12]. Nleya *et al.* (2015) [20] reviewed the available technology to recover sulfuric acid from AMD to evaluate the possibility of recovering and upgrading the acid contained up to technical grades. This study concluded that it is technically feasible to produce acid of technical grade (>95%) from AMD; however, the authors recognized that economic and environmental benefits of such recovery will largely depend on optimizing methods to purify crystals and save energy cost. Yan *et al.* (2015) [21] evaluated the extraction of metals (i.e., Fe, Cu, Zn, and Mn) from AMD (300 m^3/day) during neutralization. The resulting precipitates contained up to 42%, 12%, 31%, and 18% for Fe, Cu, Zn, and Mn, respectively, while the recovery rates were between 79% and 83%. These authors state that the refinement for Fe, Cu, and Zn could be performed by roasting and flotation. The benefits obtained from these metals offset the cost of chemical reagents used. The recovery of elements contained in AMD based on the difference between the solubility products of target elements is not a novel issue. For example, Chen *et al.* (2014) [22] studied the sequential extraction of some metals such as Fe, Cu, Zn, and Mn from AMD effluents at the Dabaoshan Mine region (China). They obtained, after adding progressively different reagents (i.e., H_2O_2, $Ca(OH)_2$, and NaS_2), a Fe-rich (85%) concentrate for use as pigment and Cu (36%) and Zn (55%) concentrates, which could be subsequently recovered. Due to the low content of Mn (7%) in the final sludge, the authors propose its use as replacement of cement materials.

However, new approaches are progressively being explored to recover CRMs from AMD beyond the classical chemical methods. For instance, Park *et al.* (2015) [23] studied the selective recovery of dissolved metals from AMD effluents by electrochemistry. This study highlighted the potential of electrochemical methods to recover valuable metals, reducing the amount of chemicals commonly used during redox and neutralizing reactions. Based on the high costs associated to chemical extraction of elements and the huge volume of sludge generated, López *et al.* (2019) [24] proposed the recovery of sulfuric acid and valuable elements such as Zn, Cu, and REEs by nanofiltration (NF)-based treatment. These authors consider that this technique is a good option to obtain elements of economic interest due to the reduction in chemicals consumption. To recover these metals, Fe should be removed before adding a Ca source (i.e., gypsum or limestone) to recover REEs by phosphate precipitation. However, this study does not provide information on energy consumption and process costs. Choi *et al.* (2019) [25] applied volume-retarded osmosis and a low-pressure membrane (VRO-LPM) process to recover Mn, Fe, Cu, Zn, As, Cd, and Pb. Crane and Sapsford 2018 [26] studied the synthesis of Cu nanoparticles from AMD using zerovalent Fe nanoparticles as a useful mechanism for the valorization of Cu-bearing AMD effluents. They affirm that it is possible to selectively elaborate Cu bearing nanoparticles from these effluents to be used in a wide range of products such as catalyzers, optical, and electronic materials or the manufacture of antifungal/antibacterial agents.

8.3 Upscaling of Successful Technologies and Economic Suitability

A critical point during the application of new technologies is the upscaling processes; many technologies failed to be implemented during their upscaling. Some reasons behind this inability are the huge investment required to settle a pilot or industrial scale project, or an inadequate or inexistent plan for the execution of the pilot program. Some technologies have been successfully applied to recover metals of economic interest at pilot or industrial scale, as previously reported in Simate and Ndlovu (2014) [27]. One of these examples is the BioSulfide process [28]. Initially, sulfur-reducing bacteria produce hydrogen sulfide gas (H_2S) inside an anaerobic bioreactor, and then, the H_2S fills a reactor tank where it reacts with the raw water to selectively precipitate metals as sulfides. The products of this reaction (i.e., metal-rich sludge and purified waters) are carried to a refine vessel where the clean water is removed from the sludge and can be recycled. Finally, the excess water is removed by filtration, producing a purified metal product which could follow a final refinement [28]. However, this process is very dependent on the price of the chemical energy sources (i.e., ethanol and butanol) used to produce sulfides. This limitation exerted by the price of organic substrates can be overcome if natural materials are used. This is the approach followed by the Rhodes BioSURE process at Grootvlei by East Rand Water Care Company (ERWAT) [29]. This process is quite similar to that of BioSulfide but using organic wastes such as wastewater treatment sludge, instead of expensive commercial carbon sources, like ethanol. Another upscaled metal recovery process is the high-pressure reverse osmosis based HiPRO process, developed by Aveng Water and able to recover more than 97% of water, obtaining a solid with commercial grades of calcium sulfate and less pure Ca and other metal sulfates [30].

Economic factors may control the interest in AMD waste recycling as secondary sources of CRMs. The first action before making a great investment to create and run a pilot or industrial plant for metal recovery is to estimate the total valuation of the elements of economic interest contained in the waste streams. An important question arises when facing the critical issue of estimating the valuation of CRMs contained in AMD effluents or sludge. The price of metals on the market varies substantially depending on purity, quantity, and other factors. The worldwide valuation of metal resources is routinely published by organizations such as the London Metal Exchange (LME) or Shanghai Metal Market (SMM). These figures are referred to high-quality products; thus, the estimation of the total valuation of raw materials contained in AMD effluents and sludge is not possible. Although case-by-case studies are needed to estimate the potential value of elements or products contained in these waste streams, a rough approximation was highlighted by Smith *et al.* (2013) [31]. These authors reported a case study of Zn selective recovery from sludge generated by an AMD treatment plant in the Wellington Oro Mine (Colorado, US). The resulting product, containing around 60 wt % of Zn, was sold to a smelter at a price of 0.33 USD/kg of Zn content, as well as the shipping costs (0.31 USD/kg of Zn). Thus, the final valuation of the product obtained from the sludge was around 30% of the average 2010 LME price. However, this valuation may vary depending on the pureness of the final product. Consequently, greater efforts are needed to obtain sustainable and cost-effective selective methods of CRMs from AMD.

Another important question to be solved is which total cost should not surpass the recovery procedure in order to be a cost-effective activity? Total costs for each recovery approach must be assessed in advance, considering all steps, including those related to raw material processing, personnel and reagents costs, energy or maintenance, and others related to put the product on the market (i.e., financial costs, infrastructure, permits, fees, or shipping [32]).

8.4 Coupling Environmental and Resources Policy: The EIT-Funded MORECOVERY Project

In January 2019, the three-year MORECOVERY (Modular recovery process services for hydrometallurgy and water treatment) project it was launched, funded by the European Union (EU) through the European Institute of Innovation and Technology (EIT)-Raw Materials. This project is executed by several EIT RawMaterials partners related to industry, SME, and research centers, namely, the Geological Survey of Finland (GTK), the Finnish Minerals Group (FMG), Keliber, LTU Business, Savonia University of Applied Science, the University of Eastern Finland (UEF), the University of Huelva (UHU) and the Spanish National Research Council (CSIC).

The main goal of this project is to guarantee the access to CRMs for the EU, by developing a service package that enables producers to carry out a non-aligned technical and economic analysis of the potential for recovery of these materials. To address this issue, the existing water treatment and hydrometallurgical pilot set-up co-owned by GTK and Savonia University of Applied Science has been improved and the piloting capacity will be combined with the laboratory-scale investigations performed by the different partners, creating a sound service package for hydrometallurgy and water treatment. To validate the service package, potential extractive waste facilities will be screened and tested in Finland and Spain, while the launch to the market will be performed by LTU Business, which will investigate business opportunities for the MORECOVERY service package. As a result of the project, the use of secondary materials and side streams in raw material production will be enhanced, while environmental goals will be achieved through cleaner mine waters and decreased amounts of extractive waste containing potentially harmful elements.

The MORECOVERY project tries to fulfil the objectives of environmental and resources policy. One of the action lines of the project is to extract CRMs from AMD using the IPB as a case study. The IPB is a large metallogenic sulfide province located in the SW of the Iberian Peninsula, where intensive mining activity has historically been performed, generating a large amount of wastes whose oxidative dissolution produces acid leachates rich in sulfate, metals, and trace elements. These AMD discharges annually transport a large amount of metals and metalloids to the Gulf of Cadiz; between 35.7–183 kt/a of sulfate, 1.7–7.9 kt/a of Fe, 2.45–5.8 kt/a of Al, 0.65–3.5 kt/a of Zn, 0.4–1.7 kt/a of Cu, 0.3–1.6 kt/a of Mn, and minor amounts of As, Cd, Co, Ni, and Pb (1–36, 3–11, 16–71, 5–36, and 9–27 t/a, respectively; [7]). The importance of the IPB on the metal delivery to oceans is observed in its contribution to the total riverine global fluxes to the oceans for some of these elements such as Zn (around 2.8–15% of Zn delivered to the oceans) and Cu (0.8–2.8%) [7].

The extremely high acidity and metal concentrations in AMD from the IPB pose a great problem for water management. As can be seen in Figure 8.3, the acidity of typical AMD from the IPB exceeds notably the maximum acidity loads commonly neutralized by traditional AMD treatment systems such as anaerobic wetlands (AnW), vertical flow wetlands (VFW), anoxic limestone drains (ALD), and oxic limestone channels (OLC).

As recently reported by Ayora *et al.* (2013) [34], these traditional systems were previously tested in the IPB. For example, limestone traps based on ALD systems commonly used in coal mines were settled in the IPB during the 1990s. These traps consisted of gravel-sized fragments of limestone which were used to fill mine shafts and galleries. This technology turned to be effective only during a few months when iron precipitates coated the limestone fragments, and the carbonate material lost its reactivity. The acidity concentrations of the AMD typically found are much higher than the threshold limit of 2 mg/L of trivalent metals (i.e., Fe(III) and Al) proposed for ALD systems [41].

Considering the problems associated to coating observed during the application of ALDs in the IPB, an experimental RAPS system was implemented in Mina Esperanza, a typical AMD from the IPB. The system was composed of a concrete tank (480 m^3) filled with calcite

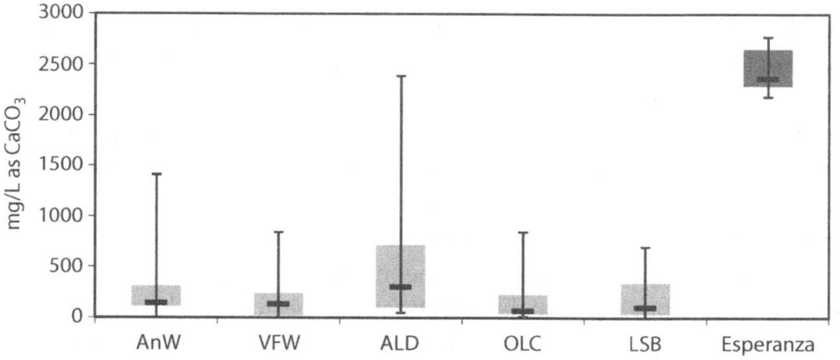

Figure 8.3 Net acidity (median, percentiles 25th and 75th) of mine waters commonly treated with traditional passive treatment technologies ($n = 80$) and typical AMD in the IPB [42]. AnW anaerobic wetlands, VFW vertical flow wetlands (RAPS), ALD anoxic limestone drains, OLC oxic limestone channels, and LSB limestone leach beds [33].

fragments and organic matter (i.e., cattle manures, sewage water, and wastes from the dairy manufacturing sector) for the treatment of around 1 L/s of AMD, allowing a residence time of around 5 days. After an irregular operation, the system finally collapsed after 86 days due to the loss of reactivity and hydraulic conductivity problems. The reasons behind these inefficiencies may be related to an insufficient residence time to enhance sulfate reduction and the high concentrations for some metals, which could be toxic for sulfate-reducing bacteria [34].

In order to overcome the coating problem associated to such metal-rich and acidic waters, a new technology, the so-called Disperse Alkaline Substrate (DAS) was developed [35]. This technology involves the mixing of a fine-grained neutralizing substance (usually calcite) with a non-reactive and granular material with high surface (commonly wood chips). Thus, the surface of the inert is to a certain degree covered by the alkaline substance. A greater reactive surface is achieved using this grain size of the alkaline material fragments compared to the same amount of coarse materials, thus increasing the dissolution rate. Consequently, a higher percentage of alkaline material is dissolved before coating, which would limit further dissolution. The addition of coarse fragments of wood shavings improves the permeability of the system, delaying the clogging problems, and increasing the efficiency of the system.

This technology was tested both at laboratory and pilot scale and proved to be successful in both cases with an efficient removal of metals and acidity [35, 36]. Rötting *et al.* (2008) [35] tested the efficiency of the system using the DAS technology by column experiments, passing AMD through columns filled with limestone, wood shavings, and sand over time (Figure 8.4). These authors reported that Al, Fe(III), Cu, and Pb were removed from the waters irrespective of the flow entering the system, while lower removal of Zn, Ni, and Cd was observed at low flow rates. The removal of most part of acidity is enhanced by the metal accumulation in the DAS, which precipitation favoring further calcite dissolution. For a 15-month period, the DAS operated successfully, avoiding the common clogging problems in Fe-rich waters, with a rate of 120 g of acidity/m^2·day, four times the loading rate advised for traditional passive treatments [35].

Once the laboratory tests were successful, the technology was upscaled to a pilot plant composed of two 3 m^3 tanks loaded with calcite and wood chips, and one 1 m^3 tank filled

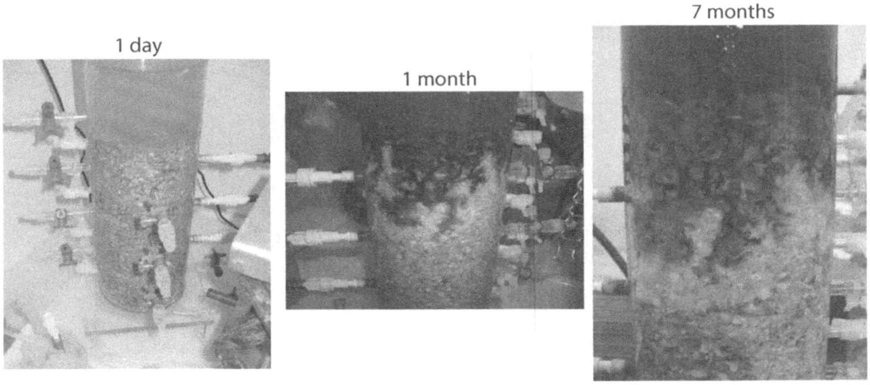

Figure 8.4 Evolution of the column experiments based on DAS technology, showing the progressive precipitation of minerals over time.

with MgO powder and wood chips, split by several chute aerators and decantation pools (Figure 8.5). Caraballo *et al.* (2009) [36] reported a high efficiency of the system after 9 months of performance, with a complete elimination of metals and metalloids such as Al, Cu, As, Pb, more than 70% of Fe, about 25% of Zn and 80% of net acidity. A detailed analysis of the main mineral phases precipitated in the limestone tanks identified goethite, schwertmannite, hydrobasaluminite, amorphous $Al(OH)_3$, and gypsum. Divalent metals such as Fe (II), Zn, Co, and Ni were almost completely removed in the third tank of MgO. These authors concluded that this treatment design is a straightforward, efficient, and long-lasting remediation approach to treat highly polluted AMD effluents like those of the IPB. For these reasons, several industrial treatment plants have been built in the recent years within the Odiel river basin in order to improve the water quality of this river.

The first AMD passive full-scale treatment plant (20,000 m^2 total surface area) was settled at Mina Concepcion, located in the IPB. The plant is fed by two different AMD sources (Figure 8.6), from (i) a pit lake overflow (MC1) and (ii) a partially restored waste dump (MC2). The MC1 has a high acidity and metal concentrations with a pH of 2.66, 512 mg/L of Fe, 100 mg/L of Al, 35 mg/L of Zn, and lesser amounts of other trace metals and metalloids (e.g., Cd, Co, and As; Figure 8.1). The MC2 AMD has lower Fe concentration (24 mg/L) and metals and metalloids (e.g., Zn, Cd, As, and Co) concentrations except for Al (119 mg/L; Figure 8.6). The average flow is around 0.8 L/s, although progressively decreases to 0.1 L/s during the dry season. The treatment plant was initially composed of a natural Fe-oxidizing lagoon (NFOL; [37]) of around 100 m^2, which favors iron oxidation and the partial removal of Fe and As. After the NFOL, the water runs into the first alkaline reactive treatment (composed of two tanks filled with limestone-DAS, RT1 and RT2, of 960 m^3 and 720 m^3, respectively), which removes the trivalent metals. After the limestone tanks, two settling ponds (S1 and S2, of 100 m^2 each; Figure 8.6) are located in the plant, which promotes the sedimentation of Fe and Al. The outflows from these ponds enter to the MgO-DAS reactive tank (RT3, 400 m^3) where pH increases up to 7.2 and divalent metals are completely or almost removed. Output waters are finally discharged into the receiving water bodies with a good final quality.

One of the most striking results obtained from the DAS technology, and directly connected to the MORECOVERY project, is the selective precipitation of CRMs in the reactive

Figure 8.5 Image of the pilot plant based on DAS technology settled at the exit of the Monterromero mine adit (IPB, SW Spain). This system is composed of two 3 m^3 tanks loaded with calcite and wood chips, and a 1 m^3 tank filled with MgO powder and wood chips, split by different chute aerators and decantation pools.

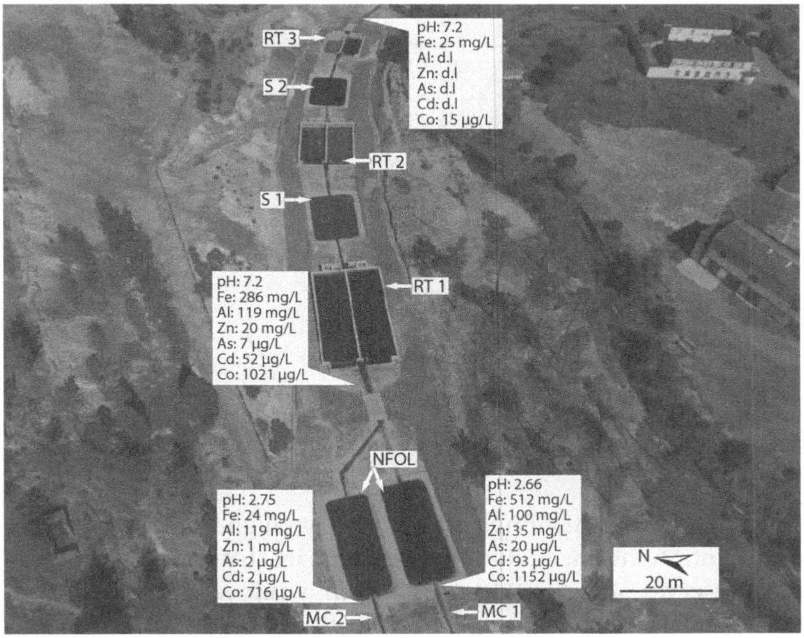

Figure 8.6 Image of the industrial AMD treatment plant based on DAS technology constructed at Mina Concepción (IPB, SW Spain), indicating the AMD effluents (MC1 and MC2), the natural Fe oxidizing lagoon, reactive tanks and settling ponds (modified from Martínez *et al.* (2019) [38]; see text for explanation). d.l: below the detection limit.

tanks of the plant. Mineral formation takes place in two different fronts; the first front corresponds to the Fe-rich front (Figure 8.7), mainly formed by schwertmannite precipitation at pH values from 2.7 to 4.0, which mainly scavenge other trace metals and metalloids contained in AMD such as As, Cr, Sb, and Mo. The second front corresponds with the Al front (Figure 8.7), formed by precipitation of basaluminite at pH values higher than 4, which can remove other metals such as Cu, Zn, Cd, Co, Ni, REEs, and yttrium (Y). This capacity of scavenging REE and Y has been recently reported by Ayora *et al.* (2016) and point at AMD treatment with DAS technology as a modest but suitable REE+Y (REY) source in sulfide mining areas. In addition to the environmental benefits associated to its treatment, the longevity of AMD guarantee that the total reserves are practically unlimited.

These facts justify the exploration of these waters as source of CRMs within the MORECOVERY project. A rough estimation of REY reserves can be done in the IPB considering the volume of AMD generated and the metal concentrations. More than one hundred AMD sources have been described along the IPB, with very variable flows from 0.1 L/s to more than 100 L/s. Considering the characteristics typical of the Mediterranean (long dry periods followed by quick and intense rainy events), AMD flows may be also multiplied several times from the dry to the rainy season. Concerning REY concentrations in mine waters of the IPB, a systematic survey in around 40 AMD locations indicates a high variability, reaching concentrations from 0.07 to 13 mg/L, with a mean concentration of 2.3 mg/L of total REY. Thus, considering an average AMD flow of about 1 m^3/s and an average concentration of 2.3 mg/L of total REY, total reserves of 80 t REY/a (100 t REY oxides/a)

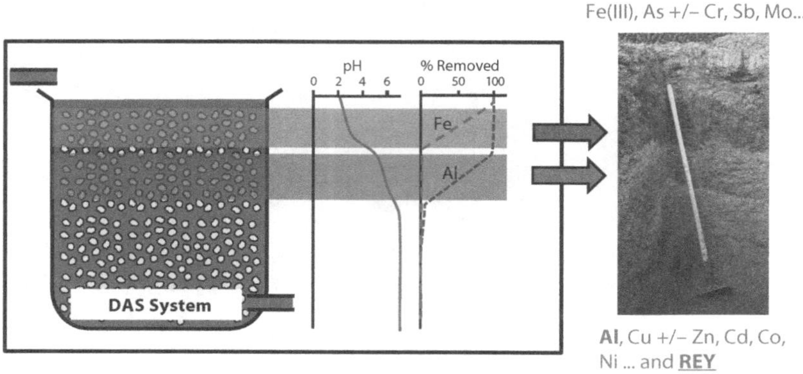

Figure 8.7 Schematic representation of the mineral precipitation fronts in reactive tanks of the DAS technology-based treatment plant and metal as well as metalloid partitioning among these fronts. REY: REE+Y.

can be estimated in the IPB. This annual inventory has a high variability depending on flow changes and concentrations uncertainties. If we compare this annual accumulation of REY with the annual reserves of the ore deposits currently mined or explored it is very small (Figure 8.8a), as pointed out in chapter "Rare earth elements – A treasure locked in AMD?". Considering the little data available on REY concentrations, and assuming that 100% of the REY will be concentrated in basaluminite, the REY rates in these precipitates would range between 0.03 and 8% REY_2O_3 (Figure 8.8b; [39]), which are similar to those of currently investigated deposits (Figure 8.8b). However, although the annual reserves are two orders of magnitude lower than current targets of exploration, the longevity of AMD, which guarantees the release of REY for a long time to come [1] with no energy investment, makes AMD a considerable resource of REY.

Although the AMD pool in the IPB exhibits lower grades of metals than in conventional deposits, these metal reserves could be extracted from the AMD in a cost-effective way if selective methods of recovery are created. One of the works scheduled within the

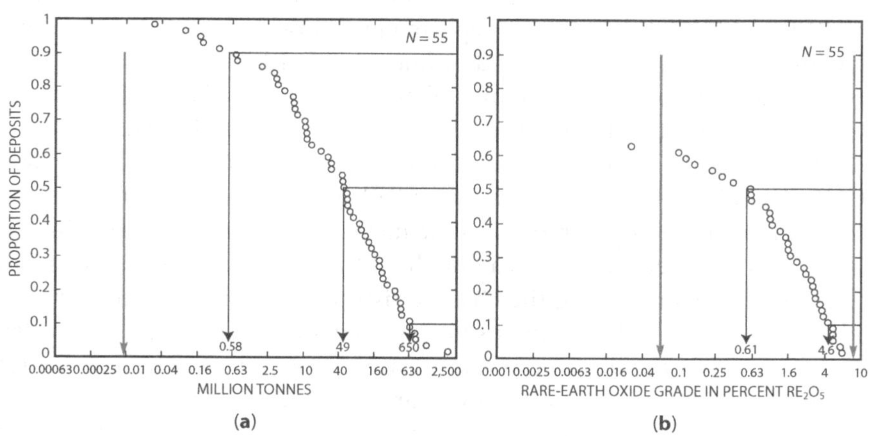

Figure 8.8 Accumulative frequency of reserves (a) and metal grade (as oxide) (b) from conventional REY deposits (modified from [40]). Red arrows indicate the REE richer and poorer AMD monitored to date.

MORECOVERY project is the concentration and selective extraction of CRMs from AMD and sludge from the DAS treatment plants. Thus, a sustainable, environmentally friendly, and economic source of CRMs from AMD effluents could be achieved in the near future.

Acknowledgements

This work was funded by the H2020 EIT Raw Materials program through the project MORECOVERY "Modular recovery services for hydrometallurgy and water treatment (H2020-EIT-PN 18190)" and by the Spanish Ministry of Science, Innovation and Universities through the SCYRE project (CGL2016–78783-C2-R).

References

1. Younger, P.L., The longevity of mine water pollution: A basis for decision-making. *Sci. Total. Environ.*, 194–195, 457–466, 1997.
2. Johnson, D.B. and Hallberg, K.B., Acid mine drainage remediation options: A review. *Sci Total Environ.*, 338, 3–14, 2005. https://doi.org/10.1016/j.scitotenv.2004.09.002.
3. Goix, S., Maurice, L., Laffont, L., Rinaldo, R., Lagane, C., Chmeleff, J., Menges, J., Heimbr̃ger, L.-E., Maury-Brachet, R., Sonke, J.E., Quantifying the impacts of artisanal gold mining on a tropical river system using mercury isotopes. *Chemosphere*, 219, 684–694, 2019. https://doi.org/10.1016/j.chemosphere.2018.12.036.
4. Younger, P.L. and Wolkersdorfer, C., Mining impacts on the fresh water environment: Technical and managerial guidelines for catchment scale management. *Mine Water Environ.*, 23, s2–s80, 2004. https://doi.org/10.1007/s10230-004-0028-0.
5. Nordstrom, D.K. and Alpers, C.N., Negative pH, efflorescent mineralogy, and consequences for environmental restoration at the Iron Mountain Superfund site, California. *Proc. Natl. Acad. Sci.*, USA, Colloquium Paper, Vol. 96, pp. 3455–3462, 1999.
6. Moreno-González, R., Cánovas, C.R., Olías, M., Macías, F., Seasonal variability of extremely metal rich acid mine drainage waters from the Tharsis mines (SW Spain). *Environ. Pollut.*, 259, 113829, 2019. https://doi.org/10.1016/j.envpol.2019.113829.
7. Nieto, J.M., Sarmiento, A.M., Cánovas, C.R., Olías, M., Ayora, C., Acid mine drainage in the iberian pyrite Belt: 1. Hydrochemical characteristics and pollutant load of the Tinto and Odiel rivers. *Environ. Sci. Pollut. Res.*, 20, 7509–7519, 2013. https://doi.org/10.1007/s11356-013-1634-9.
8. Cobelo-García, A., Filella, M., Croot, P., Frazzoli, C., Du Laing, G., Ospina-Alvarez, N., Rauch, S., Salaun, P., Schäfer, J., Zimmermann, S., COST action TD1407: Network on technology-critical elements (NOTICE) – from environmental processes to human health threats. *Environ. Sci. Pollut. Res.*, 22, 15188–15194, 2015. https://doi.org/10.1007/S11356-015-5221-0.
9. Cánovas, C.R., Macías, F., Olías, M., Basallote, M.D., Perez-Lopez, R., Ayora, C., Nieto, J.M., Release of technology critical metals during sulfide oxidation processes: The case of the Poderosa sulfide mine (south-west Spain). *Environ. Chem.*, 17, 93–104, 2020.
10. Wierzbicka-Wieczorek, M., Lottermoser, B.G., Kiefer, S., Sindern, S., Gronen, L.C., Hensler, A.S., Indium distribution in metalliferous mine wastes of the Iberian Pyrite Belt, Spain–Portugal. *Environ. Earth. Sci.*, 78, 253, 2019. https://doi.org/10.1007/s12665-019-8263-7.
11. EU, Study on the review of the list of critical raw materials, in: *Criticality assessments*, 2017, https://doi.org/10.2873/876644.
12. Zinck, J. and Griffith, W., *Review of acidic drainage treatment and sludge management operations*, MEND Report 3.43.1. CANMET- MMSL, 101p., Mining Association of Canada, Montreal, Canada. 2013.

13. Macías, F., Pérez-López, R., Caraballo, M.A., Cánovas, C.R., Nieto, J.M., Management strategies and valorization for waste sludge from active treatment of extremely metal-polluted acid mine drainage: a contribution for sustainable mining. *J. Clean. Prod.*, 141, 1057–1066, 2017. https://doi.org/10.1016/j.jclepro.2016.09.181.
14. Rakotonimaro, T.V., Neculita, C.M., Bussiere, B., Benzaazoua, M., Zagury, G.J., Recovery and reuse of sludge from active and passive treatment of mine drainage-impacted waters: A review. *Environ. Sci. Pollut. Res.*, 24, 73–91, 2017. https://doi.org/10.1007/s11356-016-7733-7.
15. Marcello, R.R., Galato, S., Peterson, M., Riella, H.G., Bernardin, A.M., Inorganic pigments made from the recycling of coal mine drainage treatment sludge. *J. Environ. Manag.*, 88, 1280–1284, 2008. http://doi.org/10.1016/j.jenvman.2007.07.005.
16. Hedin, R.S., Recovery of marketable iron oxide from mine drainage in the USA. *Land Contam. Reclamat.*, 11, 93–97, 2003.
17. Silva, R.A., Secco, M.P., Lermen, R.T., Schneider, I.A.H., Navarro Hidalgo, G.E., Sampaio, C.H., Optimizing the selective precipitation of iron to produce yellow pigment from acid mine drainage. *Min. Eng.*, 135, 111–117, 2019. https://doi.org/10.1016/j.mineng.2019.02.040.
18. Ryan, M.J., Kney, A.D., Carley, T.L., A study of selective precipitation techniques used to recover refined iron oxide pigments for the production of paint from a synthetic acid mine drainage solution. *Appl. Geochem.*, 79, 27–35, 2017. https://doi.org/10.1016/j.apgeochem.2017.01.019.
19. Liu, M., Iizuka, A., Shibata, E., Acid mine drainage sludge as an alternative raw material for M-type hexaferrite preparation. *J. Clean. Prod.*, 224, 284–291, 2019. https://doi.org/10.1016/j.jclepro.2019.03.224.
20. Nleya, Y., Simate, G.S., Ndlovu, S., Sustainability assessment of the recovery and utilisation of acid from acid mine drainage. *J. Clean. Prod.*, 6959, 17–27, 2015. http://doi.org/10.1016/j.jclepro.2015.11.005.
21. Yan, B., Mai, G., Chen, T., Lei, C., Xiao, X., Pilot test of pollution control and metal resource recovery for acid mine drainage. *Water Sci. Technol.*, 72, 12, 2308–2317, 2015. https://doi.org/10.2166/wst.2015.429.
22. Chen, T., Yan, B., Lei, C., Xiao, X., Pollution control and metal resource recovery for acid mine drainage. *Hydrometallurgy*, 147-148, 112–119, 2014. http://doi.org/10.1016/j.hydromet.2014.04.024.
23. Park, S.M., Shin, S.Y., Yang, J.S., Ji, S.W., Baek, K., Selective recovery of dissolved metals from mine drainage using electrochemical reactions. *Electrochim. Acta*, 181, 248–254, 2015. http://doi.org/10.1016/j.electacta.2015.03.085.
24. López, J., Reig, M., Gibert, O., Cortina, J.L., Recovery of sulphuric acid and added value metals (Zn, Cu and rare earths) from acidic mine waters using nanofiltration membranes. *Sep. Purif. Technol.*, 212, 2019, 180–190, 2019. https://doi.org/10.1016/j.seppur.2018.11.022.
25. Choi, J., Im, S.J., Am, J., Application of volume retarded osmosis – Low pressure membrane hybrid process for recovery of heavy metals in acid mine drainage. *Chemosphere*, 232, 264–272, 2019. https://doi.org/10.1016/j.chemosphere.2019.05.209.
26. Crane, R.A. and Sapsford, D.J., Selective formation of copper nanoparticles from acid mine drainage using nanoscale zerovalent iron particles. *J. Hazard Mater.*, 347, 252–265, 2018. https://doi.org/10.1016/j.jhazmat.2017.12.014.
27. Simate, G.S. and Ndlovu, S., Acid mine drainage: Challenges and opportunities. *J. Environ. Chem. Eng.*, 2, 3, 1785–1803, 2014. http://doi.org/10.1016/j.jece.2014.07.021.
28. Gleason, W., Metals recovered from mine water provide new revenue stream. *Min. Eng.*, 61, 22–24, 2009.
29. Neba, A., *The Rhodes BioSURE Process and the Use of Sustainability Indicators in the Development of Biological Mine Water Treatment*, PhD Thesis, Rhodes University, Grahamstown, 2007.

30. Aveng Water Blueprint for Treatment of AMD 2009; Available at http://www.avengwater.co.za/news-room/press-releases/emalahleni-blueprint-treat-ment-amd (accessed: 10/2019).
31. Smith, K.L., Figueroa, L.A., Plumlee, G.S., Can treatment and disposal costs be reduced through metal recovery? In: Wolkersdorfer, Brown, Figueroa (Eds.), *Proceedings of the Annual International Mine Water Association Conference-Reliable Mine Water Technology*, Golden, Colorado (USA), 2013.
32. Cánovas, C., Pérez-López, R., Macías, F., Chapron, S., Nieto, J.M., Pellet-Rostaing, S., Exploration of fertilizer industry wastes as potential source of critical raw materials. *J. Clean. Prod.*, 143, 497–505, 2017. https://doi.org/10.1016/j.jclepro.2016.12.083.
33. Ziemkiewicz, P.F., Skousen, J.G., Simmons, J., Long-term performance of passive acid mine drainage treatment systems. *Mine Water Environ.*, 22, 118–129, 2003. https://doi.org/10.1007/s10230-003-0012-0.
34. Ayora, C., Caraballo, M.A., Macías, F., Rötting, T.S., Carrera, J., Nieto, J.M., Acid mine drainage in the Iberian Pyrite Belt: 2. Lessons learned from recent passive remediation experiences. *Environ. Sci. Pollut. Res.*, 20, 7837–7853, 2013. https://doi.org/10.1007/s11356-013-1479-2.
35. Rötting, T.S., Thomas, R.C., Ayora, C., Carrera, J., Passive treatment of acid mine drainage with high metal concentrations using dispersed alkaline substrate. *J. Environ. Qual.*, 37, 1741–1751, 2008. https://doi.org/10.2134/jeq2007.0517.
36. Caraballo, M.A., Rötting, T.S., Macías, F., Nieto, J.M., Ayora, C., Field multi-step calcite and MgO passive system to treat acid mine drainage with high metal concentration. *Appl. Geochem.*, 24, 2301–11, 2009. https://doi.org/10.1016/j.apgeochem.2009.09.007.
37. Macías, F., Caraballo, M.A., Nieto, J.M., Rötting, T.S., Ayora, C., Natural pretreatment and passive remediation of highly polluted acid mine drainage. *J. Environ. Manag.*, 104, 93–100, 2012. https://doi.org/10.1016/j.jenvman.2012.03.027.
38. Martínez, N.M., Basallote, M.D., Meyer, A., Cánovas, C.R., Macías, F., Schneider, P., Life cycle assessment of a passive remediation system for acid mine drainage: towards more sustainable mining activity. *J. Clean. Prod.*, 211, 1100–1111, 2019. https://doi.org/10.1016/j.jclepro.2018.11.224.
39. Ayora, C., Macías, F., Torres, E., Lozano, A., Carrero, S., Nieto, J.M., Pérez-López, R., Fernández-Martínez, A., Castillo-Michel, H., Recovery of rare earth elements and yttrium from passive-remediation systems of acid mine drainage. *Environ. Sci. Technol.*, 50, 8255–8262, 2016. https://doi.org/10.1021/acs.est.6b02084.
40. Berger, V.I., Singer, D.A., Orris, G.J., Carbonatites of the World: explored deposits of Nb and REY, in: *Database and grade and tonnage models*. USGS Open-File Report 2009–1139, 2009.
41. Younger, P.L., Banwart, S.A., Hedin, R.S., *Mine water. Hydrology, Pollution, Remediation*. Kluwer Academic Publishers, 2002.
42. Caraballo, M.A., Macias, F., Rotting, T.S., Nieto, J.M., Ayora, C., Long term remediation of highly polluted acid mine drainage: a sustainable approach to restore the environmental quality of the Odiel river basin. *Environ. Pollut.*, 159, 12, 3613e3619, 2011.

9

Deriving Value from Acid Mine Drainage

M. van Rooyen* and P.J. van Staden

Hydrometallurgy Division, Mintek, Randburg, South Africa

Abstract

Acid mine drainage (AMD) is a worldwide problem, and the harmful effects of AMD on the environment and infrastructure are unquestionable. While precipitation of dissolved metals and neutralization of acidity from AMD is accomplished relatively easily with lime addition, the reduction of the remaining sulfate content to obtain water suitable for reuse, along with deriving value from the precipitated metals, requires further processing. Reverse osmosis, ettringite precipitation, barium sulfate precipitation, as well as biological sulfate reduction are identified as possible processes to be considered for removal of the residual sulfate that remains after neutralization. Each sulfate removal process offers benefits and limitations, implying that the process needs to be selected bearing in mind both the AMD composition and the ultimate intended use of the treated water. Water is the major product to be derived from AMD treatment and the extent of treatment required would depend on its intended use. A tabulation is provided of further successive treatments that might be required to render the treated water fit for one of a number of possible uses. Depending on the composition of the raw AMD, base, and (semi-)metals, as well as radionuclides, such as uranium, could be recovered from either the AMD or the neutralization sludge. Since the metals become concentrated by a factor of about 100 upon being precipitated from the aqueous AMD to the solid neutralization sludge, the blending of the sludge with the feed to a leaching plant would offer an alternative means of metals recovery from AMD. The manner in which sulfur, magnetite, gypsum, and hydrogen could be recovered from AMD is briefly discussed, but its economics would require more detailed case-specific studies.

Keywords: Acid mine drainage, deriving value, AMD treatment, fit-for-use, sulfate removal, by-products from AMD

9.1 Introduction

Mining affected water, including acid mine drainage (AMD), is a worldwide problem, leading to ecological destruction in water courses if left untreated. The harmful effects of AMD on the environment and infrastructure are undisputed so that no further discussion needs to be given here to justify efforts toward its mitigation and, where possible, even derive value from it (Figures 9.1 and 9.2). AMD from a case study in the Witwatersrand (South

**Corresponding author*: MichellevR@mintek.co.za [ORCID:000-0003-1563-5108]
[ORCID: van Staden, 000-0001-8347-3935]

Elvis Fosso-Kankeu, Christian Wolkersdorfer and Jo Burgess (eds.) Recovery of Byproducts from Acid Mine Drainage Treatment, (235–262) © 2020 Scrivener Publishing LLC

Figure 9.1 AMD at a Coal mine near eMalahleni, Mpumalanga Province, South Africa (published with permission from H. Cornelissen).

Figure 9.2 Discharge of AMD into the Brugspruit near KwaGuqa village, Mpumalanga Province, South Africa and efflorescent salts (published with permission from H. Cornelissen).

Africa) shows relatively low pH and high concentrations of total dissolved solids (TDSs), which result from relatively high element concentrations, especially Fe (Table 9.1). In addition, this AMD contains other elements including Al, Cu, Mg, Ni, U, and Zn.

The consequence of leaving AMD untreated is two-fold, namely, (a) this water would be potentially harmful if ingested or released into rivers and streams, and (b) immense volumes of water would be "lost" which, if treated further, could be of considerable value for water reuse. In the rest of this chapter, the formation of AMD and options for its remediation are briefly discussed and relevant references cited. Greater focus is placed on the value to be derived from AMD and the processing implications for doing so.

Table 9.1 Composition of AMD from a South African case study.

Parameter	Typical concentrations, given in mg/L
pH	2.5–6.5
TDS	2,000–7,000
Sulfate	2,200–5,000
Aluminium	13
Calcium	60
Cobalt	0.80
Copper	400
Iron	110–1,500
Magnesium	44
Manganese	142
Nickel	0.80
Uranium	85
Zinc	32

9.2 AMD Formation

AMD, also referred to as acid rock drainage (ARD), is formed wherever mining into sulfide-bearing orebodies has occurred, because of exposure of the (di-)sulfide minerals to oxygen and water. Depending on the mineralogical composition and the occurrence of water and oxygen ingress, AMD can also form as a natural process [1, 2]. However, mining can accelerate the formation of AMD by exposing (di)sulfide in mine wastes and tailings heaps, as well as mining-related structures such as pits and underground workings [3]. AMD formation can continue for long after mining activities have ceased. Since mineralogy and other factors affecting AMD formation are site-specific, it is difficult to predict (1) the potential for AMD formation and (2) the composition of the AMD if it will form. There are, however, certain mineralogically based tools that have been developed to assist in the prediction of AMD formation and are the subject of another chapter in this book ("AMD generation and prediction").

AMD is defined as mine water with a pH below 5.6. It usually has high electrical conductivity and often bears high concentrations of iron and metals such as aluminium and manganese. It is also common to encounter certain concentrations of radionuclides such as uranium. AMD is formed by the oxidation of di-sulfides, such as pyrite (FeS_2). Its oxidation occurs as a two-step process; the first step (Eq. (9.1)) produces ferrous sulfate and sulfuric acid, the latter causing dissolution of other metals from surrounding surfaces.

$$2\,FeS_{2(s)} + 7\,O_{2(g)} + 2\,H_2O_{(l)} \rightarrow 2\,Fe^{2+}_{(aq)} + 4\,SO^{2-}_{4(aq)} + 4\,H^{+}_{(aq)} \tag{9.1}$$

The dissolution of Fe^{2+}, SO_4^{2-}, and H^+ causes the TDSs and acid concentration of the water to increase, along with a commensurate decrease in pH.

In the second (acid consuming) step, the ferrous iron is oxidized to ferric iron (Eq. (9.2)). Oxidation kinetics is dependent on the oxygen concentration, pH, and bacterial activity in the environment [3].

$$4\,Fe^{2+}_{(aq)} + O_{2(g)} + 4\,H^{+}_{(aq)} \rightarrow 4\,Fe^{3+}_{(aq)} + 2\,H_2O_{(l)} \tag{9.2}$$

The ferric cations produced can oxidize additional pyrite, reducing to ferrous ions and producing more sulfuric acid (Eq. (9.3)) [3, 4]:

$$FeS_{2(s)} + 14\,Fe^{3+}_{(aq)} + 8\,H_2O_{(l)} \rightarrow 15\,Fe^{2+}_{(aq)} + 2\,SO^{2-}_{4(aq)} + 16\,H^{+}_{(aq)} \tag{9.3}$$

The net result of reactions (9.2) and (9.3) is net acid producing, although some of the acid could be partially neutralized by buffering minerals, such as calcite or dolomite, that might co-occur with the di-sulfides. The acidity gets buffered at around 2 where the ferric iron concentration reaches its natural limit of solubility, resulting in precipitation according to (Eq. (9.4)) which can be seen to produce more acidity that can counter the effect of acid-consuming gangue.

$$2\,Fe^{3+}_{(aq)} + 6\,H_2O_{(l)} \rightarrow 2\,Fe(OH)_{3(s)} + 6\,H^{+} \tag{9.4}$$

While AMD is typically characterized as having high acidity and low pH, alkaline mine water can also be found, particularly in carbonate-rich environments. Alkaline mine water is not as environmentally destructive as AMD and often contains iron in ferrous form (Fe^{2+}). However, the pH of the water can be lowered considerably by the combination of iron oxidation (according to reaction (9.2)) and hydrolysis (according to reaction (9.4)), with the net result being acidifying the effluent [1, 2].

9.3 AMD Treatment Options

9.3.1 General Philosophy

The negative effects of AMD can be alleviated by three interventions, namely [1, 2]:

1. *Primary* prevention of the AMD-generation process;
2. *Secondary* control, by preventing the migration of AMD beyond the point of formation; and
3. *Tertiary* control, involving the collection and treatment of AMD.

The conventional means of primary prevention to date has been to raise the pH by alkali addition sufficiently to become unfavorable for microbial activity [1, 2]. Secondary control would logically involve some form of civil construction to contain the AMD.

In this chapter, some details are provided for methods of tertiary control, namely, that of lime neutralization followed by removal of the residual sulfate content of which four options are discussed. Each sulfate removal technology produces a different quality of product water and is subject to its own set of constraints that might favor one over the other for a specific combination of location and AMD composition. An informative book by Alley [5] describes several processes for the removal of various pollutants from wastewater, a summary of which is provided in the Appendix. In summary, the work of Alley has illustrated that there is no single "best" process for the treatment of polluted water. The nature and concentration of the pollutants, the intended use of the treated water combined with the applicable regulations, costs, and available infrastructure determine the treatment technology (or combination of technologies) to be selected.

9.3.2 High-Density Sludge Neutralization of AMD

The chemicals generally applied for the neutralization of AMD include limestone ($CaCO_3$), hydrated lime [$Ca(OH)_2$], soda ash (Na_2CO_3), caustic soda (NaOH), ammonia (NH_3), calcium peroxide (CaO_2), kiln dust, and fly ash [1, 2]. Any one of these neutralizing agents should be used only with full consideration of the additional soluble elements it may contribute to the water. The various chemicals would also affect process parameters differently, such as the rheology of the sludges produced which would, in turn, affect the ease with which they could be dewatered and stacked.

The neutralization of AMD with limestone and lime or one of them is often applied in the mining industry [6–8], using the high-density sludge (HDS) process. This involves the dilution of the neutralizing agent (lime or limestone) with recycled precipitate before it is added to the AMD, resulting in the formation of a higher-density precipitate than when the neutralizing agent is added directly to the AMD [9, 10].

When ferrous iron (Fe^{2+}) concentrations in the influent AMD are below 50 mg/L, the neutralization step is usually performed at a pH of between 7 and 8, whereafter the sludge is settled [1, 2]. When the ferrous iron concentration in the AMD exceeds 50 mg/L, higher pH values for the neutralization step are required, along with aeration to oxidize the ferrous iron to ferric iron (Fe^{3+}), to be precipitated in the hydroxide form [1, 2] (Figure 9.3).

AMD neutralization can be performed using only hydrated lime [$Ca(OH)_2$], or in two stages whereby limestone ($CaCO_3$) is used in the first neutralization step to raise the pH to about 6, followed by the addition of lime $Ca(OH)_2$ to raise the pH higher if required (the selection of the neutralizing agent is subject to confirmation of adequate efficiency of precipitation). Although $CaCO_3$ is generally cheaper than $Ca(OH)_2$, and its application produces a lower volume of sludge, the use of $CaCO_3$ as neutralization reagent for AMD is not widely applied, primarily because its chemical reaction produces carbon dioxide (CO_2), which buffers the pH of the slurry at around 6 [1, 2]. Furthermore, the evolution of CO_2 causes foaming which needs to be controlled. Either way, the aqueous discharge from the HDS process has concentrations of calcium and sulfate consistent with the solubility of gypsum.

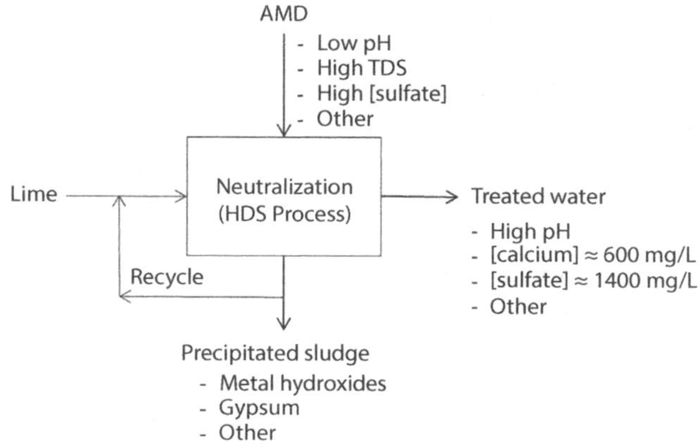

Figure 9.3 Simplified representation of HDS neutralization of AMD.

A study by the International Human Rights Commission [6] confirmed that AMD neutralization does not offer a complete solution and that discharging high concentrations of sulfate into natural water sources can have severe negative effects on health and result in water that is unsuitable for drinking, domestic use, recreation, and irrigation. It is therefore obvious that AMD neutralization is only a partial step toward AMD remediation, and further processing must render the water product suitable for use or release.

Of primary concern is the high sulfate concentration that remains in the treated water after HDS neutralization [6], ranging from 1,500 mg/L up to about 2,500 mg/L corresponding to gypsum-saturation. Such sulfate concentrations can have health effects on water users. Even higher sulfate concentrations are possible with high concentrations of highly soluble species, such as magnesium and sodium. Magnesium, at least, can be removed if the pH of the neutralization step is raised to 12. A study by the World Health Organization [11] indicates that the lowest taste threshold concentration for sulfate is approximately 250 mg/L. According to the Water Research Commission Sulfate Guidelines (summarized in Table 9.2), this is within the recommended range for sulfate concentrations of drinking water [6, 12].

The effects of the various sulfate concentrations shown in Table 9.2 suggest that concentrations below 400 mg/L are safe for most applications, including drinking water, and that only moderate corrosion might be experienced when exposed to pipes and domestic infrastructure. Sulfate concentrations exceeding 400 mg/L might become unacceptable, with no human adaptation to concentrations exceeding 1,000 mg/L. This highlights the critical need for adequate treatment of AMD, particularly the removal of sulfate from AMD and the treated water post AMD neutralization.

9.3.3 Sulfate Removal Options

9.3.3.1 Reverse Osmosis

Reverse osmosis (RO) is the most widely applied desalination technology globally and has been used for the removal of impurities from solutions since the 1950s [13]. Membrane filtration, involving microfiltration (MF) or ultrafiltration (UF), are typically applied in steps

Table 9.2 Water research commission sulfate guidelines [12].

Sulfate range (mg/L)	Drinking Health	Aesthetic	Food preparation	Bathing	Laundry
<100	No effects	No effects	No effects	No effects	No effects
100–200	No effects	No effects	No effects	No effects	Slight corrosion possible
200–400	Insignificant Health effects	Insignificant Health effects	Insignificant Health effects	No effects	Moderate Corrosion
400–600	Slight chance of initial diarrhoea, in sensitive groups, but disappears with adaptation	Slight, bitter taste	Slight chance of initial diarrhoea, in sensitive groups, but disappears with adaptation	Slight chance of diarrhoea if water swallowed, e.g., infants	Increasingly corrosive
600–1,000	Possibility of diarrhoea. Poor adaptation in sensitive individuals	Bitter taste	Possibility of diarrhoea. Poor adaptation in sensitive individuals	Increasing chance of diarrhoea if water is swallowed, e.g., in infants	Very corrosive
>1,000	High chance of diarrhoea. No adaptation	Very bitter and salty taste	High chance of diarrhoea. No adaptation	Possibility of diarrhoea if water is swallowed	Extremely corrosive

Shade indicators:

	Ideal
	Good
	Marginal
	Poor
	Completely unacceptable

preceding RO. These membrane filtration units remove some of the dissolved particulates prior to RO; however, fouling is a considerable problem within membranes [14].

A notable example of AMD treatment by RO is the High Recovery Precipitating Reverse Osmosis (HiPRO®) process [15] that is operational at two mine sites, namely, the (1) eMalahleni Water Reclamation Plant (EWRP) (25,000 m^3/day product water), and (2) Optimum Coal Mine Water Reclamation Plant (OWRP) (15,000 m^3/day product water), both in Mpumalanga, South Africa. The HiPRO process produces drinking water seeping from coal dumps. The HiPRO process involves several stages of membrane systems, incorporating UF and RO. From each of the membrane systems, a concentrated brine stream is produced, and salts are removed from these streams by means of precipitation with lime [15].

The OWRP flowsheet (Figure 9.4) differs slightly from that of the EWRP, in that the pH of the mine water at the Optimum Coal mine is circumneutral and contains iron and manganese, which is not the case for the mine water at the eMalahleni plant [15, 16]. Consequently, the OWRP incorporates the addition of ozone in an initial stage to decrease the iron and manganese to suitable concentrations prior to UF and RO, whereas the EWRP incorporates lime neutralization instead of ozonation as the first step for the removal of iron and manganese [15, 16].

For the OWRP, the clarifier feed water emanating from ozonation is dosed with an anionic polymer flocculent to promote solids agglomeration. Anti-scalant is added to minimize precipitation behind the membranes from the brines which become super-saturated in calcium-sulfate and other metal ions. The clarified water from Stage 1 is passed through a sand filter prior to further purification through the UF [15, 17].

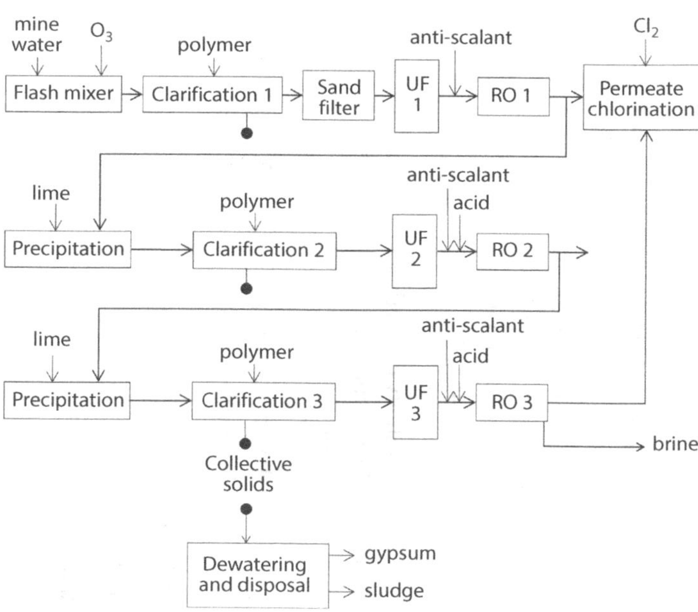

Figure 9.4 Simplified block flow diagram of the Optimum Coal Mine Water Reclamation Plant, adapted from Karakatsanis *et al.* [15].

Thereafter, the super-saturated brines from the preceding stages are fed to Stages 2 and 3. Precipitation of salts from the super-saturated brines is initiated by the addition of lime in the agitated precipitation reactors and hydro-cyclones are used to classify the slurry. Subsequently, the overflow from the hydro-cyclone, containing fine particles, enters the Stage 2 or 3 clarifier. The hydro-cyclone underflow contains the larger gypsum particles, and these are dewatered by vacuum belt filtration [15, 17]. UF is the final solids-removal process before each subsequent RO step. Sulfuric acid is also added after UF for pH correction ahead of RO, which is the final treatment step [15]. The membrane-based separation would also remove monovalent ions such as Na^+ and Cl^-.

Membranes are prone to fouling, especially if the pre-treatment steps leading up to the membrane units were performed inadequately, and this has the potential of increasing the maintenance and operating cost of this technology [18]. Water recovery in excess of 95% can reportedly be achieved with this process, and three waste streams are produced, namely, mixed sludge, gypsum, and brine [15]. Comment on the possible further treatment of these wastes is given in Section 9.4.2.1.

9.3.3.2 *Ettringite Precipitation*

Ettringite is a Ca-Al-SO_4 mineral with a very low solubility constant, which can be formed by chemical reaction between aqueous $CaSO_4$ (such as would occur in the solution phase in equilibrium with precipitated gypsum) and aqueous Al (which can be represented by $Al(OH)_3$). Hence, by adding a water-soluble aluminium salt to sulfate-saturated water, the sulfate concentration can be reduced to below permissible discharge concentration. A process utilizing ettringite precipitation for sulfate removal is described by Nevatalo *et al.* [19]. A more sophisticated ettringite-based process is SAVMIN®, with the distinguishing feature that the aluminium is recovered from the ettringite precipitate for reuse [4, 20] (Figure 9.5).

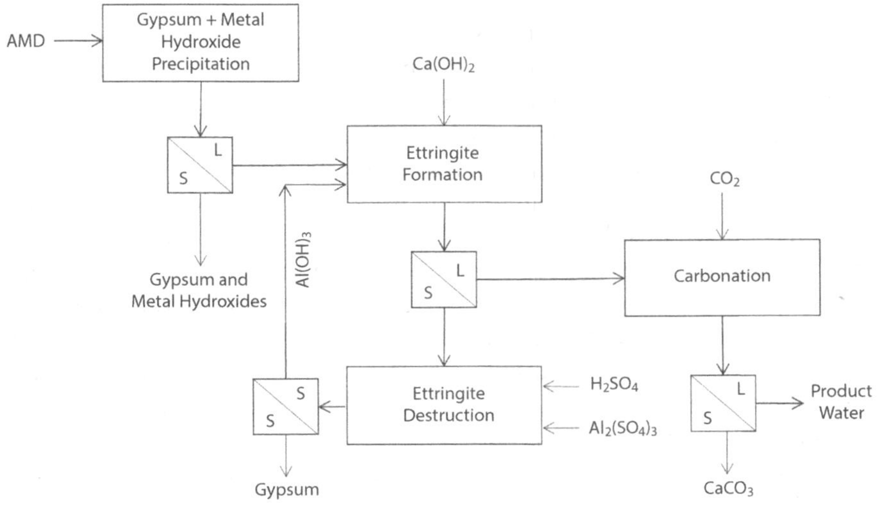

Figure 9.5 Block flow diagram showing the concept of the SAVMIN process [4].

In the "Gypsum and Metal Hydroxide Precipitation" step, the AMD is first treated with lime to immobilize all sulfate as gypsum and precipitate divalent (Me^{2+} in Eq. (9.5)) and trivalent (Me^{3+} in Eq. (9.6)) metal hydroxides.

$$MeSO_4 + Ca(OH)_2 + 2\ H_2O \rightarrow CaSO_4{\cdot}2\ H_2O + Me(OH)_2 \tag{9.5}$$

$$Me_2(SO_4)_3 + 3\ Ca(OH)_2 + 6\ H_2O \rightarrow 3(CaSO_4{\cdot}2\ H_2O) + 2\ Me(OH)_3 \tag{9.6}$$

The aqueous overflow contains $CaSO_4$ at a concentration commensurate with the solubility of gypsum. In the "Ettringite Precipitation" step, $Al(OH)_3$ is introduced (as a recycle from the "Ettringite Destruction" stage), which incorporates both the calcium and sulfate in solution into solid-phase ettringite (Eq. (9.7)).

$$2\ Al(OH)_3 + 3\ CaSO_4 + 3\ Ca(OH)_2 + 26\ H_2O \leftrightarrow 3\ CaO{\cdot}3\ CaSO_4{\cdot}Al_2O_3{\cdot}32\ H_2O \tag{9.7}$$

The discharge from the ettringite precipitation stage contains only residual $Ca(OH)_2$ alkalinity which is removed by carbonation with CO_2 to precipitate $CaCO_3$ (Eq. (9.8)) that can be filtered out (and reused as neutralizing agent) to yield the treated water product.

$$Ca(OH)_{2(aq)} + CO_2 \rightarrow CaCO_{3(s)} + H_2O \tag{9.8}$$

The aluminium is recovered from the ettringite by acidification which decomposes the ettringite to $Al(OH)_3$ and gypsum, according to Equation (9.9), which is effectively the reverse of the ettringite formation reaction, Equation (9.7). If the $CaSO_4$ on the right-hand side of Equation (9.9) were to become saturated in solution it would start precipitating as gypsum.

$$3\ CaO{\cdot}3\ CaSO_4{\cdot}Al_2O_3{\cdot}32\ H_2O + 3\ H_2SO_4 \leftrightarrow 2\ Al(OH)_3 + 6\ CaSO_4 + 32\ H_2O \tag{9.9}$$

The key to the implementation of the SAVMIN process has been to achieve the separation of the $Al(OH)_3$ and gypsum. There is not much relative density difference to rely on, being around 2.5 g/cm^3 for $Al(OH)_3$ and about 2.3 g/cm^3 for gypsum. Therefore, the separation needs to rely on differences in particle shape and size. A multitude of conventional and relatively novel separation apparatus have been tested with various degrees of success, but ultimately hydro-cyclones of suitable design were found to yield an efficient and reliable separation. Interestingly, it is the slightly higher-density $Al(OH)_3$ that reports to the overflow to be recycled, due to the $Al(OH)_3$ possessing a finer particle size than the gypsum that reports to the underflow to be discarded. Any gypsum that gets entrained in the $Al(OH)_3$ recycle detracts from the effective aluminium recycle efficiency since the gypsum adds to the sulfate-load of the "Ettringite Precipitation" step. During a recent piloting campaign, it has been found possible to restrict the gypsum entrainment in the recycle to <10% of what enters the solid-solid separation step.

Yet, any aluminium exiting with the gypsum discard represents a loss that needs to be replenished, which has been found to be achieved most effectively by the addition of $Al_2(SO_4)_3$ to the "Ettringite Destruction" stage since it yields $Al(OH)_3$ that is chemically reactive and does not introduce monovalent elements such as Na^+ or Cl^- into solution. Indications to date have been that aluminium losses can be restricted to <20% of what enters the solid-solid separation step. A logistical matter to be managed in the SAVMIN process is that all solutions throughout the process are of a scaling nature and the formation of gypsum or $CaCO_3$ on surfaces is inevitable. This requires sufficient redundancy and duplication of equipment and piping to permit descaling without interruption of the process.

Note further that ettringite precipitation would not remove highly soluble monovalent species such as Na^+ or Cl^-. The AMD encountered to date exhibited Na^+ concentrations of 100–200 mg/L which, according to Holmes [21], poses no health effects, although it imparts a faintly salty taste as its concentration approximates 200 mg/L. While Cl^- accelerates metal corrosion, the AMD exhibited moderate Cl^- concentrations of around 100 mg/L.

9.3.3.3 *Barium Carbonate Addition*

Sulfate removal from AMD can occur by means of a barium-carbonate ($BaCO_3$) process, suggested by Hlabela *et al.* [22] (Figure 9.6). Influent mine water is treated with lime ($Ca(OH)_2$), in the "Neutralization" step, to precipitate metals as hydroxides and gypsum. Hlabela *et al.* [22] specifically mention that the removal of magnesium is also targeted in the "Neutralization" step, which suggests that this step is operated at a high pH, supposedly around 12 [23] to yield magnesium-hydroxide [$Mg(OH)_2$]. The removal of the metal hydroxide and gypsum sludge is performed subsequently in a solid-liquid separation step, where after $BaCO_3$ is added to the solution, effecting the removal of sulfate from the water as the highly insoluble $BaSO_4$. $BaCO_3$ is a very expensive reagent, and hence, it needs to be regenerated from the $BaSO_4$, toward which the major part of the process is dedicated.

Barium-sulfate ($BaSO_4$) recovered from the precipitation step is mixed with carbon and reacted in a furnace at 1,050 °C to form barium-sulfide (BaS) and carbon dioxide (CO_2), Eq. (9.10). At the same time, accompanying calcium-carbonate ($CaCO_3$, formed in the

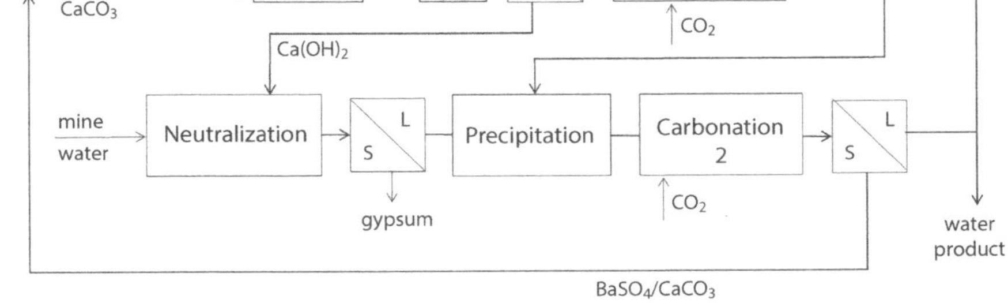

Figure 9.6 Simplified block flow diagram of the integrated barium process based on [24].

Carbonation-2 step) is calcined to produce calcium-oxide (CaO) and more CO_2, Eq. (9.11), [22, 24].

$$BaSO_4 + 2\,C \rightarrow BaS + 2\,CO_2 \tag{9.10}$$

$$CaCO_3 \rightarrow CaO + CO_2 \tag{9.11}$$

When water is added to this BaS/CaO product from the furnace, the BaS dissolves and the $Ca(OH)_2$ can be separated from it to be used in the "Neutralization" step for metals, gypsum and magnesium precipitation. Carbonation of the BaS solution with CO_2 (in the "Carbonation 1" step) strips the sulfur out of solution as H_2S to convert the BaS to $BaCO_3$ (Eq. (9.12)).

$$BaS + H_2O + CO_2 \rightarrow BaCO_3 + H_2S \tag{9.12}$$

From the H_2S/CO_2 gas mixture emanating from the "Carbonation 1" step, elemental sulfur is produced according to the "Pipco" process. This process is reported to involve the burning of a portion of the H_2S in air to form SO_2, which is reacted with the balance of the H_2S to yield elemental sulfur (S^0) and water [25].

Reliance is placed on the extremely low solubility product of $BaSO_4$ to maintain low concentrations of Ba^{2+} and SO_4^{2-} in solution. The aqueous product of the "Precipitation" step is carbonized with CO_2 (in the "Carbonation 2" step) to convert the residual $Ca(OH)_2$ in it to $CaCO_3$. The mixture of $BaSO_4$ and $CaCO_3$ is separated from the suspension emanating from the "Carbonation 2" step and recycled to the furnace. A small portion of the final water product is recycled for suspending the BaS/CaO solid phase product of the furnace.

Our own calculations suggest that a stoichiometric amount of 0.54 kg carbon (C) is required per m^3 of typical AMD (with composition as per Table 9.1) treated for conducting the chemical reactions in the furnace. These reactions withdraw oxygen from the $BaSO_4$ and $CaCO_3$ to form CO_2. However, another 1.67 kg/m^3 of carbon is required to be reacted with oxygen (introduced in the form of air) to achieve the reaction temperature of 1,050 °C. That includes the assumption that the off-gas from the furnace is used to pre-heat the air supply to about 900 °C to economize the heating requirement. In the paper by Hlabela *et al.* [22], the CO_2 emanating from the furnace is shown as being used in the two carbonation steps. However, achieving that in practice seems complex, since the CO_2 generated in the furnace will have been diluted with nitrogen, it will be contaminated with dust and water vapor and will be evolved at about ambient pressure. It would practically be more feasible to supply the two carbonation steps with pressurized CO_2 from a tank, but of course, it would add to the cost and carbon footprint of the process.

Barium carbonate is highly toxic and tragic consequences would result if any unreacted barium carbonate were to pass into the final water product by overdosing the "Precipitation" step with $BaCO_3$. As with the ettringite precipitation processes, precipitation with Ba would not remove monovalent species such as Na^+ or Cl^-.

9.3.3.4 *Biological Sulfate Reduction*

Biological treatment, using sulfate reducing bacteria, can be used to purify sulfate-bearing effluent streams. Biological sulfate reduction can be conducted in either passive (such as a constructed wetland) or active systems (such as the THIOPAQ® [26] or BioSURE® [27] processes). Passive systems maintain a large working inventory and operate slowly but require low maintenance, low operational intervention, and relies on the remedial actions of natural plant and algal growth and the formation of reducing conditions below the mud-level [28]. In contrast, the active treatment proceeds rapidly, holding a smaller inventory but requiring frequent human intervention for maintenance and monitoring of the constructed reactors, and it further requires external sources of chemicals, growth-substrate, energy (electrical power), and labor [28], usually along with a higher capital cost for infrastructure development. The active biological method offers more consistent performance while the possibility for the recovery of metals and sulfur as by-products can offset the operating costs.

A reductant and organic carbon source is needed for microbially mediated sulfate reduction, which is represented by $CH_2O_{(aq)}$. The reaction proceeds as shown in Eq. (9.13), [29].

$$2\,CH_2O_{(aq)} + SO_4^{2-} + 2\,H^+ \rightarrow H_2S + 2\,H_2CO_3 \tag{9.13}$$

Any Cd, Cu, Fe, Pb, Hg, Ni, and Zn present in the AMD will precipitate as metal (Me) sulfides according to Eq. (9.14).

$$H_2S + Me^{2+} \rightarrow MeS + 2\,H^+ \tag{9.14}$$

Manganese, Fe, Ni, Cu, Zn, Cd, Hg, and Pb can be co-precipitated partially with other metal sulfides. Sulfate reducing bacterial species can reduce metals to a more insoluble form, for instance, the reduction of U(VI) to U(IV). Sulfate reduction also assists in the consumption of acidity and thus raises the pH of the solution, which facilitates both sulfide and hydroxide precipitation reactions [17, 29]. Residual H_2S can be eliminated by its oxidation to bio-sulfur. The disposal of or value-creation from metal sulfides would require further research, probably on an individual case-by-case basis. This process has been demonstrated over an 18-month period at a mine site in Mpumalanga in South Africa, and a larger-scale demonstration plant is currently being designed.

9.4 Deriving Value from AMD

9.4.1 Fit-for-Use Water

9.4.1.1 The Cascade Model

Mining operations have the potential to cause severe pollution to natural water resources; hence, to promote a sustainable alternative, governments have initiated strict environmental regulations and effluent discharge limitations to address the harmful effects of AMD [30]. A cascade model is proposed by Van Zyl *et al.* [31], where the quality of the water source is linked to an appropriate subsequent use, and the available water is used for the most valuable applications as the purity of the water improves after each treatment or reuse [32]. For example, the cascaded treatment of AMD, with each subsequent treatment process yielding lower sulfate concentrations, would yield firstly water for irrigation (at 2,000 mg/L sulfate), coal processing (at 1,000 mg/L sulfate), general

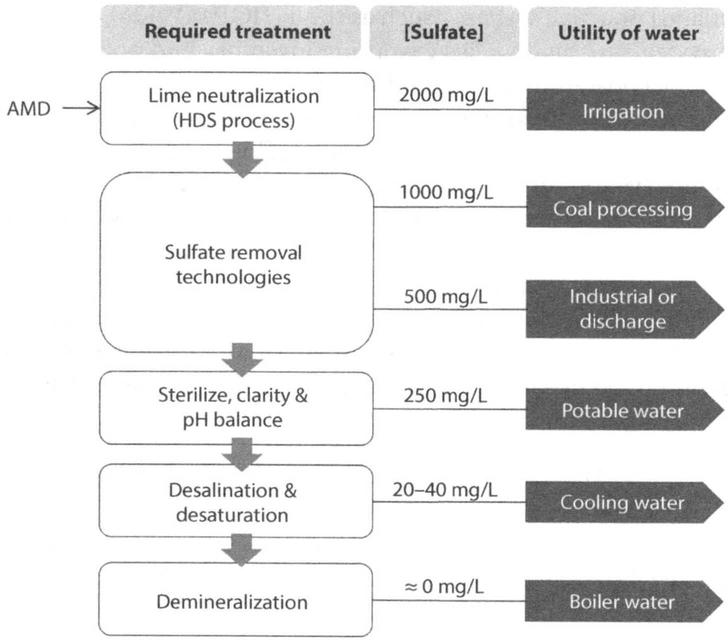

Figure 9.7 Utility of treated water cascade model, adapted from Van Zyl *et al.* [31].

industrial use and/or discharge to public streams (at 500 mg/L sulfate), potable water (at 250 mg/L sulfate), cooling water (at 20–40 mg/L sulfate), and finally, demineralized water suitable to be used in boilers (Figure 9.7).

To prevent the unnecessary use of potable water where a lower quality water could be suitable, the treatment of water for reuse should focus primarily on the selection of appropriate technologies for the production of water of the targeted application. The suitability of water for an application can be determined from the relevant water quality guidelines, which may be particular to a specific country or district [32]. A few possible uses for treated water are considered more closely in the sub-sections that follow.

9.4.1.2 Water Suitable for Irrigation

Resulting from the cascade model, water for irrigation requires the least stringent treatment of AMD to be of use. Partially treated mine water could be suitable for use as irrigation water when the majority of dissolved metals have been removed, along with a reduction of the sulfate concentration of the water to less than 2,000 mg/L [31].

In the case of Resolution Copper, a joint venture between Rio Tinto and BHP, an underground mine, was developed in Arizona, United States of America, for the recovery of copper [30]. This underground mine, called a "wet mine", was 1.5-km deep, and about $9 \cdot 10^6$ m^3 of water that had collected in the mine void had to be removed [30]. The water was pumped to the surface, treated and used for the irrigation of local produce, for which groundwater and other valuable water sources had been used before [30].

Another noteworthy example is Xstrata Coal's Ulan mine in New South Wales, Australia. This mine was a "water surplus mine" site, with more water being pumped from underground than what was consumed by the mine operations [30]. The mine's water license stipulated zero discharge of mine water, and consequently, an irrigation scheme was implemented whereby the water was pumped to several points to irrigate more than 240 ha of beef and cattle pastures [30].

9.4.1.3 *Water Suitable for Industrial Use*

Depending on the location of the source of AMD and its proximity to an industrial operation, the AMD might either in treated or untreated form be suitable for industrial use as wash water or process make-up water. The contained acid and ferric iron might even be useful reagents for dissolution reactions. The applicability of such instances would be very case specific. A case worth mentioning is the Brazilian mining company Vale, which increased its overall water reuse and recycling to 77% in the year 2012, as a direct consequence of the company's water system automation initiatives [30].

9.4.1.4 *Water Suitable for Environmental Discharge*

The standard to which AMD would need to be treated before it could be discharged to receiving water bodies would be governed by locally prevailing legislation and guidelines. Its legal release would swell the water supply for the purposes of any or all of irrigation, domestic use, and recreation.

9.4.1.5 *Water Suitable for Sanitation*

Potable water is still widely used for sanitation, such as flushing toilets, where water of potable standard is not necessarily required. Dual reticulation systems, applied in countries such as Japan and Australia, make it possible to use lower quality water for sanitation, while potable water is reserved for drinking, cooking and other household use that requires water of potable quality [32].

9.4.1.6 *Potable Water*

Of all the sulfate removal technologies discussed above, RO is the only technology capable of producing drinking water from AMD, especially when the AMD contains monovalent species such as Na^+ and Cl^-, which the precipitation-based or biological-based technologies do not remove from solution.

The production of potable water from AMD by RO is also the technologically most advanced, having been in commercial application for several years at two sites in South Africa. Over the years, the cost of municipal potable water has increased considerably, resulting in a RO becoming a cost-effective solution for the production of potable water from low- to medium-quality mine water sources.

9.4.1.7 *Cooling Water*

Air-conditioning systems rely mostly on wet or evaporative cooling towers [33]. Manufacturing and processing plants adopt similar cooling towers to meet the cooling demand for their processes [33]. However, suitably high-quality water must ensure proper operation of the cooling towers. Of particular importance to cooling towers is the inhibition of corrosion, scale formation, and microbial fouling [33]. The quality of the available water source would determine the treatment required to render it suitable as cooling water.

9.4.1.8 Boiler Water

Boiler feed water should essentially be free of dissolved solids, suspended solids and organic material to prevent scaling, corrosion, and fouling of the boiler system. A typical boiler water treatment system would include units such as filtration and UF, and this is typically followed by an ion exchange circuit or membrane technologies such as nano-filtration or RO [34]. These treatment steps are usually followed by deaeration or degasification and, finally, coagulation [34]. A listing of contaminants, their potential effects on the boiler and applicable treatment steps for their removal are provided in Table 9.3.

Table 9.3 Boiler feed water contaminants and their removal, adapted from [34].

Problematic material	Harmful effects on boiler system	Treatment
Iron, soluble or insoluble	Can deposit on boiler parts and tubes, damaging equipment.	Chemical precipitation, filtration, ultrafiltration, or ion exchange.
Copper	Can cause deposits to settle in high-pressure turbines, decreasing efficiency, and requiring costly cleaning or repairs.	Chemical precipitation, filtration, ultrafiltration, or ion exchange.
Silica, requires removal to very low concentrations	Can cause very hard scaling.	Chemical precipitation, filtration, or ultrafiltration.
Calcium	Scaling	Ion exchange
Magnesium	If combined with phosphate, can stick to boiler interior and coat internals.	Chemical precipitation, filtration, ultrafiltration, or ion exchange.
Aluminium	Can deposit as scale on boiler interior. Can also react with silica to increase likelihood of scaling.	Chemical precipitation, filtration, ultrafiltration, or ion exchange.
Hardness	Can cause deposits and scale on boiler internals.	Ultrafiltration or ion exchange.
Dissolved gasses	Chemical reactions with dissolved oxygen and carbon dioxide, causing corrosion.	Deaeration or degasification

9.4.2 By-Products from AMD Treatment Processes

9.4.2.1 Overview

Comparing the initial AMD (Table 9.1) with the calculated precipitate from lime neutralization (Table 9.4) shows a concentration by a factor of about 100. Calcium (as gypsum) and iron (as ferric-oxy-hydroxides) contribute the largest proportions of the composition and would therefore be obvious candidates for recovery as saleable products. The U concentration is with 10,800 mg/kg U_3O_8 very high compared to typical U ores. Furthermore, the concentration of copper is comparable to that of relatively high-grade ores with the concentrations of cobalt, nickel, and zinc corresponding to relatively low-grade ores of those metals.

With most soluble species (apart from calcium and sulfate) having been removed from AMD during neutralization, sulfate-removal processes that succeed a neutralization step can be expected to produce relatively clean gypsum or elemental sulfur.

9.4.2.2 Gypsum Containing Products

Hydrometallurgical processes generally involve the production of gypsum, which can often constitute substantial volumes of process streams that require disposal. Treatment of these streams, for the generation of value and the minimization of the volumes requiring actual disposal, is therefore of considerable interest.

Gypsum is used in a multitude of industrial and agricultural applications. Gypsum has fire-resisting, heat insulating, and sound absorption qualities [35] and is therefore used extensively in applications related to construction, such as low-cost building materials, ceilings, drywalls, and plasterboards. Furthermore, gypsum is easily converted into cementitious material, is quick setting, and is used widely as primary constituent of cement and bricks [35]. An interesting case is discussed by Sampson [35], whereby gypsum-containing

Table 9.4 Calculated composition of neutralization precipitate.

Element	Concentration, % (m/m)
Aluminium	0.14
Calcium	14
Cobalt	0.080
Copper	4.3
Iron	13
Magnesium	0.48
Manganese	1.5
Nickel	0.080
Uranium	0.92 (10,800 mg/kg U_3O_8)
Zinc	0.35

waste from an ilmenite facility was treated successfully to generate material suitable to be used as cement for construction. Of particular interest is the fact that the material contained U and Th albeit at sufficiently low concentrations that it did not pose health or environmental threats [35]. However, it highlights (a) the importance of adequate removal of impurities from the material for adherence to relevant regulations and standards, and (b) some of the challenges associated with the treatment of waste from AMD remediation to generate value.

Drywalls and plasterboards, with dimensions around 1,200 × 2,500 mm, and a thickness of about 6 mm, can be bought from local hardware stores at between R 150 and R 200 which, at the time of writing, equates to less than US$15 per board or about US$0.4 per kg. Therefore, the all-inclusive cost of erecting and operating a processing plant for rendering contaminated gypsum suitable for construction materials, plus the cost of marketing and distribution would need to be less than that.

9.4.2.3 High-Value Iron-Bearing Products

Iron is typically a major constituent of AMD (Table 9.1) and is precipitated during the neutralization stage where limestone or lime is added to the AMD. It is therefore an obvious element to consider for recovery from the AMD treatment process. Iron-bearing precipitate could not compete with the economics of scale of iron ore for bulk applications such as steel smelting. However, it has been investigated as a possible means for production of somewhat higher-value iron-bearing chemicals such as pigment and heavy media.

Magnetite, Fe_3O_4 or more correctly the mixed oxide $Fe^{II}Fe_2^{III}O_4$ [36], along with titanomagnetite is responsible for the magnetic properties of rocks [36]. It also finds use in heavy medium separation, as a pigment, polishing agent, as a chemical filler, in cosmetics and other applications. The development of a process for magnetite production from AMD collected at a coal mine is briefly summarized here, based on [37]. The AMD had predominantly iron, aluminium and magnesium in sulfate medium.

The AMD, which in this case contained about 3.5 g/L Fe, is neutralized with lime, similar to the process described in Section 9.3.2, followed by the rejection of excess gypsum by physical separation (Figure 9.8). The resultant slurry, containing jarosite in sulfate-saturated solution, is then subjected to solid-liquid separation following which the solution can undergo sulfate removal *via* any of the processes described in Section 9.3.3.

The jarosite is converted to magnetite by the addition of magnesium oxide (MgO) and ferrous iron at 80 °C. The slurry is subjected to a solid-liquid separation step and the magnesium-sulfate bearing solution can be neutralized and discharged to an effluent pond. Finally, the magnetite is recovered from the solids by magnetic separation.

Another process for the production of magnetite from AMD is proposed by Akinwekomi, Maree, Zvinowanda, and Masindi [38], which is based upon the principle of selective precipitation and uses sodium carbonate as primary reagent. Considering that mineable quantities of magnetite occur naturally, any process that produces magnetite from AMD has the potential to be seen as relatively complex and may likely not be competitive. Such processes might be complicated further by the presence of other AMD contaminants such as U, which would require additional purification steps prior to the magnetite formation step. Both processes mentioned have been proven to be technically sound, and demonstration of these processes at larger scale would provide valuable operational insight regarding industrialization viability.

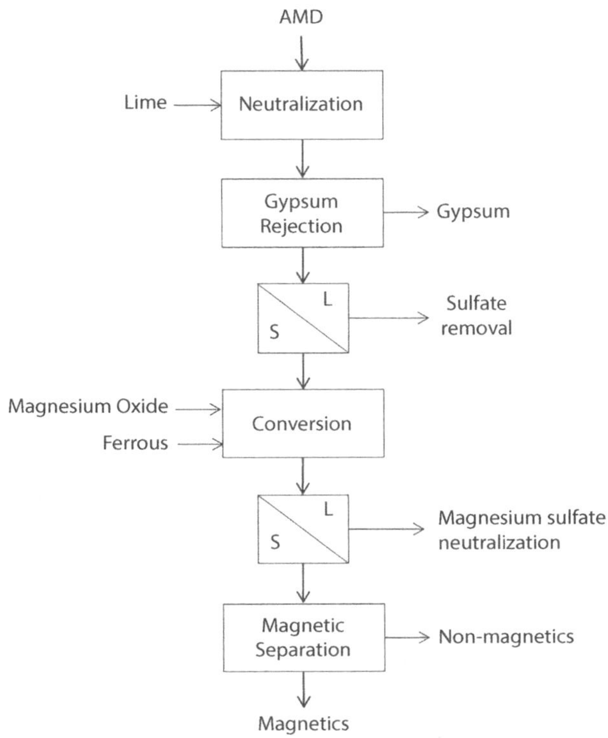

Figure 9.8 Block flow diagram for production of magnetite from AMD after [37].

9.4.2.4 *Uranium and Base Metals*

Along with base metals, such as Cu, Co, Zn, and Ma, AMD can also contain U, such as the AMD emanating from South African gold deposits [39] and other deposits globally [40, 41]. The concentration of U in AMD can vary considerably, from a few μg/L [42], up to several mg/L [40]. The World Health Organization (WHO) issued a provisional guideline value for U not to exceed a concentration of 30 μg/L in drinking water [42]. Consequently, whenever U is present in the AMD, and the water is treated for reuse, close attention should be paid on ensuring the removal of U to permissible concentrations.

For the recovery of U from AMD, several approaches are possible. The first involves the removal of the U before the neutralization step (Figure 9.9). This can be done with ion exchange [2, 43], while the remaining constituents pass to limestone/lime neutralization where the majority of the base metals and gypsum are precipitated. An implication of this approach is that the ion exchange equipment treats the entire AMD stream, which is often multiple tens of millions of litres per day, resulting in the requirement for large equipment. Furthermore, the influent AMD typically contains suspended solids as well as high concentrations of dissolved solids, which would result in frequent fouling of the resin inventory, increasing the need for resin regeneration and the associated operating cost and operational complexity.

Base metals can then be extracted by re-leaching of the precipitated neutralization sludge. Secondly, it could be considered recovering the U from the precipitated sludge produced by

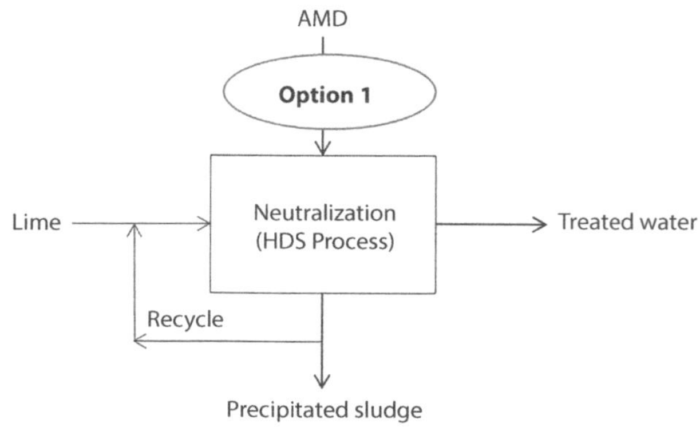

Figure 9.9 Recovery of U from AMD prior to neutralization.

the neutralization of AMD with limestone/lime (Figure 9.10). This might be achieved by a resin-in-pulp process which might simultaneously recover accompanying base metals. The concentration of U in either of the precipitated sludges might still be too high for environmental disposal.

Extracting U and other elements from neutralization residue would probably not offer the economics of scale that ore bodies do. However, the blending of such residue with the feed to a leaching plant for these elements might prove a concept worth considering. The choice of option 1 or 2 might therefore be influenced by whether the limiting design criterion is safe sludge disposal or optimal economics of the downstream processes for metals recovery from the sludge.

A third, seemingly attractive option for the removal of U from mine water, was developed by Earth, and discussed by Howard, Grobler, Robinson, and Cole [44]. This process uses a proprietary ion exchange technique to remove U from mine water to permissible

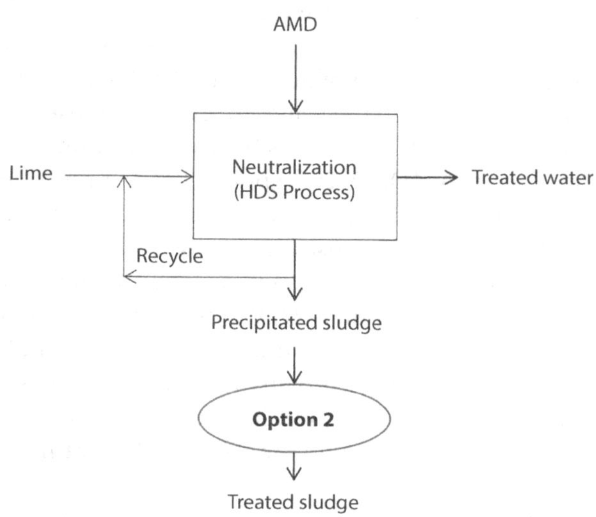

Figure 9.10 Recovery of U from the precipitated sludge post lime neutralization.

concentrations, while generating saleable products and minimal waste. Biological sulfate reduction would precipitate base and (semi-)metals as their respective sulfides, along with the production of elemental sulfur, which would be saleable products.

9.4.2.5 *Hydrogen*

The pursuit by a South African company to produce hydrogen from AMD toward the emerging hydrogen economy was reported in 2019 [45]. Very little technical detail has been provided regarding the hydrolysis technology to be used and of the benefits or disadvantages that the properties of AMD might hold for it. Water suitable for the production of hydrogen from AMD will likely follow the cascade model (Figure 9.7) up to the production of demineralized water fit for boiler water. However, electrolysis to produce hydrogen would require good conductivity of the water which, in turn, would require the addition of high-purity caustic or sulfuric acid. For the purposes of this chapter, this interesting application will just be noted.

9.5 Synopsis

9.5.1 AMD Remediation

The potentially negative effects of AMD on the environment and infrastructure are undisputed. While these effects of AMD can be alleviated either by prevention of its formation, by prevention of its spread or by its collection and treatment once formed, the focus in this chapter has been on the latter. Furthermore, the extent to which value could be derived from its treatment has been explored.

Lime neutralization can neutralize the acidity and precipitate most commonly occurring metals from solution. However, it leaves residual calcium sulfate in solution commensurate with the solubility of gypsum in water, making it unsuitable for most uses and causing it to exceed the guidelines for release to the environment. This frames the last step in AMD treatment as that of reduction of its sulfate content to fall within the limits for the intended use of the treated water. For the purpose of illustration and model calculations where necessary, data derived from a case study on the treatment of AMD from the Witwatersrand in South Africa has been used. RO is a physical treatment means which produces very clean product water, but the example cited in this chapter also leaves three waste streams, namely, mixed sludge, gypsum, and brine.

Sulfate removal can also be achieved by precipitation, and the two processes mentioned in this chapter rely on the very low solubilities of respectively ettringite and barium sulfate, with the sulfate ultimately being rejected as gypsum. The economics of both precipitation processes benefit from the regeneration and recycle of the key chemicals, namely, aluminium hydroxide for ettringite formation and barium carbonate for barium sulfate formation. The recycle of those reagents contribute substantially to the complexity of the flowsheets. The ettringite precipitation process needs to contend with gypsum scaling throughout most of the unit operations. It further relies on the separation of two solids (gypsum and aluminium hydroxide) from suspension to achieve the aluminium hydroxide recycle, which is arguably its greatest complication. With the barium sulfate process, great care is required to prevent over-dosing with the highly toxic barium carbonate so that it does not report to the product water. The requirement for a pyrometallurgical step for barium carbonate regeneration seems to be a particularly complicating factor, and our own calculations suggest it may possess a relatively large carbon footprint due to carbon dioxide emissions.

Biological sulfate reduction has been demonstrated to alleviating sulfate content by reducing it to sulfide. That can precipitate base and (semi-)metals as their respective sulfides and the reducing conditions can also alter higher-valence metals to their lower-valence form of lower solubility. Either passive or active systems can be considered, with the latter offering greater control and consistency of performance. An important requirement for this process is a carbon source for microbial growth (which might also act as reductant) which is of reasonable cost yet relatively easy to metabolize.

9.5.2 Deriving Value From AMD

Water that has been made suitable for use has been identified as the most important potential product to be derived from AMD treatment. With increasing intensities of treatment, AMD could be made suitable for irrigation (at 2,000 mg/L sulfate), coal processing (at 1,000 mg/L sulfate), general industrial use or discharge (at 500 mg/L sulfate), potable water (at 250 mg/L sulfate), cooling water (at 20–40 mg/L sulfate), and demineralized water as boiler feed.

RO is the only technology capable of producing potable water from highly mineralized AMD, particularly when the AMD contains monovalent species such as Na^+ and Cl^-, which the precipitation-based or biological-based technologies do not remove from solution. The treatment steps required to render the AMD suitable for an intended use would therefore depend on the specific contaminants occurring in the AMD. A guide obtained from literature has been reproduced in the Appendix for ease of reference. RO and the ettringite precipitation process for sulfate removal yield gypsum which could be considered as construction material. If the gypsum should be contaminated with any of the elements present in the AMD, additional decontamination plant and process would be required, the extent of which would be case-specific.

Two processes for the production of magnetite from AMD have been mentioned. While both have been proven to be technically feasible, demonstration at larger scale would be advantageous and provide valuable operational validity and insight. U and base metals can be recovered around the lime neutralization step, either from solution (prior to neutralization) or by re-mobilization from the gypsum sludge (after neutralization). In both cases, it is envisaged that ion exchange resin would be used for the selective concentration of the valuable metals for recovery.

According to the case study presented here, the metals initially dissolved in the AMD are concentrated by a factor of around 100 when precipitated in the gypsum sludge. Therefore, should a leaching plant be located in the vicinity of the AMD treatment plant, blending of the neutralization sludge with the feed to the leaching plant could be considered as an alternative means of recovering the value-metals. Biological sulfate reduction or barium sulfate precipitation yield elemental sulfur as a potentially saleable product; however, the disposal of, or value-creation from metal sulfides would require further research.

Appendix: Recommended wastewater treatment for various pollutants [5].

Pollutant	Treatment technology	Achievable concentration (mg/L)
Ammonia	Aerobic biological	0.05–0.5
	Air stripping	0.3–0.5
	Steam stripping	0.1–0.3
Arsenic	Carbon adsorption	0.6
	Charcoal precipitation	0.06
	Chemical oxidation	0.2
	Ferric chloride precipitation	0.05–0.13
	Ferric sulfate precipitation	0.003–5
	Lime/Alum precipitation	0.003–0.2
	Lime/ferric hydroxide precipitation	0.005–0.2
	Lime softening	0.03
	Sulfide precipitation with filtration	0.05
Biodegradable organics	Aerobic fixed growth	0.03
	Aerobic suspended growth	0.03
	Anaerobic fixed growth	1.0
	Anaerobic suspended growth	1.0
	Activated carbon	0.03
	Chemical oxidation	0.03
	Land treatment	1.0
	Membrane filtration	0.001
Boron	Ion exchange	0.001
	Reverse osmosis	0.14
Cadmium	Hydrogen precipitation	0.003 at pH 11.0
	Lime/hydroxide precipitation	0.04 at pH 10.0
	Sulfide precipitation	0.0001 at pH 11.0
Carbon monoxide	Air stripping	–
Chloride	Electro-dialysis	–
	Reverse osmosis	15
Chromium^{6+}	Activated carbon adsorption	0.02
	Bisulfite	0.05
	Ion exchange	0.02
	Meta-bisulfite	0.001
	Sulfur dioxide	0.01
Chromium^{3+}	Hydroxide precipitation	0.02 at pH 8.5
Copper	Hydroxide precipitation	0.02 at pH 9.0
	Sulfide precipitation	0.01 at pH 9.0
	Ion exchange	0.03

(Continued)

Appendix: Recommended wastewater treatment for various pollutants [5]. (*Continued*)

Pollutant	Treatment technology	Achievable concentration (mg/L)
Cyanide	2-stage alkaline oxidation	0–0.4
	Electro-dialysis	0.1
	Ozonation	0
Dissolved solids	Electro-dialysis	–
	Ion exchange	–
	Reverse osmosis	–
Fluoride	Alum precipitation	0.6
	Lime precipitation	6
$Iron^{2+}$	Aeration	0.1
	Chlorination	0.05
$Iron^{3+}$	Hydroxide precipitation	1.0 at pH 4.0
	Lime precipitation	1.0
Lead	Hydroxide precipitation	0.015 at pH 10
	Sulfide precipitation	0.3
$Manganese^{2+}$	Chlorination	0.05
	Permanganate oxidation	0.05
$Manganese^{3+}$	Hydroxide precipitation	0.05 at pH 9.5
Mercury	Activated carbon	0.001
	Alum precipitation	0.01 at pH 7
	Ion exchange	0.001
	Lime/hydroxide precipitation	0.001
	Sulfide precipitation	0.01 at pH 10
Nickel	Hydroxide precipitation	0.01 at pH 10.5
	Lime precipitation	0.01 at pH 10
	Reverse osmosis	0.01 at pH 8
	Sulfide precipitation	
Nitrogen Total Oil, including fat and grease emulsion breaking	Anoxic biological denitrification	0.5
	Acid cracking	–
Oil, including fat and grease removal	API separator	15
	Dissolved air flotation	10
	Diatomaceous earth filtration	5
	Membrane filtration	5
	Sand filtration	10

(*Continued*)

Appendix: Recommended wastewater treatment for various pollutants [5]. (*Continued*)

Pollutant	Treatment technology	Achievable concentration (mg/L)
Pesticides	Activated carbon adsorption	0.01
	Aerobic biological treatment	–
	Anaerobic biological treatment	–
	Resin adsorption	0.001
pH	Base addition	–
	CO_2 addition	–
	Strong acid addition	–
	Weak acid addition	–
Phenol	Activated carbon adsorption	–
	Aerobic biological treatment	–
	Chemical oxidation with lime precipitation	–
	Ozonation	–
Phosphorous	Alum coagulation	–
Selenium	Ferric sulfate coagulation	0.05
	Ion exchange	–
Settleable solids	Comminution	–
	Flotation	–
	Screening	–
	Sedimentation (grit chambers)	–
Sulfide	Activated carbon adsorption	–
	Air stripping	–
	Aerobic biological treatment	–
	Anaerobic biological treatment	–
Zinc	Carbonate precipitation	0.3 at pH 9.0
	Hydroxide precipitation	0.07 at pH 10.0
	Sulfide precipitation	0.001 at pH 9.0

Source: Table 12.1 in Alley [5].

References

1. Wolkersdorfer, C., *Water Management at Abandoned Flooded Underground Mines: Fundamentals, Tracer Tests, Modelling, Water Treatment*, p. 466, Springer, Heidelberg, 2008.
2. Younger, P.L., Banwart, S.A., Hedin, R.S., *Mine Water: Hydrology, Pollution, Remediation*, vol. 5, p. 464, Springer, Dordrecht (Kluwer), 2002.
3. Blodau, C., A review of acidity generation and consumption in acidic coal mine lakes and their watersheds. *Sci. Total Environ.*, 369, 1–3, 307, 2006.
4. Van Rooyen, M., Acid Mine Drainage Water Treatment using the SAVMIN Process, in: *SAIMM Hydro 2016*, pp. 52–60, South African Institute of Mining and Metallurgy, Cape Town, South Africa, 2016.

5. Alley, E.R., *Water quality control handbook*, vol. 2, pp. 459–460, The McGraw-Hill Companies, Inc., New York, 2007.
6. Harvard Law School, *The Cost of Gold: Environmental, Health, and Human Rights Consequences of Gold Mining in South Africa's West and Central Rand*, International Human Rights Clinic, Harvard College, Cambridge, Massachusetts, 2016.
7. Horak, S., *Background information document for the activities associated with the immediate and short term interventions regarding acid mine drainage in the Western, Central and Eastern Witwatersrand Basins*, Digby Wells, Randburg, 2012.
8. Ramontja, T., *Mine Water Management in the Witwatersrand Gold Fields with special emphasis on Acid Mine Drainage*, Council for Geoscience, Johannesburg, 2010.
9. Bosman, D.J., The improved densification of sludge from neutralised acid mine drainage. *J. South. Afr. Inst. Min. Metall.*, 9, 340, 1974.
10. Coulton, R., Bullen, C., Williams, C., Williams, K., The formation of high density sludge from mine water with low iron concentrations., in: *International Mine Water Association Symposium 1*, Newcastle Upon Tyne, University of Newcastle, 2004.
11. Fawell, J., Mascarenhas, R., *Sulfate in Drinking Water - Background Document for Development of WHO Guidelines for Drinking Water Quality*, World Health Organisation (Ed.), WHO Press, Geneve, Switzerland, 2004.
12. Water Research Commission, *Quality of Domestic Water Supplies, Volume 1: Assessment Guide*, Department of Water Affairs and Forestry, and Water Research Commission (Ed.), Water Research Commission, Pretoria, p. 93, 1998.
13. Lee, K.P., Arnot, T.C., Mattia, D., A review of reverse osmosis membrane materials for desalination—Development to date and future potential. *J. Membr. Sci.*, 370, 1–2, 1, 2011.
14. Howe, K.J. and Clark, M.M., Fouling of microfiltration and ultrafiltration membranes by natural waters. *Environ. Sci. Technol.*, 36, 16, 3571, 2002.
15. Karakatsanis, E. and Cogh, V.E., Drinking Water from Mine Water using the HiPRO Process - Optimum Coal Mine Water Reclamation Plant, in *International Mine Water Conference 2010*, Nova Scotia, Canada, 2010.
16. Hutton, B., Kahan, I., Naidu, T., Gunther, P., Operating and Maintenance Experience at the Emalahleni Water Reclamation Plant, in *International Mine Water Conference 2009*, Pretoria, South Africa, 2009.
17. Mottay, R. and Van Staden, P.J., *Positioning of the SAVMIN Process*, Mintek, Randburg, South Africa, 2018.
18. Goosen, M., Sablani, S., Al-Hinai, H., Al-Obeidani, S., Al-Belushi, R., Jackson, A., Fouling of reverse osmosis and ultrafiltration membranes: A critical review. *Sep. Sci. Technol.*, 39, 10, 2261, 2005.
19. Nevatalo, L., Van Der Meer, T., Kerstiens, B., Method for removing sulphate, calcium and/or other soluble metals from wastewater, PCT Patent Number PCT/FI2013/050816, assigned to Outotec, 2014.
20. Van Rooyen, M., *Acid Mine Drainage Treatment using the SAVMIN Process, in SAIMM Hydro 2016*, South African Institute of Mining and Metallurgy (Ed.), Cape Town, South Africa, 2016.
21. South African Water Quality Guidelines, Volume 1: Domestic Water Use, http://www.dwaf.gov.za/IWQS/wq_guide/domestic.pdf, 1996.
22. Hlabela, P., Maree, J., Bruinsma, D., Barium carbonate process for sulphate and metal removal from mine water. *Mine Water Environ.*, 26, 1, 14, 2007.
23. Tolonen, E.-T., Rämö, J., Lassi, U., The effect of magnesium on partial sulphate removal from mine water as gypsum. *J. Environ. Manage.*, 159, 143, 2015.
24. Mulopo, J., Continuous pilot scale assessment of the alkaline barium-calcium desalination process for acid mine drainage treatment. *J. Environ. Chem. Eng.*, 3, 1295, 2015.

25. Maree, J., Theron, D., Nengovhela, R., Hlabela, P., Sulphur from smelter gases and sulphate-rich effluents. *J. South. Afr. Inst. Min. Metall.*, 105, 641, 2005.
26. Cline, C., Hoksberg, A., Abry, R., Janssen, A., Biological Process for H_2S Removal from Gas Streams: The Shell-Paques/THIOPAQ® Gas Desulfurization Process. in *Proceedings of the Laurance Reid Gas Conditioning Conference*, Norman, Oklahoma, pp. 1–18, 2003.
27. Rose, P., Long-term sustainability in the management of acid mine drainage wastewaters – development of the Rhodes BioSURE Process. *Water SA.*, 39, 5, 582, 2013.
28. Grewar, T. and Singh, E., *Active Biological Sulphate Reduction*, Mintek, Randburg, 2015.
29. Doshi, S.M., Bioremediation of acid mine drainage using sulphate reducing bacteria., *in National Network of Environmental Management Studies*, USEPA (Ed.), University of Indiana, Washington DC, 2006.
30. Toledano, P., Roorda, C., *Leveraging mining investments in water infrastructure for broad economic development: Models, Opportunities and Challenges*, in CCSI Policy Paper, Columbia Center on Sustainable Investment (Ed.), pp. 1–30, Columbia Center on Sustainable Investment, Columbia, 2014.
31. Van Zyl, H., Maree, J., Van Niekerk, A., Van Tonder, G., Naidoo, C., Collection, treatment and re-use of mine water in the Olifants River Catchment. *J. South. Afr. Inst. Min. Metall.*, 101, 1, 41, 2001.
32. Grewar, T., *Fitness for use of treated water sources*, pp. 3–33, Mintek, Randburg, 2017.
33. Singapore National Water Agency, *Technical Reference for Water Conservation in Cooling Towers*, Singapore National Water Agency, Singapore, 2017.
34. SAMCO, What is a boiler feed water treatment system and how does it work?, https://www.samcotech.com/what-is-a-boiler-feed-water-treatment-system-how-does-it-work/, 2016
35. Sampson, D.H., *Chemical Engineering Methods and Technology. Gypsum: Properties, Production and Applications*, pp. 35–68, Nova Science Publishers, Inc, New York, 2011.
36. Cornell, R.M. and Schwertmann, U., *The iron oxides: Structure, properties, reactions, occurrences and uses*, pp. 9–37, WILEY-VCH Verlag GmbH & Co. KGaA, Weinheim, 2003.
37. Muller, H., Mokoena, S., Neale, J., Gericke, M., *Mine Water as a Resource (MIWARE) Report 4*, Mintek, Randburg, 2017.
38. Akinwekomi, V., Maree, J., Zvinowanda, C., Masindi, V., Synthesis of magnetite from iron-rich mine water using sodium carbonate. *J. Environ. Chem. Eng.*, 5, 3, 2699, 2017.
39. Mokwana, C., *Application of the RIP technology for the removal of uranium from Robinson Lake*, Mintek, Randburg, South Africa, 2018.
40. Groudev, S., Georgiev, P., Spasova, I., Nicolova, M., Bioremediation of acid mine drainage in a uranium deposit. *Hydrometallurgy*, 94, 1–4, 93, 2008.
41. Nascimento, M., Fatibello-Filho, O., Teixeira, L., Recovery of uranium from acid mine drainage waters by ion exchange. *Miner. Process. Extr. Metall. Rev.*, 25, 2, 129, 2004.
42. Hirose, A., Fawell, J.K., Uranium in Drinking-water, in: *Background document for development of WHO Guidelines for Drinking-water Quality*, World Health Organisation (Ed.), WHO Press, Geneve, Switzerland, 2012.
43. Senes Consultants Limited, *Acid Mine Drainage – Status of Chemical Treatment and Sludge Management Practices. MEND Report, vol 3.32.1.*, The Mine Environment Neutral Drainage [MEND] Program, Richmond Hill, 1994.
44. Howard, D., Grobler, C., Robinson, R., Cole, P., Sustainable Purification of Mine Water using Ion Exchange Technology. in *International Mine Water Conference 2009*, Pretoria, South Africa, 2009.
45. Creamer, M., On-the-Air, in: *Engineering News*, Creamer Media, Johannesburg, 2019.

10

Rare Earth Elements—A Treasure Locked in AMD?

Leon Krüger

Hydrometallurgy Division, Mintek, Randburg, South Africa

Abstract
A Rare Earth Element (REE) is described by the International Union of Pure and Applied Chemistry as a member of the 17 elements in the Periodic Table of Elements which includes the elements of the lanthanide series as well as scandium and yttrium. Since the lanthanides are characterized by the filling of the 4*f* orbitals, these elements display unique chemical properties such as high coordination numbers, large magnetic moments, high magnetic anisotropy, and very sharp spectral lines. This leads to their use in many technological applications where they can have a profound effect on how complicated devices or systems function. These uses can essentially be divided into two broad groups, namely, process enablers and technology components. The first can include catalytic applications, while the latter can include energy generation and storage as well as optical applications. The latest developments in these areas will be discussed in this chapter. Although there are many conventional terrestrial sources of REEs globally, an unusual source could be from the treatment of Acid Mine Drainage (AMD) as a byproduct of value extraction processes. The recovery of REEs from AMD as well as the techno economics of the process will also be discussed in this chapter.

Keywords: Rare earth elements, REE, REO, uses, applications, AMD and recovery

10.1 AMD—Annoyance or Resource

Acid Mine Drainage is a very topical subject in mining and environmental management activities and has often only been regarded as a waste stream that has to be treated to minimize the legacy of the mining operation or the effect that it may have on the environment. Over the last number of years, the mind-set around AMD has shifted markedly as more and more people started recognizing its value as a potential resource. Since AMD is inherently acidic, it has the potential to leach metals from rock or soil material that it comes into contact with. This natural leaching process can therefore lead to the accumulation of a range of elements in the AMD, some of low value, but also some that may be in high demand. Although base metals such as copper and cobalt, as well as more controversial value metals such as uranium, are quite common in AMD streams and will certainly be considered for recovery, developments around more exotic metals such as rare earths have aroused the interest of the mining industry in AMD as a potential source. This interest is

Email: leonk@mintek.co.za [ORCID: 0000-0001-7603-6120]

Elvis Fosso-Kankeu, Christian Wolkersdorfer and Jo Burgess (eds.) Recovery of Byproducts from Acid Mine Drainage Treatment, (263–314) © 2020 Scrivener Publishing LLC

probably fueled further by the fact that relatively standard metallurgical processes such as precipitation or ion exchange would be sufficient to recover the value metals from the AMD. For example, a common process step during the treatment of AMD is the removal of metal ions by precipitation at elevated pH to produce an alkaline precipitate in which the metals are concentrated as their hydroxides. The full value from these hydroxides can then be unlocked by relatively uncomplicated further processing. An important point that has to be made, however, is that the technical feasibility of recovering metals from AMD does not necessarily lead to a good business case for doing so. The best chance for a viable recovery of value metals from AMD is if they are either very valuable such as gold or platinum group metals or that they are regarded as critical metals for industrial or military applications. One such group of metals may be the rare earth elements. The question is, however, whether there is enough value in the rare earths to consider a case for their recovery from AMD. In order to answer this question, several aspects will have to be considered, such as whether rare earths are sought after as value metals and if so, whether there are actually enough of them in AMD to make it viable to recover them.

10.2 Rare Earths—The Almost Forgotten Elements!

> *"The rare earth elements perplex us in our researches, baffle us in our speculations, and haunt us in our very dreams. They stretch like an unknown sea before us mocking, mystifying and murmuring strange revelations and possibilities."*

Sir William Crookes
16 February 1887 [1]

The REEs are defined by the International Union of Pure and Applied Chemistry (IUPAC) as the set of 17 elements in the Periodic Table of Elements, including the 15 lanthanides plus scandium (Sc) and yttrium (Y). Although Y and Sc possess different electronic and magnetic properties, they exhibit similar chemical properties to the lanthanide elements. In addition, they are also found with the lanthanide series of elements in nature and are therefore classified as REEs. An unusual member of the REEs is promethium (Pm), since it is a radioactive element and is only found naturally in very small amounts in uranium ores as a result of natural nuclear fission. Considered from an industrial perspective, however, the term REEs usually refers to the remaining 15 elements only (i.e., excluding Sc and Pm).

The history of the rare earths can probably be traced back to 1751, when Swedish mineralogist and chemist Axel Fredrik Cronstedt, found a reddish-brown mineral in the quarry of Båstnäs near Riddarhyttan in Sweden. It fascinated him since it appeared to be unusually heavy, but after it had been proclaimed to be nothing more than an iron aluminium silicate by Swedish chemist Carl Wilhelm Scheele, the world lost interest for almost 40 years [2]. This all changed in 1788, when a Swedish soldier and chemist, Lieutenant Carl Axel Arrhenius, found a very dense black mineral near the village of Ytterby. A closer inspection by the Finnish chemist Johan Gadolin led him to the conclusion that it was a new metal oxide that he called ytterbia. After this discovery, the heavy mineral from Båstnäs was revisited and it became clear that it was very similar to the mineral from Ytterby, but yet distinct. They were classified as a new kind of "earth" from the archaic term for acid soluble elements [3]. At that time, these elements had not been found anywhere else, and therefore, they were assumed to be uncommon or rare; hence the name rare earths.

In 1885, Austrian scientist and inventor, Dr. Auer von Welsbach, announced his invention of incandescent mantles using mixed or simply separated rare earths. This led to a great demand for rare earths and agents from the Welsbach Company visited all the main mining centers of Europe and America to try and source raw materials to satisfy this demand [4]. During these visits, it soon became apparent that the so-called "rare earths" occurred relatively widely in nature and were certainly not rare. In spite of this, the misnomer remained, and to today, they are still called REEs.

10.3 Characteristics—What is with the *f*-Orbitals?

Through the years, the REEs have found more and more technological applications, to the point that they are currently declared critical elements by the European Union, United States, as well as Japan. This growth in technological applications is driven by the unique properties that the REEs possess. A clue to these unique properties of the REEs can be found under the surface in the electron configuration and the nature of the orbitals that are filled as the REE is formed. In order to discuss the electronic behavior and perhaps link that to the unique properties of the rare earths, it would probably be useful to take a step back and consider what an orbital is and why they are important in the nature of bonding between atoms as well as how they influence the properties of elements and chemical complexes.

Orbitals are not containers containing electrons, but rather a region in the space around the nucleus of an atom where they are likely to be found. Classical physics explains matter and energy on a scale of normal human observation well, but shows severe limitations in describing matter and its interactions with energy when the scale is reduced to that of atoms and below. In order to overcome these limitations, so that the behavior of electrons could be described, a new kind of physics called quantum theory was developed. From quantum theory, a quantum mechanical model was developed that describes orbitals or regions where electrons may be found by a set of quantum numbers that are related to the energy (n), angular momentum (l), and a vector component of the angular momentum (m), the so-called magnetic quantum number of the electrons in this region. The final quantum number completing the description of the state of an electron is the spin quantum number (s). This number has been the subject of many theories and speculations, but probably one of the best ways to visualize it was advanced by George Eugene Uhlenbeck who proposed that this number may be viewed as arising from the electron generating its own magnetic field like a spinning ball of charge [5].

The principal quantum number (n) describes the size of the orbital. Since an electron is attracted by the opposite charge of a nucleus in an atom, the distance that it can move away from the nucleus is determined by the nuclear charge that it is exposed to and the energy that it possesses. This means that the principal quantum number also describes the energy of an orbital. The second quantum number (l) describes the shape of the orbital since the angular momentum determines where the electron can move and therefore where it is likely to be found. The magnetic quantum number will determine the orientation of an orbital with respect to a given direction, usually that of a strong magnetic field and has no effect on the shape of the orbital or on the energy of an electron. Figure 10.1 gives a pictorial representation of a p_z orbital showing the shape and likelihood of finding an electron in space around the nucleus.

Quantum mechanics further states that no two electrons can occupy the same quantum state simultaneously (Pauli Exclusion Principle) [6]. This is quite important, since it means that only two electrons can occupy an orbital. Since two electrons with the same spin orientation in an orbital will violate Pauli's Rule, it means that if one is +½, the other must be −½, and no more can be added because a third would have to adopt the same quantum state as that of one of the first two, again violating Pauli's Rule [6].

The *Aufbau* principle was initially proposed by the Danish physicist Niels Bohr, who used quantum mechanics to study and describe atomic structure [6]. According to the *Aufbau* principle, the elements are formed by adding electrons to available orbitals of the lowest energy as protons are added to an atom. Atomic orbitals can therefore be considered as the basic building blocks of the atomic orbital model (a modern basis for describing the behavior of electrons in matter) [6]. Following the *Aufbau* principle, one eventually arrives at a point where the next electron has to occupy an orbital described by the principle quantum number $n = 4$ and angular momentum quantum number $l = 3$. Since the shape of the orbital is determined by the angular momentum quantum number, this combination describes a new kind of orbital that does not exist in the shells below. The reason for this is that the quantum mechanical model states that the highest value of l possible for each principal quantum number is equal to $n - 1$. This means that $l = 3$ can only exist from principal quantum number $n = 4$ onward. When one considers the filling of these new orbitals, two things become immediately apparent. Firstly, the number of orbitals (and therefore electrons), and secondly, the nature of the orbitals, which are clearly different to that found in the transition metals. Since these orbitals are unique, the question to consider is whether it is also linked to the unique properties of REEs as applied in so many important technologies.

Although an atom can consist of several layers of orbitals, only electrons in the outermost, or highest in energy, orbitals are normally available for bonding. The properties of the orbitals that they occupy are therefore important in determining the chemical properties of

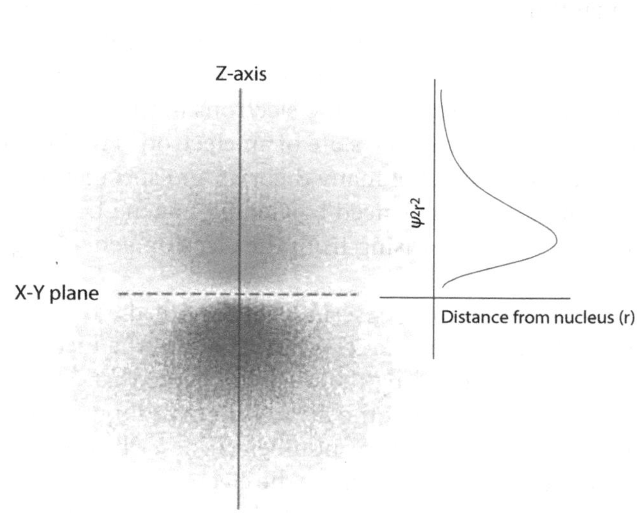

Figure 10.1 Probability of finding an electron occupying one of the *p*-orbitals.

an element. A useful way of representing the distribution of electrons in an atom or molecule is through the electron configuration notation. For example, the electron configuration of Fe can be written as:

$$1s^2 2s^2 2p^6 3s^2 3p^6 3d^6 4s^2$$

and can be depicted as in Figure 10.2.

Since the outermost electrons are of particular interest for bonding, a common way to simplify this notation is by using the so-called noble gas shortcut. For example, the noble gas Ar can be written as:

$$1s^2 2s^2 2p^6 3s^2 3p^6$$

The electron configuration for iron can therefore be written simply as: [Ar] $3d^6 4s^2$.

In the same way, the electron configuration of the common REE oxidation states can be written using the noble gas Xe as base (Table 10.1).

Whereas the number of electrons in the outer orbitals is certainly an important factor in determining the characteristics of a REE, the nature of the orbitals may be even more important and therefore warrants a closer look. Since there are seven possible magnetic quantum numbers in the *4f* subshell, there are seven *f*-orbitals and their shapes will be more complex than the underlying orbitals due to their higher angular momentum. Figure 10.3 shows the complex shapes of the *4f* orbitals [8].

As there can only be two electrons in each orbital, there may be up to 14 *f*-electrons in the different rare earths as shown by the example of Yb (Figure 10.4).

As new electrons are added to an atom, they tend to shield one another from the nuclear charge that they are all subjected to and so reduce the effective nuclear charge that each electron experiences. The more directional an orbital is, the less the shielding effect. For example,

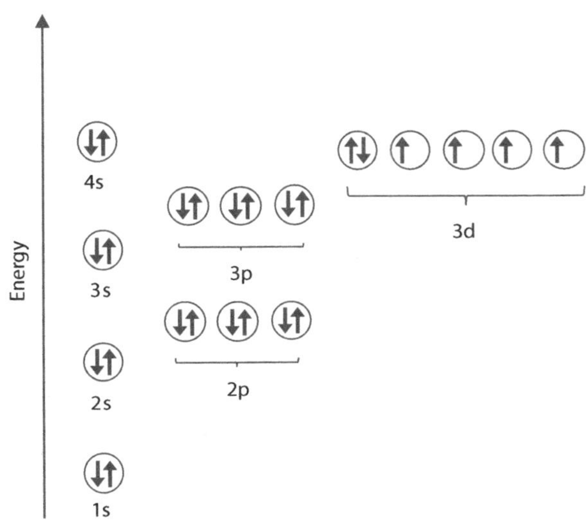

Figure 10.2 Electron configuration of Fe.

Table 10.1 Electron configurations and some properties of common REE atoms and ions [7].

Element	Symbol	Atomic number	Atomic (0)	2+	3+	4+	E°, V	Radius M^{3+}, Å
Lanthanum	La	57	$[Xe]4f^{0}5d^{1}6s^{2}$		$[Xe]4f^{0}$		−2.37	1.061
Cerium	Ce	58	$[Xe]4f^{1}5d^{1}6s^{2}$		$[Xe]4f^{1}$	$[Xe]4f^{0}$	−2.34	1.034
Praseodymium	Pr	59	$[Xe]4f^{3}6s^{2}$		$[Xe]4f^{2}$	$[Xe]4f^{1}$	−2.35	1.013
Neodymium	Nd	60	$[Xe]4f^{4}6s^{2}$		$[Xe]4f^{3}$	$[Xe]4f^{2}$	−2.32	0.995
Promethium	Pm	61	$[Xe]4f^{5}6s^{2}$		$[Xe]4f^{4}$		−2.29	0.979
Samarium	Sm	62	$[Xe]4f^{6}6s^{2}$	$[Xe]4f^{6}$	$[Xe]4f^{5}$		−2.30	0.964
Europium	Eu	63	$[Xe]4f^{7}6s^{2}$	$[Xe]4f^{7}$	$[Xe]4f^{6}$		−1.99	0.950
Gadolinium	Gd	64	$[Xe]4f^{7}5d^{1}6s^{2}$		$[Xe]4f^{7}$		−2.29	0.938
Terbium	Tb	65	$[Xe]4f^{9}6s^{2}$		$[Xe]4f^{8}$	$[Xe]4f^{7}$	−2.30	0.923
Dysprosium	Dy	66	$[Xe]4f^{10}6s^{2}$		$[Xe]4f^{9}$	$[Xe]4f^{8}$	−2.29	0.908
Holmium	Ho	67	$[Xe]4f^{11}6s^{2}$		$[Xe]4f^{10}$		−2.33	0.894
Erbium	Er	68	$[Xe]4f^{12}6s^{2}$		$[Xe]4f^{11}$		−2.31	0.881
Thulium	Tm	69	$[Xe]4f^{13}6s^{2}$	$[Xe]4f^{13}$	$[Xe]4f^{12}$		−2.31	0.869
Ytterbium	Yb	70	$[Xe]4f^{14}6s^{2}$	$[Xe]4f^{14}$	$[Xe]4f^{13}$		−2.22	0.858
Lutetium	Lu	71	$[Xe]4f^{14}5d^{1}6s^{2}$		$[Xe]4f^{14}$		−2.30	0.848

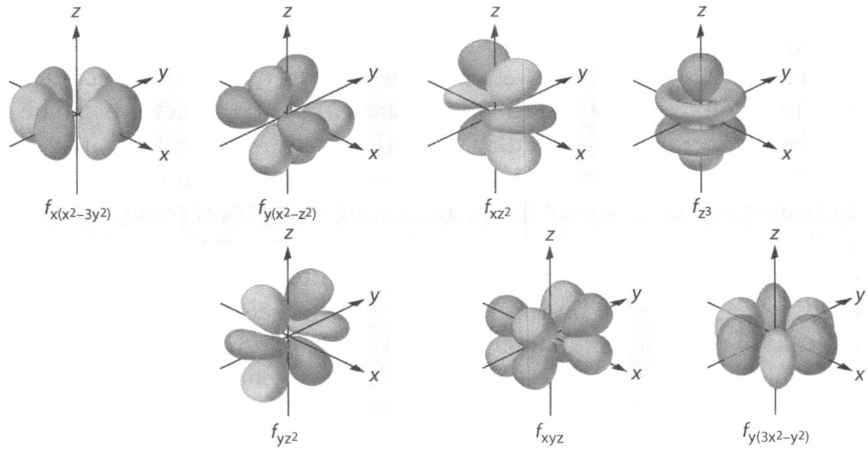

Figure 10.3 Shapes of the 4*f* orbitals [8].

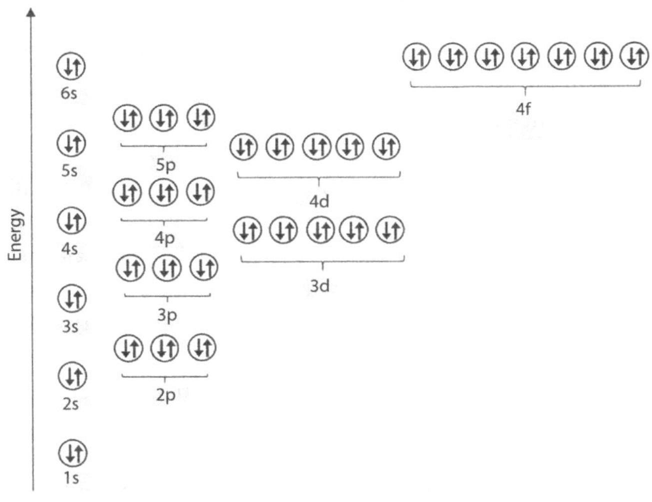

Figure 10.4 Electron configuration of Yb.

an *s*-orbital is spherical in shape, and since it is not directional at all, *s*-electrons shield very well against a nuclear charge. The opposite is true of *f*-orbitals (very directional), and therefore, *f*-electrons shield poorly. The order of shielding power can be given as follows:

$$S(s) > S(p) > S(d) > S(f)$$

with S representing the shielding power of the electrons.

If the *f*-electrons shield one another poorly, then it follows that a considerable part of their wave functions are within the Xe core and that these electrons will only have a limited effect on bonding. This also means that the entire electron cloud shrinks as more *f*-electrons (and therefore also more protons in the nucleus) are added. This phenomenon is called the lanthanide contraction and results in the atoms becoming smaller rather than larger as is the normal pattern when one moves across a period in the Periodic Table (Table 10.1).

Another consequence of the penetration of the *f*-electrons into the inner electron core of the atom is the stabilization of the +3 valence state. The two 6*s* electrons can be removed relatively easily due to the good shielding from the nuclear charge that these *s*-electrons experience. One of the *f*-electrons can also be removed, but this then results in an increased effective nuclear charge experienced by the remaining *f*-electrons, resulting in them being pulled even closer to the nucleus. The energy required for removing any additional 4-*f* electrons then becomes so large that it will not happen under normal conditions, hence the unusual stability of the +3 state. In this state, none of the remaining *f*-electrons are available for any kind of covalent bonding and they are only perturbed very slightly by ligand fields. The chemical interactions of the REEs in the +3 valence state are therefore almost purely ionic in nature [9]. In fact, their thermodynamic properties as well as their electrode potentials can be correlated very well with interactions based solely on the charge and size effects [7].

Although all this appears to point to a rather uninteresting chemical behavior of the REEs, with the *f*-electrons exerting only a secondary influence on the chemical properties of its compounds, it is the physics where the *f*-electrons promise to make a real difference. There are two physical properties of REEs that are very important for many technological applications of these elements. It is especially their magnetic and spectroscopic properties that show profound differences from the *d*-block elements.

The magnetic properties of the REEs are characterized by strong Russel-Saunders coupling between the spin and orbital angular momenta of electrons in unfilled 4*f* shells (J values) [1]. It should be noted that such magnetic behavior is qualitatively different to elements depending on the spin only (S values), which is the case with most of the *d*-block elements. Only for the empty, half-filled and completely filled *f*-orbitals, where the sum of the *ml* values is zero (i.e., L = 0 and therefore J = S), do the magnetic properties depend on spin only as with most of the *d*-block elements [7]. In addition to this, an indirect coupling of angular momenta of neighboring ions can also take place *via* conducting electrons, resulting in very large magnetic moments. Although the magnitude of the magnetic moment is important for many applications, the magnetic anisotropy is equally important since it relates to the retention of magnetization. Strong coupling of the angular momentum (L) to the large 4*f* spin (S), in the presence of an electrostatic field, can lead to an enormous magnetic anisotropy of electrostatic origin in most of the REEs [1]. The reason for this is that spin-orbit coupling essentially results in the charge distribution of the *f*-orbital to be pulled along with the spin as the magnetization direction is changed. Much higher electrostatic energy is therefore required from the crystal field in order to effect this change. However, as temperature increases, the anisotropy strongly decreases and finally vanishes at the ordering temperature; in the case of REEs, below room temperature. Although this magnetic anisotropy is temperature dependent, the moments can be confined further by structural effects in some REEs such as Tb and Dy, which leads to a reduced temperature dependence of the anisotropy in these elements, allowing for higher temperature applications.

Since the *f*-electrons are shielded almost perfectly from external fields by the overlying 5*s* and 5*p* shells, the quantum states arising from the different $4f^n$ configurations therefore tend to be nearly constant for a given ion. Since the spin-orbit coupling constants for *f*-electrons can be rather large (ca. 1,000 cm^{-1}), almost all rare earth ions have ground states with a single well-defined value of total angular momentum (J). The difference between the ground state and the next J-state is at an energy of several times kT and would therefore be almost unpopulated. Another consequence of the good shielding of the *f*-electrons from external fields is that these fields split the states due to the $4f^n$ configurations by very little (only around 100 cm^{-1}). This means that the *f-f* electronic transitions lead to absorption bands that are extremely sharp, distinct from the broad bands that are observed for *d-d* transitions. Almost all of the absorption bands of trivalent rare earth ions found in the visible and near UV spectra display this unusual line like character [7].

10.4 Applications—Sweating the Unique Characteristics

10.4.1 Introduction

REEs are used in a range of chemical forms and in a large variety of industrial applications. Some are simple such as polishing agents while others are used in much more complex applications such as military range finders. According to Gareth Hatch from Technology Metals Research (TMR), applications of REEs can be divided into essentially two broad categories, namely, process enablers and technology building blocks. The process enablers are cases where the REEs are used somewhere in the life cycle of materials or components, but does not remain part of the final product. In these applications, properties of relatively simple REE compounds are often used to achieve a certain outcome, such as highly polished surfaces or crude oil conversion into lighter fractions [10]. Where REEs are used as technology building blocks, the REE compounds or engineered components are generally more complex and the unique properties of the REE are critical to the functioning of the larger engineered system.

10.4.2 Rare Earths as Process Enablers

Although they may be vulnerable to substitution, simple rare earth compounds remain essential to many medical, chemical, and manufacturing processes. Almost 60% of global consumption of REEs is accounted for by applications in these sectors and four major uses can be identified:

- catalysis in oil refining, emission control, or energy generation
- metallurgical processes
- glass and ceramics industry
- therapeutic application in medicine

10.4.2.1 Catalysis

Fluid catalytic cracking (FCC) is a very important technology, used widely in the petroleum industry to convert the hydrocarbon chains in high boiling, high molecular mass crude oils into more valuable fractions such as light automotive fuels and olefinic gases. Catalytic cracking processes has all but replaced the thermal cracking process since it can produce fuels of higher octane ratings as well as more olefinic gases than thermal cracking. A typical FCC catalyst consists of four major components [11]:

- crystalline zeolite
- matrix
- binder
- filler

Zeolites can be described as crystalline aluminosilicates containing cations of alkaline elements, REEs or other metals of varying valence states depending on the charge balance requirements [11]. They are the active components in FCC catalysts and make up about 15 to 50% of the mass of the catalyst [12]. The binder and filler have a structural function since they provide the mechanical strength and competence of the catalyst. The binder is usually a silica sol while the filler is normally a low shrink-swell capacity clay such as kaolin to ensure stability. Zeolites are not soluble in the reaction medium and can therefore be categorized as heterogeneous catalysts. One of the most important properties of zeolites in a catalytic application is their surface acidity, since it can facilitate the formation of carbocations on its surface. These carbocations can then undergo a number of rearrangements, ultimately leading to the breaking of C–C bonds, yielding shorter chain hydrocarbons as shown in Figure 10.5.

A good FCC catalyst should have the following characteristics:

- Good thermal and hydrothermal stability
- High activity
- Large pores
- Good structural strength
- Low production of carbonaceous compounds during the dehydrogenation reaction

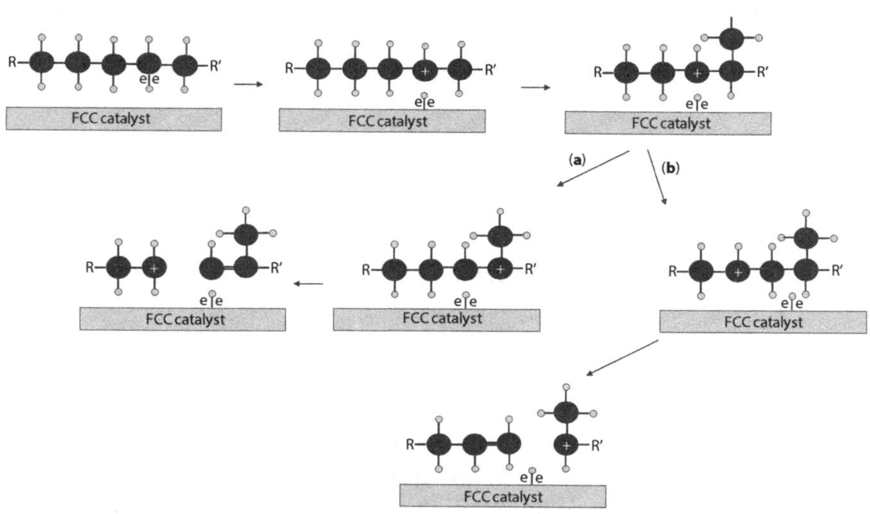

Figure 10.5 FCC catalytic cracking of hydrocarbons *via* the formation of carbocations.

The use of rare earth containing zeolites in catalysts for the cracking of hydrocarbons was first described by Plank and Rosinski in the 1960s [13]. REEs, and in particular La, the largest trivalent ion, makes critical contributions to the optimal functioning of an FCC catalyst. During the synthesis of a REE zeolite, the normal Na^+-cation in the Y-zeolite is replaced by La^{3+} through an ion exchange reaction, followed by migration to smaller cages upon calcination [14]. The current view is that once the La^{3+}-ions are located in these cages, they form coordination bonds to the oxygen atoms of the zeolite framework. These bonds improve the mechanical stability of the zeolite by connecting the tetrahedral building blocks of the zeolite through bridging bonds. This enhanced stability imparts high resistance to failure due to severe hydrothermal conditions [14].

Both Brønsted and Lewis acid sites occur in zeolites. Brønsted sites are almost exclusively generated as a result of the following reactions [15]:

$$NH_4^+ + Na^+Z^- \xrightarrow[-Na^+]{} NH_4^+Z^- \xrightarrow[-NH_3]{\approx 300-400^\circ C} H^+Z^- \tag{10.1}$$

$$[La(H_2O)_n]^{3+} + 3Na^+Z^- \xrightarrow[-3Na^+]{} [La(H_2O)_n]^{3+}(Z^-)_3 \xrightarrow[-(n-2)H_2O]{\approx 300^\circ C} [LaOH)(H_2O)]^{2+}H^+(Z^-)_3 \rightarrow [La(OH)_2]^+(H^+)_2(Z^-)_3 \tag{10.2}$$

The reaction given in Equation 10.2 is normally referred to as the Hirschler-Plank scheme. Since the La^{3+} has to neutralize three negative charges in the zeolite framework, strong electrostatic fields build up after most of the water had been removed from the La ion. The remaining water molecules then dissociate under the influence of the electrostatic field, yielding a proton and an OH group. From Equation 10.2, it can be seen that each La^{3+} ion present in the zeolite will result in two Brønsted acid sites, contributing substantially to the activity of the zeolite as catalyst for the cracking of hydrocarbons. It has been shown that the degree of hydrolysis is correlated to an increase in ionic size, hence La^{3+} being the REE of choice in FCC catalysts [15].

Vanadium is one of the most potent poisons for FCC catalysts since it can destroy the zeolite structure and also increase the production of coke and light gases. If vanadium is present, it can be hydrolyzed to vanadic acid which, in turn, can lead to the hydrolysis of the SiO_2/Al_2O_3 framework in the zeolite, leading to its destruction. Since the REE oxides are basic, they can react with the vanadic acid, trapping it and so preventing reaction with the zeolite [16]. A possible explanation may be the ability of REEs to form vanadates, so that the vanadium is kept in the pentavalent state ($REEVO_4$).

The REE therefore appears to perform two critical roles in the FCC catalyst. They do not only control the stability, activity, and selectivity of the zeolite but also acts as a vanadium-trap, improving the coke selectivity even more and also improving the activity retention of the catalyst. REEs, with La and Ce in particular, are therefore indispensable components of modern FCC catalysts, accounting for about 22% of REE consumption [17].

Another catalytic application of REEs is in automotive catalysts used to reduce atmospheric emissions from internal combustion engines. REEs, and in particular Ce, is used as part of the so-called "washcoat" in an automotive catalytic converter. The washcoat is a thin layer on the substrate or catalytic support and normally serves as a carrier for the PGM catalytic particles as well as a dispersion medium to enlarge the surface area. The function of the Ce in this layer is, however, not restricted to mere physical support, since the unique properties of the CeO_2 lattice provides valuable oxygen storage capacity for the conversion of pollutants by the PGM particles [18]. When the CeO_2 is in a reducing atmosphere at high temperature, it forms a range of oxygen deficient CeO_{2-x} units while retaining the fluorite crystal structure. This structure is maintained even when substantial loss of oxygen had taken place, resulting in a large number of oxygen vacancies. Since the reaction is reversible, the CeO_{2-x} oxides will be re-oxidized readily to CeO_2 when the environment turns oxidizing. This means that CeO_2 can act as an O_2 buffer because oxygen can either be stored or released, exploiting the Ce^{3+}/Ce^{4+} redox couple in the following reversible reaction [19].

$$CeO_2 \underset{H_2O/CO_2}{\overset{H_2/CO}{\longleftrightarrow}} CeO_{2-x} + \frac{x}{2}O_2 \tag{10.3}$$

This is particularly useful in enhancing the efficiency of the catalysts during rich (net reducing) or lean (net oxidizing) air-to-fuel ratios [18].

A logical progression in the use of the exceptional oxygen buffering capacity of CeO_2 or doped CeO_2 in automotive catalytic systems is to add these particles directly to the fuel as a water emulsion of CeO_2 nanoparticles. This approach is currently a field of considerable research and development and very promising results had been reported [20–24]. The cerium oxide nanoparticles present in the fuel stimulate more efficient combustion when compared to fuel without the addition of the nanoparticles by lowering the carbon combustion activation temperature. The reason for this appears to be the ability of CeO_2 to release and store oxygen depending on the partial pressure of oxygen [25]. Ce oxide can therefore supply the oxygen for the oxidation of hydrocarbons in the fuel as well as any soot particles formed during combustion, while being converted to Ce_2O_3 according to the following reactions [25]:

Combustion of hydrocarbons

$$(2x+y)CeO_2 + CxHy \rightarrow \left[\frac{(2x+y)}{2}\right] Ce_2O_3 + xCO + \frac{y}{2} H_2O \tag{10.4}$$

Burning of soot particles

$$4\ CeO_2 + C(soot) \rightarrow 2\ Ce_2O_3 + CO_2 \tag{10.5}$$

Other important pollutant molecules produced during the combustion of diesel fuel are the various nitrogen oxide gases (NO_x). A useful way of eliminating the harmful effects of NO_x gases, formed during the combustion process, is to reduce them to nitrogen. This is where the oxygen buffer capacity of cerium oxides can make an important contribution. Ce_2O_3 formed after the oxidation of the diesel molecules and soot particles can accept oxygen from the NO_x molecules leading to the Ce_2O_3 being re-oxidized to CeO_2, while the nitrogen oxide is reduced by the following reaction [25]:

$$Ce_2O_3 + NO \rightarrow 2\ CeO_2 + \frac{1}{2}\ N_2 \tag{10.6}$$

This potential application appears to be a very elegant approach to enhance the performance of compression ignition engines, and therefore, considerable research is currently underway to increase the understanding of the physics and chemistry as well as to resolve practical difficulties with the application of such an in-fuel catalytic process.

While the unusual oxygen storage capacity of ceria is applied very effectively in a number of hydrocarbon catalytic technologies, the unique properties of the CeO_2 lattice and the presence of the Ce^{3+}/Ce^{4+} redox couple also opens up other exciting catalytic applications in energy production. One such application is the use of doped CeO_2 as electrolyte in Solid Oxide Fuel Cells (SOFC). These fuel cells have been attracting much attention in base load stationary applications, recently. Although it has limited load following capability, which limits its general application, it has a number of advantages such as high temperature operation which improves efficiency, long term stability, high tolerance to impurities in fuels, low emissions, and competitive cost [26]. Electrolytes in fuel cells are required to meet very specific requirements and in the case of solid electrolytes, there are several critical property criteria that such a substance should conform to. One of the most important is that it has to have a high oxygen conductivity over a wide range of oxygen partial pressures. This is made possible by the cubic fluorite crystal structure of CeO_2, which has fairly large open spaces between the ions in the structure, enabling rapid diffusion of oxygen ions inside the lattice (Figure 10.6). In addition to this, it should also be non-porous so that cross diffusion of fuel and oxidant will not take place. Since the electrolyte is in contact with both air and fuel electrodes as well as the sealing materials, it should not only be compatible with these materials but also have good mechanical properties [26].

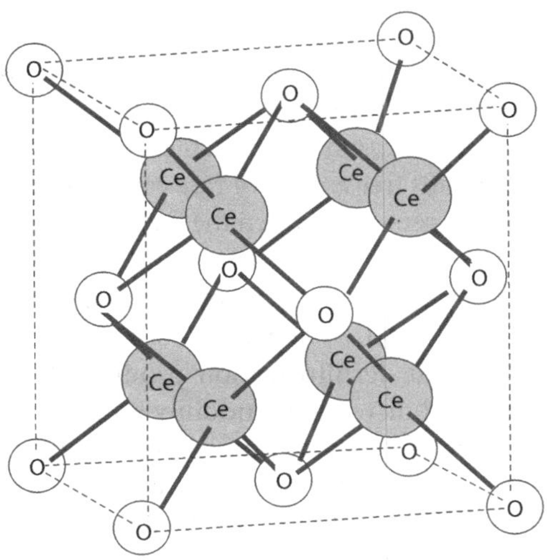

Figure 10.6 The cubic fluorite structure of CeO_2.

Oxygen-ion transfer in CeO_2 takes place by means of a process of vacancy diffusion. When an oxidizing environment is present, the cubic fluorite structure of pure CeO_2 has no oxygen vacancy defects and therefore oxygen transfer is poor in pure CeO_2 under these conditions. Oxygen vacancy defects can, however, be introduced into the CeO_2 structure by creating functional point defects in the lattice. This can be done by substituting some of the Ce^{4+} cations with trivalent acceptor cations in the lattice. Since this results in a positive charge deficiency in the lattice, an oxygen vacancy is formed for each two doped cations to compensate. This altered structure is stable over a wide range of temperatures up to its melting point (≈2,400 °C) and many such centers of perturbation can be introduced into the structure. A number of properties such as ionic conduction and the thermoelectric effect can be customized by manipulating the concentration of these vacancy defects [27].

SOFCs are gaining in importance because they offer substantial advantages over conventional electrical power sources. These include higher efficiency, low emissions, and a high thermal efficiency energy output over a wide array of applications with systems of one or two kW in size having similar, or in some instances better, efficiencies than those of 200 kW in size. Although the chemistry of the REEs, and Ce in particular, appears to be one-dimensional (ionic behavior only), the unique chemical behavior of their compounds leads to a wide range of interesting applications.

10.4.2.2 Physical Metallurgy

In the quest to improve the strength of steels at room temperature under static loading, strength often came at the expense of toughness. Since modern requirements from especially the automotive industry, demand both toughness and strength, much research has been done to understand and quantify the factors influencing the properties of steel and alloys. Researchers in the field of physical metallurgy have shown that control of impurities and microstructure is particularly important in making strong steel with acceptable toughness [28, 29]. Since the early 1990s, the use of REEs for the modification of microstructures in steel has been investigated and they have since been used widely in steel metallurgy due to their remarkable ability to remove impurities, improve microstructure morphology, and refine inclusions [29].

The purification of steel requires deoxidization, desulfurization, and removal of elements with low melting points. Due to strong ionic interactions, the REEs form stable compounds with oxygen, sulfur, lead, arsenic, tin, and antimony. Since these compounds are not metals and have very high melting points, purification can be achieved by allowing them to collect in the upper slag in the furnace. The Gibbs free energy of rare earth compounds dictate that they will complex with sulfur first and then remove it, when the levels of oxygen are low enough [30, 31].

Besides removing impurities from steel by complexing with them, REEs can also improve the characteristics of the steel by controlling the, morphology, as well as the distribution of inclusions and impurities in the steel. When REE compounds are added to liquid steel, they react with deoxidization and desulfurization products to form high melting point inclusions that tend to cluster and distribute randomly around grain boundaries. Since the evenly distributed small spheres of REE inclusions have a similar thermal expansion coefficient to steel, the fatigue strength of steel is enhanced because stress concentrations are avoided during casting. This sulfide and graphite shape control to modify the inclusions remaining after solidification is one of the most important uses of REEs in metallurgy since it increases the strength of the steel.

According to Wang, magnesium alloys modified by rare earth metals can have a range of enhanced properties such as toughness, corrosion resistance, fatigue resistance, and hot ductility [32]. These rare earth containing magnesium alloys can therefore be used in more demanding applications in the automobile industry than would be possible for conventional magnesium alloys. Magnesium alloys are often not used in automotive applications, where high temperature strength and high resistance to corrosion are specified. Both these requirements are addressed effectively through alloying with rare earths. When REEs are added to Mg alloys, they facilitate the formation of stable phases at the grain boundaries. In particular, the presence of thermally stable $Al_{11}RE_3$ needle shape particles along grain boundaries result in less labile grain boundaries, leading to increased creep resistance of the alloy, which in turn, results in high thermodynamic stability up to ≈180 °C [33].

Most metals and alloys suffer from preferential corrosive attack along grain boundaries, when exposed to a corrosive environment. Grain boundaries are usually more susceptible to oxidation than the grain matrix since a depletion in protective elements had occurred in these areas. Preferential attack can therefore occur along or adjacent to grain boundaries while the matrix remains relatively inert. This type of attack can have a high rate of reaction and usually proceeds deep into the metal, reducing the overall strength. It is caused by the formation of local galvanic couples due to the presence of impurities at the grain boundaries. When REEs are added to the magnesium alloy, they trap the impurities by forming intermetallic compounds with them and so decrease the activity of the cathode [34].

Another way that the REEs can influence the microstructure and texture of steel is by micro-alloying of iron with rare earth metal. Rare earth metals dissolve in minute quantities in the steel in the form of a rather unconventional solid solution. Since the atomic radius of the REE atom is influenced by polarization, the atomic covalent diameter of La can reduce to just 5.5% larger than that of the Fe atom at a 60% degree of polarization. This means that the REE atom can actually occupy points in the iron crystal lattice and so distort the lattice. Such distortion leads to local stress fields that interact with those of the dislocations in the lattice, impeding their motion and so resulting in an increase in the yield stress of the steel.

10.4.2.3 Glass and Ceramic Industries

While the composition of optical glass is very important in optical systems, the surface finishing and accuracy of the polishing of components such as lenses can make the difference between success and failure of a system. Cerium oxide has been shown to be one of the best glass polishing agents available and is capable of providing high-quality polished surfaces. In order to achieve the desired results, it is important to match the hardness of the polishing powder to the glass being polished. For example, the hardness should be similar to that of the material polished if one wants to avoid the formation of large grooves caused by the polishing particles being forced deep into the material by the mechanical pressure of the polishing action. Table 10.2 gives a comparison of the hardness of common polishing compounds to different types of glass [35].

Cerium oxide is a very interesting polishing agent since its high polishing efficiency is not only derived from its mechanical properties but also from its chemical interactions with the substrate during polishing. Polishing of glass and other materials cannot be regarded as merely a simple mechanical process using finer and finer abrasive media, but a number of

Table 10.2 Hardness values of different polishing materials and glasses [35].

Polishing compound	Hardness (Mohs Scale)	Glass type	Hardness (Mohs Scale)
Diamond	10	Silica	7
Alumina	9	Soda lime	5.3
Zirconia	8	Borosilicate	5.8
CeO_2	7–8	Lead	4.8

other interactions between the elements of the polishing system also play an important role [36]. There are three polishing mechanisms at work during the polishing of a glass surface:

- Mechanical polishing: This involves the classical action of an abrasive, to strip the surface irregularities and expose new underlying material.
- Molecular polishing: This involves the removal of silica molecules from the surface by breaking of chemical bonds and proceeds by the following reaction:

$$=Ce-OH+HO-Si\equiv\rightarrow\ =Ce-O-Si\equiv +\ H_2O \qquad (10.7)$$

  CeO_2 in water is hydrolyzed to produce H_3O^+ ions, which in turn, undergoes an exchange reaction with Na on the glass surface, forming a Ce-O-Si molecule. Since the chemical bonds in this new molecule are stronger than those of the Si-O-Si glass network, the polishing effect can be viewed as silica molecules being removed from the surface of the glass by the formation of the more stable Ce-O-Si molecule.
- Rheological polishing: Since glass can be considered an amorphous solid, almost no plastic flow is observed at room temperature, even under load. Friction between the polishing medium and the glass is inherent to the polishing process and it generates enough heat so that the glass surface will eventually soften. This allows for some movement of molecules, resulting in a smoothing of the surface. Minute imperfections such as scratches and indentations can therefore be removed during this phase due to the heat induced plastic flow during polishing. Removing these defects eliminates high-stress points that may become the initiators for fractures. Therefore, glass, for example, in automotive and architectural applications is almost always polished after it is strengthened or laminated [36].

Pure cerium oxide is not always used in polishing applications, since in many cases the presence of other REE oxides does not make a material difference and therefore less expensive mixtures can be used. For special applications, other REE oxides (such as Pr_2O_3) are sometimes added to modify the properties of the polishing compound depending on the material that has to be polished.

Nuclear energy has been an important component of the global energy mix for many decades, and therefore, management of waste and reprocessing of fuel have also been an important consideration. Vitrification has been an attractive solution for immobilization of nuclear wastes because there is a large range of glasses that can capture and integrate a wide range of elements into the glass phase. Benefits of this process include high corrosion durability and a much reduced volume of the final waste form. Since containment of nuclear waste is such an important topic in the nuclear industry, much research has been done over the last 40 years on the best composition of high and intermediate level waste glasses, mainly borosilicate formulations, as well as their behavior over the extended storage periods required for such wastes. The main focus areas of this research relates to waste loading, practical application, physical, and chemical properties of these glasses, as liquids and solids, as well as its containment [37]. New rare earth oxide-based glasses have been developed in the last number of years that show very good promise for enhanced immobilization of post reactor radioactive wastes. The rare earth multivalent cations can stabilize the glass matrix by reducing the number of point defects from radiation induced ion implantation and the structural changes that takes place in the glass when holes and free electrons are trapped in the matrix. This results in a number of characteristics that make them particularly suitable for nuclear waste containment [38, 39]:

- fission products incorporation rates up to 25 wt%, which is an almost 40% gain compared with current glasses;
- very high corrosion resistance;
- low crystallization tendency;
- high neutron absorbing cross sections.

An area where the role of REEs is often not recognized is in structural applications, where their influence on the mechanical properties of materials leads to altered materials that are particularly useful in modern technologies. A good example is the role that rare earth oxides play in the stabilization of zirconia (ZrO_2), which is a widely used ceramic material for structural applications. It has a very high melting point (2,680 °C), which makes it an ideal material for refractory applications. Although its use was initially limited to refractories, extensive research led to its application in many other technologies such as oxygen sensing, thermally insulating and corrosion resistant coatings for advanced engines, joint replacement material for human hips, and a number of other structural and wear applications with demanding temperature specifications. In its pure state, ZrO_2 has severe limitations for use in these applications since a destructive tetragonal-to-monoclinic phase transformation can take place. This happens particularly when the ZrO_2 is exposed to a combination of temperature, humidity, and stress. Many of the attractive features of ZrO_2, especially fracture toughness and strength, are compromised by this transformation and therefore constraining it is vital for reliable use of this material. Research into the stabilization of the ZrO_2 matrix has shown that addition of certain metal oxides can prevent this inherent phase transformation if these oxides have properties satisfying a number of criteria. They are:

- a cation radius close to that of Zr^{4+} to minimize lattice distortions which may result in high shear stresses;
- a stable cation valence state of +2 or +3, which will result in oxygen vacancies in the structure;

- the oxide crystal structure should be either cubic or tetragonal;
- the oxide should have some degree of solubility in ZrO_2.

The properties of several rare earth oxides such as CeO_2 and Y_2O_3 satisfy these rules very well and have been shown to be excellent stabilizing oxides. Cerium dioxide, for example, has played a fundamental role since it stabilizes the tetragonal phase of zirconia at room temperature. This phase provides arguably the most attractive characteristics from a mechanical engineering perspective and has found a large number of technological applications [40–42].

Although stabilization of zirconia is arguably one of the most important applications of rare earths in ceramics, they are also used in many other applications (Table 10.3).

10.4.2.4 Medicine and Health Care

The rare earths display a similarity to calcium in biological systems and this has been the basis for research into potential therapeutic applications of the rare earths since the early 1900s. One of the important attributes of REE^{3+} ions is their exceptional redox stability, which makes them very useful for biological applications in environments where organic reducing agents such as ascorbate and thiols are present in a biological system. In addition to this, they also exhibit very useful luminescent properties attributable to partially allowed *f-f* transitions as well as the completely allowed *4f-5d*, charge-transfer transitions, opening up many imaging possibilities in biological systems. Other chemical properties of the rare earths such as the remarkable oxygen buffering capacity of CeO_2 may also have far reaching therapeutic effect. For example, CeO_2 may be used to prevent harmful effects of reactive oxygen species in the human body.

Recent developments in the application of the unique nuclear properties of some isotopes of rare earths such as ^{177}Lu have also generated much interest in the nuclear medicine field. The strategy for such targeted cancer therapy involves the attachment of ^{177}Lu atoms to molecules designed to show selective association with cancerous growths after injection. The radioactive properties of ^{177}Lu essentially fulfil a dual role in this treatment scheme. As the radioactive decay of ^{177}Lu emits low energy gamma photons which can travel through tissue, they can be used for visualization and targeting by external imaging techniques. At the same time, beta particles are also emitted from ^{177}Lu decay, but since they only travel short distances in tissue, the energy of these particles is primarily absorbed by the targeted tumor, killing the cancer cells. A useful practical benefit of the ^{177}Lu decay is that it has a half-life of 6.64 days, which allows it to be shipped from the reactor production facility to clinical sites all over the world. Major clinical applications using ^{177}Lu include treatment of neuroendocrine tumors and prostate cancer metastases [44].

Following the application of gadolinium(III)-containing compounds as contrasting agents in magnetic resonance imaging (MRI) during the 1980s, interest in rare earth–based molecules and nanomaterials increased and their use in exciting new therapies such as photodynamic therapy, radiation therapy, drug/gene delivery, bio-sensing, and bio-imaging have been investigated [45, 46] (Table 10.4).

Table 10.3 Use of rare earths in ceramics [40–43].

REE	Application	Rare earth property
Y	Structural stabilization of ceramics	Oxide structure
	Synthetic gems (cubic zirconia)	Oxide structure
	Oxygen sensors	Oxide structure
	Dental coloring	Oxide structure
La	Structural stabilization of ceramics	Oxide structure
	Inhibition of grain growth in ceramic capacitors, enhance dielectric constant	Oxide structure
Ce	Stabilization of colors in ceramics	Oxygen buffer
	Structural stabilization of ceramics	Oxide structure
	Inhibition of grain growth in ceramic capacitors, enhance dielectric constant	Oxide structure
	Dental coloring	Oxide structure
Pr	Coloring agents ceramics	Narrow stable spectroscopic bands
	Dental coloring	Oxide structure
Nd	Coloring agents ceramics	Narrow stable spectroscopic bands
	Inhibition of grain growth in ceramic capacitors, enhance dielectric constant	Oxide structure
	Dental coloring	Oxide structure
Sm	Coloring agents ceramics	Narrow stable spectroscopic bands
Eu	UV fluorescence ceramics	Narrow stable spectroscopic bands
Tb	UV fluorescence ceramics	Narrow stable spectroscopic bands
Dy	Coloring agents ceramics	Narrow stable spectroscopic bands
Ho	Coloring agents ceramics	Narrow stable spectroscopic bands

(*Continued*)

Table 10.3 Use of rare earths in ceramics [40–43]. (*Continued*)

REE	Application	Rare earth property
Er	Coloring agents ceramics	Narrow stable spectroscopic bands
	Inhibition of grain growth in ceramic capacitors, enhance dielectric constant	Oxide structure
	Dental coloring	Oxide structure
Tm	Coloring agents ceramics	Narrow stable spectroscopic bands

Table 10.4 Medical or therapeutic applications of rare earths [47].

REE	Medical applications
La	Lanthanum oxide nanoparticles can be used for MRI
Ce	Cerium-doped lutetium orthosilicate is a scintillator that has been mainly used for PET imaging, to display tissue and organ function
Pr	Praseodymium oxide nanoparticles have been used in radiotherapy techniques
Nd	Neodymium has been used in lasers developed for the treatment of skin cancers, as well as permanent hair removal
Sm	^{153}Sm has been used to treat severe pain from tumors that have advanced into bone tissues
Eu	The unique optical properties of europium are used in nanoprobes designed for bio-imaging
Gd	Gadolinium improves MRI images of growths, while its magnetic properties useful for radio-contrast agents in MRI scans
Tb	^{149}Tb has been used in targeted cancer therapy
Dy	^{165}Dy has been employed in the treatment of rheumatoid knee effusions
Ho	Solid-state lasers containing holmium is effective in the treatment of certain cancers or reducing the size of kidney stones
Er	Erbium-based lasers have found applications in other medical field such as dentistry
Tm	^{167}Tm has been used in power sources for mobile X-ray devices
Yb	^{176}Yb can be used to produce ^{177}Lu which is known to be a promising radioisotope for a medical application
Lu	^{177}Lu has shown much promise in new targeted radiotherapy, for treatment of, for example, cancers such as prostate cancer

10.4.3 Rare Earths as Technology Building Blocks

The second group of end uses for rare earths involves incorporating rare earths into complex alloys and compounds, which can then be used in engineered components. These components can either be employed in stand-alone applications or in sub-assemblies, which can be integrated with other components to produce a complex engineered product. In these applications, it is not always the amount of REE present in the component that is important, but rather how critical their presence is for the functionality of the ultimate end application.

10.4.3.1 Permanent Magnets

Magnetism has been known since ancient times from naturally occurring rocks containing magnetite (Fe_3O_4). Although the mysterious behavior of these rocks was not understood, they were nevertheless used in many applications. Chinese sailors used naturally occurring lodestone rocks to navigate across oceans and the material also found widespread use in Chinese and Greek medicine. As the understanding of magnetism and magnetic materials improved through the centuries, so the uses for these materials increased. Magnetism, and permanent magnets in particular, started to play a growing, but often unseen role in modern technology. Many common devices used in daily life contain permanent magnets. The growth of applications for permanent magnets resulted largely due to improvements in magnetic properties of materials which allowed engineers to design smaller, lighter, and more efficient devices [48]. The properties of magnetic materials that are of the greatest technological importance are remanence, coercivity, and maximum energy product. Magnetic remanence can be described as the residual magnetization in a ferromagnetic material after an external magnetic field had been removed, while the magnetic coercivity is a measure of the ability of the ferromagnetic material to withstand demagnetization. Both these properties are non-equilibrium and highly sensitive to structure. Coercivity is particularly sensitive to microstructure, while remanence is sensitive to texture (crystallographic alignment) [48]. The peak energy product of a magnet indicates the maximum energy available for use outside the magnet and will depend on both coercivity and remanence [49].

Even though rare earths are strongly associated with magnetism, the pure metals are not of much use as practical magnetic materials. The reason for this is that while REEs are ferromagnetic metals and can be magnetized to form permanent magnets, their Curie temperatures are lower than room temperature. At these temperatures, the thermal energy is sufficient to overcome the internal aligning forces of the microcrystals leading to a rapid decrease in remanence and coercivity. This strongly limits the use of pure rare earth metals in most applications of permanent magnets.

Modern magnetic devices require magnets possessing large coercive forces. In order to meet this requirement, magnetic materials used in these devices should possess high magnetic anisotropies as well as high levels of magnetization and magnetic ordering. Rare earths can provide high anisotropies, originating from strong coupling of the angular momentum (L) to the large 4*f* spin (S), in the presence of an electrostatic field. Since 3*d* transition metals such as Fe or Co possess relatively high Curie temperatures, it was postulated that combining REEs with 3*d* transition metals should lead to materials with favorable properties for exploitation in magnetic devices [50]. In 1966, Austrian researcher Karl Josef Strnat, working on a US Airforce Materials Laboratory programme, discovered a very good magnetocrystalline anisotropy in intermetallic compounds between Co and a number of rare earths [51, 52]. These compounds combined the higher Curie temperature and magnetization from the 3*d* elements with the larger intrinsic coercivity of 4*f* elements. Materials such as $SmCo_5$ can therefore be regarded as the first generation of unusually powerful magnets. Since then, much research has been done to develop new magnetic materials and manufacturing processes, leading to materials that became critical to a wide range of technologies. As a result of this research, even more powerful permanent magnet materials, such as the ternary intermetallic compound $Nd_2Fe_{14}B$ had been discovered, leading to considerable scientific and technological interest [50]. The magnetic energy product (BH_{max}) of these neodymium magnets is almost 20 times larger than conventional ferrite magnets by unit volume. This means that rare earth magnets can be much smaller than other magnets at the same field strength. Table 10.5 gives magnetic properties of several common magnetic materials, highlighting the extraordinary magnetic properties of rare earth–based magnetic materials [53].

Although the Nd-based rare earth magnets have magnetic properties ideal for a vast range of technological applications, they have a number of other properties that may cause some complications. For example, their Curie temperatures remain relatively low, limiting their use in applications where they are exposed to high temperatures. This necessitated the addition of small amounts of some of the heavier rare earths such Tb and Dy which introduces further confinement of the magnetic moments by structural effects. This results in these alloys maintaining their magnetic anisotropy even at substantially higher temperatures.

The high Fe content (64–68%) of these rare earth magnets makes them especially susceptible to corrosion when they are exposed to damp environments. Once again, small amounts of other REEs, such as dysprosium, are added to the NdFeB alloy to make it more resistant to corrosion, but this does not eliminate the problem completely, and therefore, rare earth magnets are routinely coated before use in practice. A triple layer of nickel-copper-nickel is very often used as a coating for permanent magnets, while other options include corrosion resistant metals such as gold, silver, tin, or zinc. As alternatives to metal coatings in less demanding applications, epoxy, Teflon, as well as a range of paints and varnishes can be used.

Since rare earth magnetic materials are formulated to optimize their magnetic properties and not their mechanical properties, all these materials are generally hard and brittle. The NdFeB alloys are also prone to thermal shock and have low resistance to crack propagation.

Table 10.5 Magnetic properties of selected magnetic materials [53].

Material	Remanent induction (T)	Intrinsic coercivity (MA/m)	Energy product (kJ/m³)
Sr Ferrite	0.43	0.2	34
Alnico 5	1.27	0.05	44
Alnico 9	1.05	0.12	84
$SmCo_5$	0.95	1.3	176
Sm_2Co_{17}	1.05	1.3	208
$Nd_2Fe_{14}B$	1.36	1.03	350

Care should therefore be taken in the design of magnetic devices so that these materials are not used as structural units.

Despite these disadvantages, NdFeB alloys have such good magnetic properties that their uses as components in technology applications have all but replaced electromagnetic systems. Rare earth magnets provide design engineers with the desirable attributes of high remanence, predictable demagnetization behavior, a high magnetic energy product which results in smaller magnets for the specified magnetic properties, and a high inherent coercivity, which reduces the risk of magnet demagnetization.

In modern developed economies, 40–45% of all electrical energy generated is consumed by electrical motors with large industrial drives responsible for the majority of this consumption [53]. Even in consumer products, such as digital cameras, domestic appliances, and power tools as well as in the automotive industry, electric motors are crucial components. In a typical car, between 70 and 150 individual rare earth magnets are used for electric motors, sensing, actuators, and loudspeakers. Huge pressure is therefore placed on the development of more efficient electric motor technology to reduce the emissions of greenhouse gasses as well as to increase the battery life of small portable devices. This is where rare earth permanent magnet brushless motors have been making a huge contribution, especially where compactness, high torque per unit volume, better dynamic response through reduced inertia of the rotor, as well as high reliability are the principal requirements [53].

Perhaps, the most known use of rare earth magnets is in a wide range of energy generation applications, wind turbines in particular. Due to their high coercivity, NdFeB magnets can convert electrical energy extremely efficiently to mechanical energy in electrical motors or mechanical energy to electricity in generation applications. Over the last number of years, the two most rapidly growing markets for NdFeB magnets have been in electricity generation from wind turbines and drivetrains for electric vehicles. Rare earth permanent magnet alternating current generators, for example, are more efficient and reliable than conventional induction generators, since they are not subject to frequent slipring/brush failures or losses. Even though the capital cost of this technology is high, primarily due to the high cost of rare earth permanent magnets, these costs can be offset against much lower operational and maintenance costs as a result of higher reliability and lower failure rate. Wind turbines containing rare earth permanent magnet generators normally offer substantially higher availability compared to alternative solutions such as doubly fed induction generators and are therefore crucial to the viable production of electricity from wind energy.

Electric and hybrid electric vehicles (EV and HEV) are becoming more common as automotive manufacturers are placed under more pressure, to reduce the overall emissions of their product ranges, by legislation as well as community pressure to improve urban air quality and to combat climate change. Since this can only be done through greatly improved fuel efficiency, electric vehicles are becoming very attractive prospects [54]. EVs and HEVs depend on electric drives for their mobility and since the dynamic performance of these vehicles is required to be comparable to their internal combustion counterparts, they have to meet a number of key requirements [55]:

- high torque and power per unit volume;
- operation over a broad speed range, comparable to that achieved with internal combustion drive trains;

- high efficiency over the operational range;
- delivery of constant power operation over a wide operational range;
- delivery of high torque for rapid acceleration and ascent of steep inclines;
- reliability and robustness suitable for the automotive industry;
- quiet operation;
- competitive cost.

With the advent of high coercivity rare earth permanent magnets, electric motors using these magnets are becoming very appealing drive technologies. In addition, the continuous development of new motor spatial design and control philosophies has made brushless permanent magnet drives even more promising to deliver the drive characteristics required by the automotive industry for EVs and HEVs [56].

The important contribution of rare earth permanent magnets to human mobility is not only limited to terrestrial vehicles, since they also play a crucial role in the aviation industry. In contrast to the automotive industry, rare earth magnets are not directly involved in aircraft propulsion, but they nevertheless play a vital role in the functioning of the aircraft. Rare earth permanent magnet motors are mainly used in aircraft for various electric actuation systems. Because of the outstanding magnetic properties of rare earth permanent magnets, a powerful magnetic field can be maintained without further energy supply after magnetization. In comparison to traditional electric motors, a rare earth permanent magnet motor offers benefits that are extremely desirable for aircraft design. These characteristics include high efficiency, simple structural design, reliability in operation, small size, and low weight. This enables design engineers to achieve a level of performance and response that conventional electric excitation motors cannot attain. In addition, it also allows them to design specialized motors to meet very specific and demanding operational specifications, such as those for actuation of flight control surfaces.

The four kinds of actuation systems used in aircraft include hydraulic, electric, pneumatic, and mechanical systems. While hydraulic actuation systems are still the most widely used, new electric actuation systems are very promising alternatives. With the development of permanent magnets and high-power microprocessors, electric actuation systems have developed to the extent that they can compete with hydraulic actuation systems in the aerospace industry. Important systems that use NdFeB permanent magnets are [56]:

- Flight Control System
  The electric actuating system used in the flight control system is also called a power telex actuator and is mainly used for the operation of wing surfaces and rudder. This system consists of an electro-hydraulic actuator and electro-mechanical actuator, both employing high voltage permanent magnet brushless DC motors. Load control is achieved by using pulse width modulation.
- Electric Environmental Control System
  An environmental control system on an aircraft controls a number of critical functions on an aircraft such as air supply, cabin pressurization, and temperature control for the comfort of all on board the aircraft. Further functions of this system include cooling of the avionic systems, detecting smoke, and suppressing fires. Such systems use variable speed, high voltage rare earth

permanent magnet brushless DC motors capable of high speeds and power to drive a compressor.

- Air Oxygen Production System
  Modern systems developed for military aviation use include a molecular sieve method that passes bleed air from turbine engines through a molecular sieve that separates the oxygen from the air for breathing. The motor used for this application drives a valve and is specified to be capable of low-speed and high-precision. As a result of their high coercivity to volume ratio, brushless rare earth permanent magnet DC motors are ideal for these applications.
- Fuel system
  In the past, most of the hydraulic pumps and valves used to be driven by conventional DC motors. These motors have now largely been substituted by rare earth permanent magnet brushless DC motors. This allows the motor and the pump to be immersed in the fuel, with the controller and the magnet rotor body installed in the same shell, completely sealed.

Although the use of rare earth permanent magnets is probably dominated by generation and drive systems, medical imaging, and therapeutic applications are also an important area where the unique properties of these materials play an important role. MRI, for example, has become almost indispensable to clinical and basic medicine, since it can provide essential diagnostic information, such as non-invasive high soft tissue contrast. MRI has become a major user of Nd-based permanent magnets and is probably placed in the top three in the international market of neodymium permanent magnets. New developments in image reconstruction have resulted in the processing of MRI signals into images without the requirement for a completely homogeneous static field. This led to the development of portable low-cost MRI systems, based on rare earth permeant magnets. Such systems may make "point-of-care" and potentially lifesaving MRI diagnosis available for routine scans, especially to people from socio-economic backgrounds for whom this kind of examination is normally too expensive [57, 58].

Although much research has been done on finding substitutes for rare earths in magnetic materials, while maintaining magnet performance, it does not appear that much progress has been made toward a viable rare earth free magnet industry. The magnetic properties of rare earth–based magnets are such that they are likely to continue to drive inventiveness and technological application not only in electromagnetic systems but also in a range of more exotic applications. Rare earth magnets and in particular neodymium-based magnets have been used as components in an enormous number of applications, and a list of some of these applications is given in Table 10.6 [59–62].

10.4.3.2 Energy Storage

Batteries are essentially small chemical reactors where a chemical reaction takes place producing energetic electrons, to power external devices. Since the chemical equilibria in these batteries are reversible, they would in principle make ideal energy storage devices. Over the last number of decades, the research and development of rechargeable batteries had been driven by the proliferation of portable devices as well as the demands of renewable energy. Although many designs were introduced during this time, commercial use of rechargeable

Table 10.6 Applications for selected rare earth magnetic materials [59–62].

Rare earth magnetic material	Key property	Application
$SmCo_5$ Sm_2Co_{17} $Sm(Fe,Mo)_{12}$ Sm_2Fe_{17}	High Curie temperature High coercivity Low corrosivity High energy product	High end electric motors Fluid couplings Traveling wave tube amplifiers for RF signals Cryogenic or high temperature magnetic systems Bench top NMR spectrometers Actuators in rotary encoders Oxygen sensors Dental coloring
$Nd_2Fe_{14}B$ $(Nd\text{-}Pr)_2Fe_{14}B$ $(Nd\text{-}Pr\text{-}Dy)_2Fe_{14}B$ $Nd(Fe,Ti)_{12}$ $Nd(Fe,Mo)_{12}$	Highest remanence High coercivity Highest energy product Higher Curie temperature	Head actuators for HDD Mechanical e-cigarette firing switches Voice coil motors High temperature motors and generators Hybrid and electric traction drives Commercial and industrial generators Wave guides: TWT, undulators, wigglers Electric bicycles Energy storage systems Magnetic braking Magnetically levitated transportation Motors, industrial, general automotive, Pipe inspection systems Relays and switches Reprographics Torque-coupled drives Wind power generators Gauges Hysteresis clutch Magnetic separation Acoustic transducers HDD, CD, DVD Magnetic refrigeration MRI Sensors Advertising Latches Toys

batteries was dominated by lead acid and NiCd batteries. As the quest for extended lifetimes and recharge cycles continued and many chemical reactions were investigated, it became clear that (NiH_2) batteries have a superior lifespan compared to conventional rechargeable battery types [63]. A major drawback of these batteries, however, is that the structure of the batteries is impractically large, because the hydrogen has to be stored in high pressure gas tanks. These practical hurdles provided impetus to the development of the metal alloy/metal hydride or nickel-metal hydride (NiMH) battery for applications where more compact and robust batteries are required. A NiMH battery uses the hydrogen redox couple at the negative electrode and the following equations describe the reactions taking place in the battery when charging and delivering electrical current [64].

At the positive electrode:

$$2\ \mathrm{NiOOH} + 2\ \mathrm{H_2O} + 2\ \mathrm{e^-} \leftrightarrow 2\ \mathrm{Ni(OH)_2} + 2\ \mathrm{OH^-}\ \ (E^0 = 0.49\,\mathrm{V}) \quad (10.8)$$

At the negative electrode:

$$\mathrm{MH_2} + 2\ \mathrm{OH^-} \leftrightarrow \mathrm{M} + 2\ \mathrm{H_2O} + 2\ \mathrm{e^-} (E^0 = -0.83\,\mathrm{V}) \quad (10.9)$$

Overall reaction:

$$\mathrm{MH_2} + 2\ \mathrm{NiOOH} \leftrightarrow \mathrm{M} + 2\ \mathrm{Ni(OH)_2} (E^0 = -1.32\,\mathrm{V}) \quad (10.10)$$

A good metal hydride hydrogen storage material should have several characteristics in order to make it suitable for practical application in a NiMH battery electrode. The alloy must react with and release hydrogen readily at moderate pressures and temperatures, while maintaining its reactivity and capacity over thousands of cycles. Extensive research has shown that some hexagonal intermetallic compounds of the composition AB_5, AB_2, and AB, where A and B are metallic elements, are good candidates for such applications [65–68]. REEs show an unusual stability of the +3 state due to their particular electronic configuration. This means that the rare earths are electropositive and will form hydrides readily. In addition to this, they can also form alloys of the AB_5 type with base metals such as Ni and Co. The most striking property of these compounds is that they can absorb and desorb large amounts of hydrogen at room temperature. Hydrogen absorption of $LaNi_5$ at 0.25 mPa at room temperature reaches over six atoms per LaN_5,which is nearly twice as high as in liquid hydrogen [65]. During absorption of hydrogen, the unit cell of $LaNi_5H_{6.7}$ expands by about 25% without leading to a change from its hexagonal symmetry due to the high coordination number of rare earth bonding. This allows the $LaNi_5$ to maintain its structural stability and reactivity and making it a very useful NiMH battery electrode material [69].

Currently, the most common use for NiMH batteries is for providing power to electric motors in hybrid car designs. Since the market for these vehicles is still growing, the demand for NiMH batteries and consequently the demand for their contained rare earths are also growing. NiMH batteries also have many other consumer uses, such as power tools emergency lighting, cordless telephones, radio controlled toys, portable printers, and GPS systems.

The main competitors for NiMH batteries are lithium-based batteries. In comparison to lithium ion batteries, NiMH batteries have several advantages such as a lower cost, better inherent safety characteristics, long operational life, outstanding over-charge/discharge characteristics, as well as good electrochemical performance over a wide temperature range. For automotive use, these advantages are vital and are more important than theoretical efficiency. This means that nickel metal hydride batteries, utilizing the exceptional hydrogen absorption capacity of the nickel-cobalt lanthanum alloy, shows great promise to be the batteries of choice in automotive applications [70].

Currently, the driving force for developments of energy storage devices is focused on the supply of high energy and power density output to portable devices (Li-ion batteries, lithium-sulfur batteries, and super capacitors), whereas the research on high capacity stationary energy storage is directed at elevated temperature batteries such as the sodium ion battery, as well as Redox Flow Batteries (RFBs). Since a critical element in all these energy storage devices is the electrode, much focus is placed by researchers on discovering and developing new electrode materials [71]. Developments of rare earth use in the portable battery electrode applications are summarized in Table 10.7 [71].

It has been apparent for many years that renewable energy, such as wind and solar, would become an important source of supply to electrical grids of the future. However, the intermittent nature of the supply from these renewable power sources would have to be managed carefully in order to maintain the stability of national grids. This has again highlighted the need for energy storage systems that can be employed for load levelling, variable load demand management, and maintaining power delivery when the renewable energy is not available. These systems are required at almost every demand scale, from domestic scale demand to huge integrated grid systems. One such storage system, showing promise for large scale energy storage is the flow battery. A flow battery is a type of electrochemical cell where electrical energy is stored or released through redox reactions taking place in two electrolytes, partly isolated from one another by a membrane. Although the membrane prevents the electrolytes from mixing, it does allow for the passage of specific ions in order to complete the redox reaction. There are generally two kinds of flow batteries, redox, and hybrid flow batteries. If one of the electroactive components in the battery is used in solid state, the battery is referred to as a hybrid flow battery. The simple operational principle of flow batteries results in low cost and easy scalability. They are also very efficient and allow for flexible application due to modular designs and easy transportability. This modularity and scalability of flow batteries allow them to span a large scope of application from the kW to MW range [72]. The high redox potential of the Ce^{3+}/Ce^{4+} redox couple makes Ce a very attractive proposition for high capacity electrical energy storage. Although there are two other REEs (Pr and Tb) with even higher standard redox potentials, it is not feasible to use them in flow batteries because almost all available electrolytes will suffer from poor oxidative stability at these high voltages. The electrochemical reaction in the Ce-Zn hybrid flow battery can be given by the following equilibria:

Negative electrode:

$$Zn^{2+} + 2\ e^- \underset{\text{discharge}}{\overset{\text{charge}}{\Longleftrightarrow}} Zn\ (-0.76\ V) \tag{10.11}$$

Positive electrode:

$$2\ Ce^{3+} \underset{\text{discharge}}{\overset{\text{charge}}{\Longleftrightarrow}} 2\ Ce^{4+} + 2\ e^- \ (1.28 - 1.72\ V) \tag{10.12}$$

Table 10.7 Use of rare earths in batteries [71].

Energy storage device	Application	Importance of REEs
Lithium/sodium ion battery	Doping of electrode material	Introduction of REEs into the crystal lattice of the electrode material will result in substantial distortion as a result of the large ionic radius and complex coordination of REEs. Since this normally results in the formation of smaller crystallites, the capacity and the rate at which the battery can deliver its charge can be improved considerable due to lower resistance to charge transfer.
	Rare earth oxide coated cathode to improve cycle stability	High voltage lithium-ion cathodes usually consist of transition metal oxides. When high voltage conditions exist, conventional electrolytes, especially ether-based electrolytes can decompose to products that are highly corrosive to the cathode material. In this case, surface coating is a better solution than doping, with thin layer coating of CeO_2 showing exceptional stability improvements.
	Rare earth oxide coated anode to maintain Coulombic efficiency	Since a CeO_2 layer improves the rate at which lithium ions diffuse, it results in a lower propensity for polarization, while charge-transfer impedance is also reduced. This results in an enhanced conductivity between particles.
	Rare earth material used as electrode material to enhance the capacity of graphite without raising its plateau.	Density functional theory calculations showed that that the hydrogen atom on graphite composite is negatively charged. This means that more lithium ions can bind on the graphite, resulting in much higher capacity.

(*Continued*)

Table 10.7 Use of rare earths in batteries [71]. (*Continued*)

Energy storage device	Application	Importance of REEs
	Rare earth alloys to enhance the stability of alloys for the maintenance of cyclic stability	The alloy-type lithium-ion anodes usually contain Si, Ge, Sn, Bi, Sb, and Pb. These materials provide distinctive advantages with respect to energy density per unit volume. Volume expansion during cycling induces permanent cracking in the material leading to poor cycle stability. The exceptional elastic properties of rare earth alloys mitigate this cracking and therefore improve the cycle stability. Rare earths can also buffer stresses during lithium alloying, resulting in greater the cycle stability.
Lithium-sulfur battery	Rare earth alloys to enhance the stability of Li-S batteries by maintaining Coulombic efficiency	Polysulfide formation during the sulfur to Li_2S reaction can lead to low Coulombic efficiency. This happens when the dissolved polysulfides moves through the separating membrane to form a deposit on the lithium anode. Since rare earths are strong Lewis acids, they show high capacity for the absorption of polysulfides. CeO_2 embedded porous carbon electrode material shows good polysulfide absorption with added CeO_2 catalyzed transformation of the polysulfides to Li_2S.
Supercapacitors	Rare earth doped/composite material for supercapacitors to enhance capacitance of electrode materials	Addition of cerium to the capacitor composite material led to the phase transition from β-MnO_2 to α-MnO_2. The alpha form shows enhanced ion transport characteristics, leading to the exposure of more manganese active sites for higher specific capacitance.
	Rare earth materials as supercapacitor electrodes.	Rare earth compounds such as oxides, sulfides, tellurides, nitrides, as well as colloidal hydroxides, exhibit supercapacitive behavior. The specific gravimetric capacitance of these rare earth compounds is lower than that of transition metal oxides such as NiO, Co_3O_4, and MnO_2, but they have a much larger volumetric capacitance since they are much denser.

Total reaction:

$$Zn^{2+} + 2\,Ce^{3+} \underset{\text{discharge}}{\overset{\text{charge}}{\Longleftrightarrow}} Zn + 2\,Ce^{4+} \tag{10.13}$$

Although flow batteries show much promise as large scale energy storage devices, there are a number of hurdles that will still have to be overcome before flow batteries can enter wide scale, general use. These include an enhancement of rate capability, a reduction in the migration of active species, as well as the development of large-scale engineering principles. This will allow for scale-up of the technology to very large installations designed for minimum maintenance and increased life [72]. Current research on the Ce^{3+}/Ce^{4+} reaction is therefore focused on these aspects and, in particular, on immobilizing the Ce on the electrode in order to eliminate the ion exchange membrane as well as the shifting of redox potential through *4f* coordination chemistry [71].

10.4.3.3 Phosphors

According to IUPAC, luminescence is defined as "the spontaneous emission of electromagnetic radiation by electronically excited species which are not in thermal equilibrium with their environment" [73]. Simply put, luminescent materials can be described as materials that emit light that does not derive energy from the temperature of the emitting material. A class of materials that have found widespread commercial use are inorganic materials doped with small amounts of rare earth or transition metal ions, called phosphors. Luminescence in these materials is initiated by the host crystal absorbing energy and then transmitting it to activator ions embedded within the crystal lattice. This energy excites the electrons within the activator ion to a higher energy level. Upon relaxing back to lower energy levels, the electrons release energy as an emission of light with a wavelength corresponding to the energy difference between the levels (Figure 10.7).

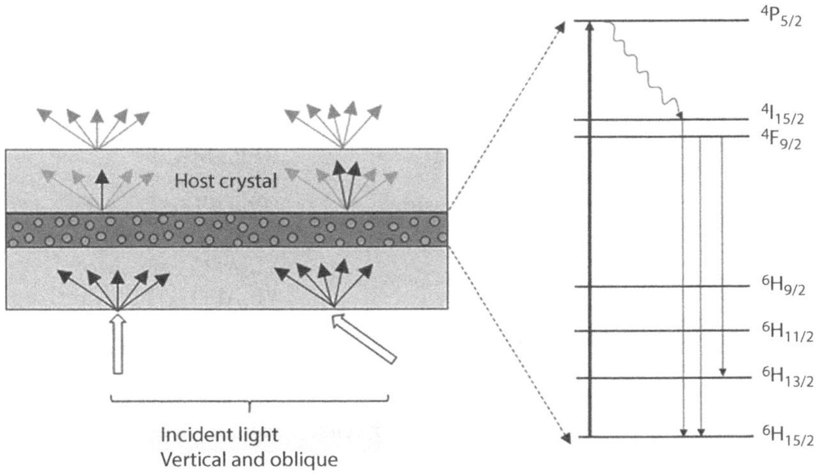

Figure 10.7 Structure and function of a REE doped phosphor.

The stable luminescence of trivalent rare earth ions makes them very attractive activators. Rare earth–based phosphors are termed triband or triphosphorous, emitting green [$LaPO_4$:Ce,Tb], red [$BaMgAl_{10}O_{17}$:Eu and (Sr, Ca, Ba)$_5$(PO_4)$_3$Cl:Eu], as well as blue [Y_2O_3:Eu] light. These phosphors are very versatile and are found in a large variety of applications [74]. Table 10.8 gives a summary of the main uses of rare earth–based phosphors.

Table 10.8 Selected properties, applications, and compositions of rare earth phosphors [75].

Property	Application	Example
RGB emission	Fluorescent lamps (CFL)	Y_2O_3:Eu^{3+} $LaPO_4$:Ce^{3+},Tb^{3+} $BaMgAl_{10}O_{17}$:Eu^{2+}
	Plasma displays	Y(P,V)O_4:Eu^{3+}, $LaPO_4$:Tm^{3+}
	LCD backlighting	(Y,Gd)BO_3:Eu^{3+}
	Biolabeling	YVO_4:Eu^{3+}, $GdVO_4$:Eu^{3+}
	Temperature sensing	Y_2SiO_5:RE^{3+}, $Y_3Al_5O_{12}$:Sm^{3+} $Y_3Al_5O_{12}$:Dy^{3+}
Band emission	LED lamps	$Y_3Al_5O_{12}$:Ce^{3+} $Tb_3Al_5O_{12}$:Ce^{3+}, SrS:Eu^{2+}, CaS:Eu^{2+},Ce^{3+} $CaAlSiN_3$:Eu^{2+}, $Sr_2Si_5N_8$:Eu^{2+}
Persistent luminescence	Security, markers	$Y_2SiO_5Eu^{3+}$ $SrA_{12}O_4$:Eu^{2+},Dy^{3+}, $Sr_2MgSi_2O_7$:Eu^{2+},Dy^{3+}
	Medical imaging	(Ca,Zn,Mg)$_2Si_2O_6$:Eu^{2+},Dy^{3+},Mn^{2+} $Zn_{2.94}Ga_{1.96}Ge2O_{10}$:$Cr^{3+}$,$Pr^{3+}$
X-ray excitability	Scintillators	(Lu,Gd)$_3Al_5O_{12:}RE^{3+}$ Gd_2SiO_5:Ce^{3+} $LiCaAlF_6$:Eu^{2+},Ce^{3+}
Upconversion	Medical imaging	REF_3, $MREF_4$
	Temperature sensing	(Pb,La)(Zr,Ti)O_3:Er^{3+} NaYF4:Yb^{3+},Er^{3+} $Yb_3Al_5O_{12}$:Er^{3+},Mo^{6+}, $Yb_3Al_5O_{12}$:Tm^{3+},Mo^{6+}
	Photovoltaics	β-$NaYF_4$:Yb^{3+},Er^{3+} LiYF4:Yb^{3+},Er^{3+}
Downconversion	Lighting imaging solar cells	$NaGdF_4$:RE^{3+}

This table is a reproduction of the table compiled by de Sousa Filho *et al.*, in a review paper on rare earths in lighting applications [75].

10.4.3.4 Glass Additives

Glass is a non-crystalline amorphous solid that is normally transparent and has a wide range of uses, technological, or decorative. The most familiar type of glass is silicate based (silicon dioxide or quartz), with the primary constituent as sand. Many of these classes of glass derive their value from its transparency and clarity. Since silica often contains noteworthy amounts of impurities such as the Fe^{2+} ion, glass made from such silica often exhibits a strong green-yellow color. Although standard mineral processing and hydrometallurgical techniques can be used to remove substantial amounts of these impurities, this is not sufficient for the manufacture of high transparency glasses. If the ratio of $Fe^{3+}/\Sigma Fe$ can be increased, then the transparency of the glass will be improved dramatically since the absorption due to Fe^{3+} is shifted more to the yellow side of the spectrum and will be faint since the Laporte forbiddeness of these electronic transitions are lifted only slightly by symmetry distortions. In order to ensure that a glass is of a low color as possible, the melt should therefore be in an oxidizing state. This can be achieved easily by the addition of an oxidizing oxide such as CeO_2. In this case, there would be two redox pairs present in the melt and the decolorization reaction can be given as:

$$Fe^{2+} + Ce^{4+} \leftrightarrow Fe^{3+} + Ce^{3+} \qquad (10.14)$$

Cerium dioxide appears to be one of the best chemical decolorizers because the iron is converted almost quantitatively to Fe_2O_3 in cerium containing glasses. Care has to be taken to control the amount of CeO_2 added for decolorization, since CeO_2 absorbs in the violet range and too much cerium could therefore result in a yellow or brown color in the glass. This decoloring of glass is referred to as chemical decoloring since a chemical reaction takes place to reduce the absorption of light. A similar effect can, however, also be achieved through a physical process of selective light absorption. This is normally done by the addition of neodymium oxide or didymium oxide which essentially compensates for the color generated by the presence of Fe^{3+} by absorbing in that wavelength.

The addition of Ce to glass has a further advantage since cerium compounds are opaque to UV radiation. This is due to intense orbitally allowed $4f^{n-1}5d$ transitions, with energies that fall in the UV wavelength region. Silicate glass can unfortunately only dissolve a limited amount of Ce before it becomes saturated and high concentrations of Ce also results in a yellow or brown color; an undesirable characteristic for applications such as windows or sunglasses. In contrast to silicate glass, phosphate glasses can dissolve large amounts of rare earth oxides. For example, a binary cerium phosphate glass can contain up to 40 mol% CeO_2. In addition to this, the UV absorption of the phosphate network is moved toward the blue side of the spectrum compared to the silicate network and this leads to a minimization of the yellow coloration of the glass by Ce [76].

When silicate glasses are exposed to high energy radiation such as gamma or X-rays, trapped electrons and holes can be created, resulting in new optical absorption that may darken the glass severely. Addition of Ce can result in resistance to such decay in the optical properties of silicate glasses, possibly by providing paths for displaced electrons to combine with holes [77]. The main applications for radiation shielding glasses are in the nuclear energy production and nuclear waste treatment fields where resistance to decay in its optical properties becomes important. Further applications are in areas such as radiation hardening of electronic devices, serving as biological shields, and preserving cultural artefacts and relics.

Of all the REEs, La_2O_3 probably plays the most critical role in the alteration of a number of physical and chemical properties of ceramics and glasses. In order to appreciate the significance of the effect of La on the properties of an optical glass, it is necessary to take a step back and consider the behavior of light when it travels from air through another medium such as glass. As light passes through glass, it slows down due to the increased density of the glass and is deflected obliquely at the interface. The reason for this is that the wave is not traveling at the same speed at all points along its front with a subsequent change in the direction of wave propagation. The relative degree of deflection is normally expressed by the Refractive Index which is calculated from the difference in speed of light through vacuum and the medium under consideration (Figure 10.8).

From an optical perspective (lens design), one of the most important aspects is to minimize the spherical aberration of a lens by reducing the curvature in the lens. This can be done by selecting a glass with a high Refractive Index since it bends light more efficiently. This also means a glass of high density. Doweidar and Saddeek investigated the influence of La_2O_3 on the structure of lead borate glasses and showed that the addition of La^{3+} to these glasses, promote the formation of non-bridging oxygen ions in the glass network by neutralizing the negative charge on the glass forming ions. Consequently, the density of the glass increases not only because heavy ions are introduced but also since the molar volume of the matrix decreases with increasing La_2O_3 concentration [78]. La_2O_3 therefore imparts excellent optical properties to glass and is used widely in many optical systems.

The Laporte optical selection rules result from the conservation of angular momentum, energy, and linear momentum. This means that electronic transitions between states of

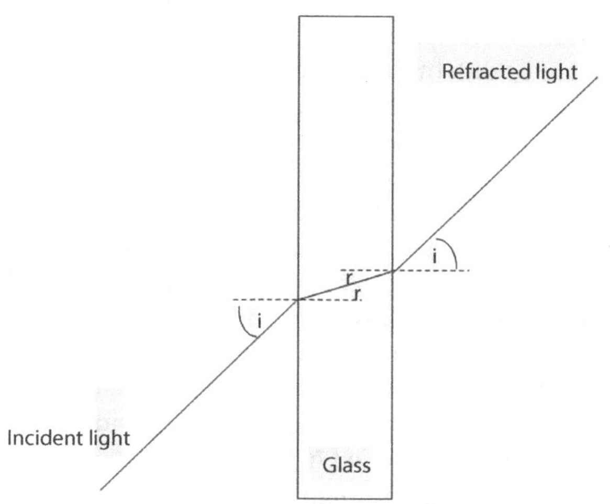

Figure 10.8 Refraction of light by glass.

the same parity are forbidden. Consequently, if a molecule is centrosymmetric, transitions within a given set of orbitals such as the *f* orbitals in a REE (redistribution of electrons within a given subshell) are forbidden. Although the 4*f*-4*f* transitions in REE ions are therefore formally forbidden by the Laporte Rule, they become somewhat allowed if the center of symmetry is disrupted by mixing of a 4*f*-orbital with a 5*d*-orbital or with the excited states of the neighboring ligands, resulting in absorption in the optical range (i.e., resulting in color) [79]. These absorption bands are relatively weak in intensity due to the nature of the transitions, but they are very narrow because the 4*f* electrons are very effectively shielded by the filled $5s^2$ and $5p^6$ shells, which have a higher energy than the 4*f* shell [1]. Although these narrow bands are broadened when the REE ions are inserted into a glass network, the colors arising from the rare earth ions remain relatively insensitive to the overall composition of the glass. Due to the stability of the trivalent state of the rare earths, these ions are likely to remain in this valence state in glasses, and therefore, the redox environment will also not affect the colors. These glasses are often dichroic, and hence, they will exhibit different colors with changes in lighting conditions. In some instances, their color is also determined by the amount of REEs in the glass as well as the thickness of the glass. Some rare earths such as Nd, Er, and Pr are used to color some special glasses due to the uniqueness of its color resulting from the narrow *f*-*f* absorption transitions.

A good example of uses for these unique narrow absorption bands in glass is the use of a mixture of Nd and Pr in blacksmithing and glassblowing safety glasses. Incorporating this mixed oxide into the glass network leads to a selective absorption of yellow light at around 589 nm. This blocks the yellow light from hot sodium in molten glass as well as the strong infrared light from a blacksmith's forge and so reduces the risk of eye damage, while preserving visibility. This mixture of REEs is also used in photographic filters as an optical band-stop filter that attenuates the orange portion of the spectrum to very low levels. This effect is especially useful for enhancing photos of autumn scenery since it will make leaves more vibrant. In the 1920s, Leo Moser developed a special kind of glass called "Heliolite" glass by adding Nd and Pr to glass at a ratio of 1:1. This glass changes color from amber to red to green depending on the light and is predominantly used for glass art [80].

Perhaps, the most direct use of materials exhibiting narrow spectroscopic absorption bands is as a spectrophotometric calibration material. When adding Nd and Pr in specific ratios to glass, it results in a reference material that has a number of benefits for spectrophotometric calibration [81]:

- Dual purpose filter
  - (i) Calibration filter in the range from 329–875 nm
  - (ii) Photometric accuracy calibration in the UV range
- Robust
- Insensitive to temperature and humidity

Similarly, holmium glass is also used extensively as a wavelength standard, since it possesses some characteristics that make it an excellent wavelength standard. It does not induce a slit positioning error as is the case with atomic emission lamps, while it is also compact and easy to use. Arguably, the most important property of these glasses and the main reason why it is so widely used is its high stability over extended periods of use [82].

10.4.3.5 *Lasers*

In 1960, Theodore Maiman demonstrated a solid-state ruby laser, featuring Cr^{3+} ions as gain centers in a sapphire matrix (Al_2O_3). A laser is a device that can convert either electrical or optical energy into light. It normally consists of a hollow space closed off with flat or spherical mirrors at the ends. This cavity is filled with a material whose atoms or molecules can be excited by electrical or optical energy to emit single wavelength light. In a solid-state laser, the gain medium is a crystalline material containing ordered and rigid bonds. Laser light is produced after being illuminated by light of a higher frequency than the desired laser frequency. One of the most successful and versatile lasing materials developed since the first ruby laser is the neodymium-doped yttrium aluminium garnet; Nd:$Y_3Al_5O_{12}$ used in Nd: YAG lasers. Due to the low crystal field effect on the *f*-orbitals of rare earth ions, gain materials containing rare earth ions provide unique versatility and are therefore very common in laser devices. Rare earth ions offer a particularly valuable technical advantage as flexible lasing materials. The reason for this is that the quantum energy levels of the lasing wavelength determining Nd-ions remain relatively insensitive to the host crystal matrix. For example, the strong fluorescence wavelength of an optically excited, free Nd^{3+} essentially does not change from 1,064 nm when introduced into a YAG or glass matrix. This is not the case with transition metal ions which will exhibit Stark shifts (the crystal field changes the energy levels of the transition metal ions) that will alter the wavelength at which the ions, introduced into the crystals, will fluoresce [83]. Nd:YAG lasing takes place when electrons transition from the $4F_{3/2}$ to the $4I_{11/2}$ atomic energy level, producing light at 1,064 nm. Starting in its $4I_{9/2}$ ground state, Nd^{3+} ions embedded in the YAG crystal are excited into a broad multiple of higher "excitation band" levels by means of optical pumping. From these levels, the ions relax rapidly through a non-radiative transition into the $4F_{3/2}$ state, from where they can transition spontaneously to the $4I_{11/2}$ state, releasing a photon. Another rapid, non-radiative relaxation from $4I_{11/2}$ back to the ground state then completes the cycle (Figure 10.9).

These photons generated in the lasing medium will bounce back and forth between the two mirrors, resulting in other electrons falling into the lower energy state by releasing photons. In this way, millions of electrons are stimulated to emit photons. The light generated within the lasing medium is reflected many times between the mirrors before it escapes through a half reflecting mirror. This is called stimulated emission and results in a power gain. The light produced from this stimulated emission becomes the optical output of the laser as shown in Figure 10.10.

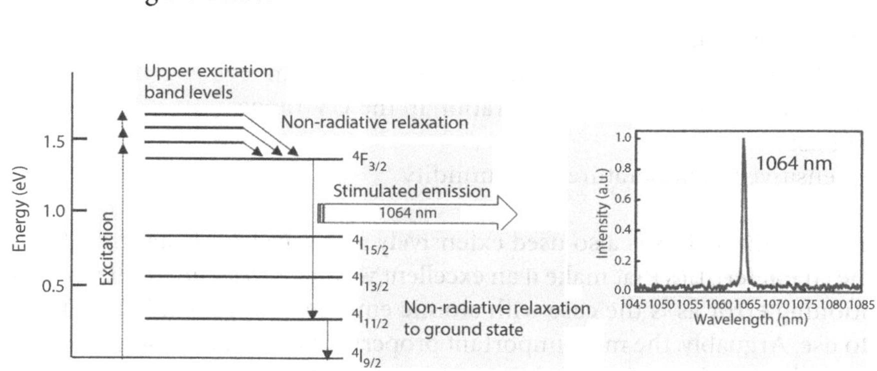

Figure 10.9 Electronic transitions in a Nd:YAG laser (courtesy of Dr. Fei Tang).

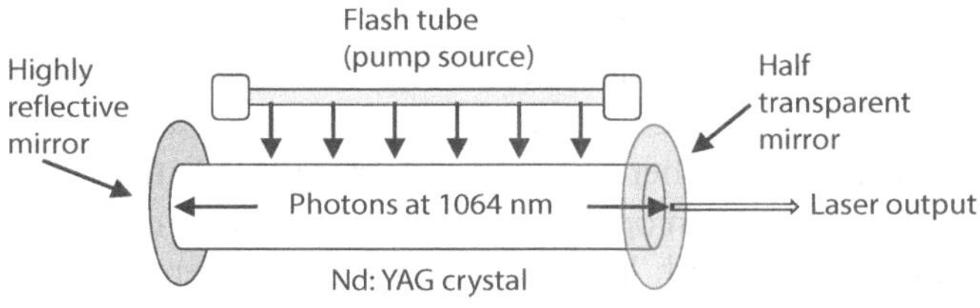

Figure 10.10 Functioning of a Nd:YAG laser.

Advantages of a Nd:YAG laser:

- Low power consumption
- High gain
- Good thermal properties
- Good mechanical properties
- The efficiency of Nd:YAG laser is very high compared to the ruby laser

These properties of Nd:YAG lasers lead to great versatility and they have found a wide array of uses ranging from medical to military applications. Nd:YAG lasers are used by eye specialists to correct complications that may occur from cataract surgery such as posterior capsular opacification and, in treatment of noncancerous thyroid nodules, cancerous growths in the liver as well as skin cancer [84]. Continuous-wave tuneable lasers have found extensive use in scientific research, especially where random wavelength access and spectral lines of extremely narrow frequency ranges are required. This is specifically true for fields such as molecular spectroscopy and atomic physics where spectral lines have to be resolved into their components and specific wavelengths can often not be produced by fixed frequency lasers [85]. Due to the concentrated power generated by these lasers, they are also often used in manufacturing applications such as glass and plastics etching as well as metal and semiconductor cutting. Military applications include rangefinders and laser target designators as well as higher power (100 kW) destructive weapon systems to defend against rockets, artillery, and mortars [86].

An interesting new development in the automotive industry is the use of small Nd:YAG lasers as fuel ignition devices instead of conventional spark plugs. Fuel ignition by laser offers several advantages over conventional electric spark ignition. Those of most interest are:

- no quenching of the combustion flame kernel;
- targeted ignition at any point in the combustion chamber;
- possibility of multiple combustion points in the combustion chamber;
- time-based ignition control.

If automotive engineers can exploit these benefits in practical designs, then engine efficiency may be improved through reliable lean-burn operation. This would lead not only to a reduction in fuel consumption but also to a reduction in harmful emissions. Laser ignition in internal combustion engines has the potential to unlock new design avenues to ensure their continued use by reducing their environmental footprint. It may therefore play a role in providing a controlled transition to electric mobility through more efficient ignition in hybrid engines and the improved combustion of innovative new fuels [87].

After Nd:YAG lasers, fiber lasers and fiber optical amplifiers are probably the next best known optical technologies containing rare earths. Fiber laser technologies are advancing at a rapid rate and are finding such a variety of practical industrial applications that they are becoming serious alternatives to other laser technologies [88]. At the heart of a modern fiber laser and many fiber amplifiers is an optical fiber doped with rare earth ions. Since rare earth ions normally form complexes with large coordination numbers, a deficit in non-bridging oxygen ions in isolated sites in a silicate network results in a grouping of rare earth ions in order to share the available oxygens. This leads to poor solubility of rare earth ions in silica glass and can even result in phase separation at higher concentrations. The most effective co-dopant for improving REE solubility in silicates is Al. Al can be incorporated in glass as a network former as well as a network modifier and therefore dissolves very well in a silicate. The 4-coordinated Al can share non-bridging oxygens with REE ions allowing them to be incorporated into the silicate network [89]. As with solid state lasers, other rare earth ions such as Er, Yb, Nd, and Tm may also find use as the active gain centers in fiber lasers. In the common "double-cladding" fiber laser, the lasing mode propagates in the rare earth doped fiber core. A pump light is located in the inner cladding, while the outer cladding provides total internal reflection for the pump in order to improve efficiency [88]. Figure 10.11 gives a diagrammatical description of the structure as well as the functioning of a fiber laser [90].

The stimulated emission from the REE doped core is guided within the core section and will therefore be amplified to high intensities before it leaves the fiber as a powerful laser beam at a wavelength of 1,060 nm. Since the core diameter can be very small (3–10 μm),

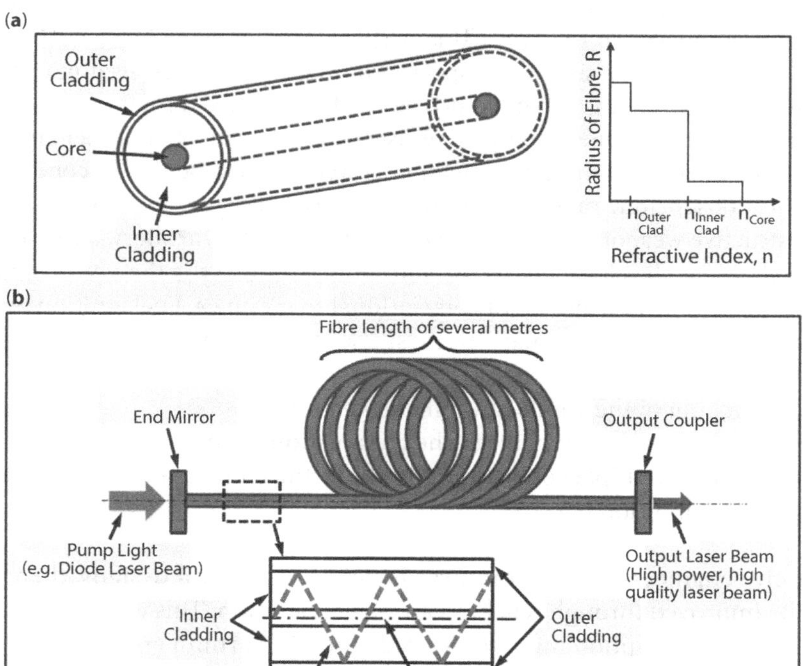

Figure 10.11 Structure and functioning of a fiber laser [90]. (a) Structure of the fiber laser; (b) Pump light propagating in the fiber.

the power density of the laser beam will be very high. This high power density produced in fiber lasers makes them ideal for material processing applications such as the cutting and welding of metal. The inherent design of fiber lasers, which allows for efficient cooling through the large surface area cladding, reduces temperature induced instability in the refractive index, and so allows for high power applications where conventional Nd:YAG lasers are plagued by instability. Yb is the preferred doping material for the lasing medium in fiber lasers since it is highly absorbing of pump radiation and produces high power laser beams in the 1,060–1,080-nm wavelength range. These lasers can produce beams in the kilowatt range, which means that they could be used for the processing of a range of materials [91]. These lasers have been shown to realize much higher cutting speeds in thin to medium section stainless steel, mild steel, and aluminium cutting when compared to a CO_2 laser. Since light from a fiber laser is already in a fiber, it is easy to direct the beam very accurately to the work zone of material processing equipment such as laser cutters and welders. The high power density that can be achieved also makes them interesting for military applications.

In order to transmit optical signals through optical fibers over long distances (> 100 km), compensation for attenuation losses within the fiber is required. The conventional way of doing this was by converting the optical signal to an electrical signal, amplifying this electrical signal and converting it back to an optical signal. Although such an approach is feasible, it is limited by the optical to electrical and electrical to optical conversions. By using lasers, amplification can be achieved without having to do this signal conversion first, since the stimulated emission in the gain medium of the fiber laser results in the amplification of incoming light. For this reason, rare earth doped fiber amplifiers are finding increasing importance in optical communications systems. Perhaps, the most deployed application is Er doped fiber amplifiers because of their ability to amplify signals at the low loss 1.55-μm wavelength range.

Rare earth–based lasers have also found application in a wide range of other fields ranging from medicine to entertainment. Table 10.9 gives a summary of some commercial application of rare earth lasers and laser systems [92–95].

When one considers all the applications of rare earths and the difficulty of substitution due to their unique properties, it becomes clear that they could offer substantial financial benefits to the global economy. An enormous number of individual products from a variety of product lines contain REEs and their economic footprint can be measured by output, payrolls, jobs, and taxes paid. According to a report compiled by the American Chemistry Council and the Rare Earth Technology Alliance, the rare earth industry is regarded as undeniably essential to the North American economy (US and Canada) [96]. The report estimated that for every direct job in the rare earth industry, five other jobs are created elsewhere in the American economy, leading to an almost two billion dollar economic footprint. In addition to this, there are enormous economic contributions from using rare earths in the creation of products, resulting in a downstream output of in excess of $300 billion. It is therefore not surprising that many nations the world over have been embracing REEs and in many cases, they are regarded as or have been declared strategic metals. Leading industrial countries or economic zones such as the United States of America, Japan, and the European Union have all formally declared rare earths as critical metals [97–99]. Rare earths are therefore clearly hugely valuable and even carry strategic importance, but does AMD contain enough of them to make their recovery viable?

Table 10.9 Applications of rare earth lasers.

Laser	Industry	Application	Effect
Er:YAG and Nd:YAG	Medical	Dermatology—skin rejuvenation, lesion treatment, tattoo removal, hair removal	Skin cell ablation, thermal fragmentation and photo thermolysis
		Dentistry	Hard tissue ablation (cavity preparation), soft tissue ablation and coagulation (surgery), selective heating (tooth whitening)
		Surgery	Soft tissue ablation and coagulation (surgery)
		Diagnostics: Optical coherence tomography Tissue characterization Porosity measurement Cavity delineation	Low coherence interferometry Fluorescence spectroscopy Gas in scattering media absorption spectroscopy
		Cancer treatment	Hyperthermia Photosensitization
Nd:YAG	Military	Laser rangefinder/laser target designator LADAR/LIDAR	Time of flight principle
		Atmospheric communication	Modulated optical pulsing
		High energy weapon	
Er:YAG and Nd:YAG	Industrial and commercial	Cutting, welding, drilling, engraving	Thermal ablation
		Accelerometers	Interferometry
		Holography	Light diffraction
		Optical tweezers	Refractive index
		3D scanners	Time of flight principle
		Additive manufacturing	Laser sintering
		Metrology	Time of flight principle
	Entertainment		Light color Diffraction Time of flight principle

10.5 Occurrence—From Magma to AMD

> *"By the year 2000, we will not be wasting our coal ash, in which geochemists have shown that there is a notable concentration of rare elements, such as germanium and rare earths. We will be recovering these elements."*
>
> **Dr. Edward Steidle**
> *16 February 1952* [100]

Contrary to what the name of this group of elements may suggest, they are not really rare and are relatively evenly distributed in the crust of the earth at a concentration similar to that of base metals such as Ni and Co (Figure 10.12).

Rare earths occur in many mineral forms, and although there are at least a dozen of these that are considered viable for extraction, the bulk of global rare earth recovery is mainly from the two minerals monazite and bastnäsite [101]. Rare earth minerals occur in diverse geological environments, but they can be divided into two categories. They can occur in primary deposits resulting from igneous and hydrothermal enrichment processes or in secondary deposits where the elemental concentration proceeded *via* sedimentary and weathering processes. Hard rock deposits such as bastnäsite are considered to be of primary origin, while the placer deposits such as monazites in beach sands are of secondary origin. Carbonatites and weathered deposits are the most dominant sources of primary rare earth deposits, while the most substantial types of secondary deposits are the placer and ion-adsorption clay deposits [101]. Rare earth ion-adsorption type deposits are enriched by

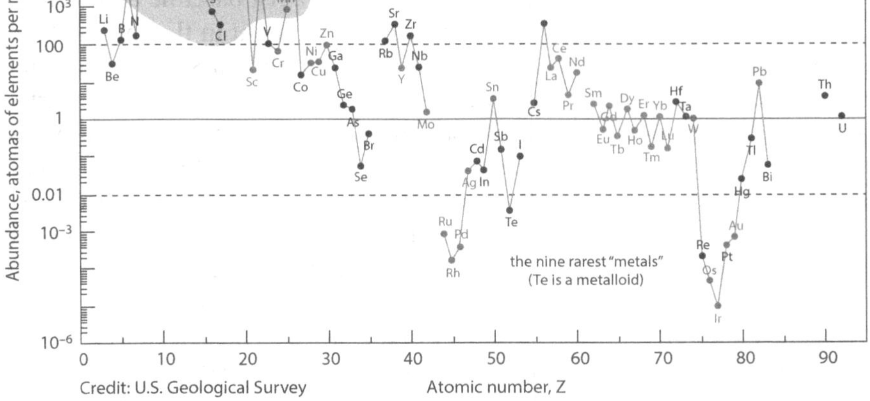

Figure 10.12 Abundance of the rare earth elements in the crust of the earth.

the weathering of rare earth mineral containing igneous rocks such as granite. As a result of surface weathering, rare earth minerals are decomposed and the liberated rare earth ions (REE^{3+}) are absorbed on clay minerals (normally kaolin and smectite groups), substituting base metal or alkali earth cations. Ion-adsorption ores generally contain higher amounts of HREE and lower amounts of Th and U than other rare earth ores. Rare earths can be recovered from REE bearing ion exchange clays by simple dilution with an electrolyte solution such as ammonium sulfate at ambient temperature [102].

Although the bulk of the actual economic extraction of rare earths may come from conventional terrestrial deposits and is likely to remain that way for the foreseeable future, there are some non-conventional sources of rare earths that may become interesting as supplementary sources, especially as the drive to cleaner mining and processing gains momentum. In such cases, the viability of rare earth recovery would then not be determined by financial considerations only but may be secondary to a primary purpose such as an environmental consideration or legislation.

Although there may be other such sources, AMD emanating from coal deposits is an interesting case that may be used to consider the potential for rare earth recovery from low pH streams associated with any mining operation. Coal can become enriched in rare earths by a variety of mechanisms such as organic complexing or fossil apatite uptake. It was shown in a number of studies that the distribution of rare earths between the organic and mineral phases or between the coals and the inorganic layers that they are associated with, varies over the coalfield and this is probably a result of a number of effects such as sedimentation, water chemistry, ion-exchange, and chelation [103–106]. The concentration of rare earths in coal deposits is generally too low to even consider direct recovery, but they can be liberated into waste streams by natural leaching processes. When pyritic waste rock material from coal mines is exposed to air and moisture, the di-sulfides are oxidized and generate sulfuric acid. This sulfuric acid leads to very low pH values in the water in contact with virgin coal deposits or coal waste repositories such as fine coal dumps or ash dumps. An effective leaching environment is therefore established, and REEs are leached into solution. Once in solution, rare earths are mostly complexed as $REESO_4^+$, which is the dominant rare earth species in AMD. This sulfate complexation tends to inhibit the sorption of the rare earths onto clays and therefore stabilizes them in solution [107]. A 2017 study on AMD from 18 coal mine discharges in the Appalachian Basin in the United States showed an enrichment in rare earths of 1–4 orders of magnitude against baseline unaffected waters in the area. Most discharges showed varying levels of enrichment in the middle REE fraction when compared to the average distribution in a North American Shale Composite, but this is unlikely to have an effect on the supply and demand balance over the rare earth series.

The rare earth content of the AMD released from the Appalachian coal mines is given in Table 10.10. The rare earth concentrations are given as flow weighted mean concentrations, while the Av AMD in the table is an estimate for the entire Appalachian discharge [108].

Clearly, a stream containing about 40 μg/L of rare earth metals does not appear to be very attractive from a standalone operational perspective, even if it can be upgraded substantially. The total REEs contained in such a stream (538 t) is also too modest a production figure for sustaining a flowsheet that is acknowledged to be one of the most complex and expensive in the minerals industry. However, financial considerations may not be the only imperatives that will determine whether rare earths are recovered from AMD or not. AMD is a potentially toxic stream that holds dangers not only for humans,

Table 10.10 Estimated rare earth content in Appalachian acid mine drainage. Concentrations in μg/L, discharge in L/s.

Study	Y	La	Ce	Pr	Nd	Sm	Eu	Gd	Tb	Dy	Ho	Er	Tm	Yb	Lu	ΣREE	Discharge volume
2008	7.8	5.1	12	1.6	6.5	1.6	0.4	1.7	0.28	1.5	0.3	0.5	0.1	0.6	0.1	40.1	13 280
2015	14	4.7	11	1.6	7.6	2.5	0.6	3.1	0.5	2.7	0.5	1.4	0.2	1.0	0.2	52.3	2 841
2017	7.0	0.8	3.7	0.5	2.3	0.8	0.2	1.0	0.2	0.9	0.2	0.4	0.1	0.3	0.1	18.3	1 048
Av AMD	8.8	4.7	11	1.5	6.4	1.7	0.4	1.9	0.3	1.7	0.3	0.9	0.1	0.7	0.1	40.8	418,000
Total flux (t/a)	116	63	148	20	85	22	5.5	25	4.1	22	4.3	12	1.6	8.9	1.4	538	

but for all living organisms on earth. Treating AMD therefore has several value propositions. These include:

- License to operate
- Positive public sentiment
- Less expensive waste deposition facilities
- Value of contained metals
- Value of new water resource

Although all these aspects may contribute to a positive business case for the recovery of the rare earths from AMD, a critical factor is access to an existing rare earth separation facility. Without purification and separation of the recovered rare earths, only a fraction of their value, at best, will be realized.

There are several processes for the treatment of AMD, but one of the simplest and least expensive processes is a chemical precipitation approach that essentially consists of two stages:

- Stage 1: raising the pH of the AMD stream to a pH of 11–12, which will result in the almost quantitative precipitation of all the (semi-)metals in the stream. The result of this is a solids stream or sludge containing all the value metals as well as an aqueous stream containing gypsum at saturation.
- Stage 2: removing the gypsum from the saturated stream from stage 1, through the formation of a very insoluble calcium-aluminium sulfate salt.

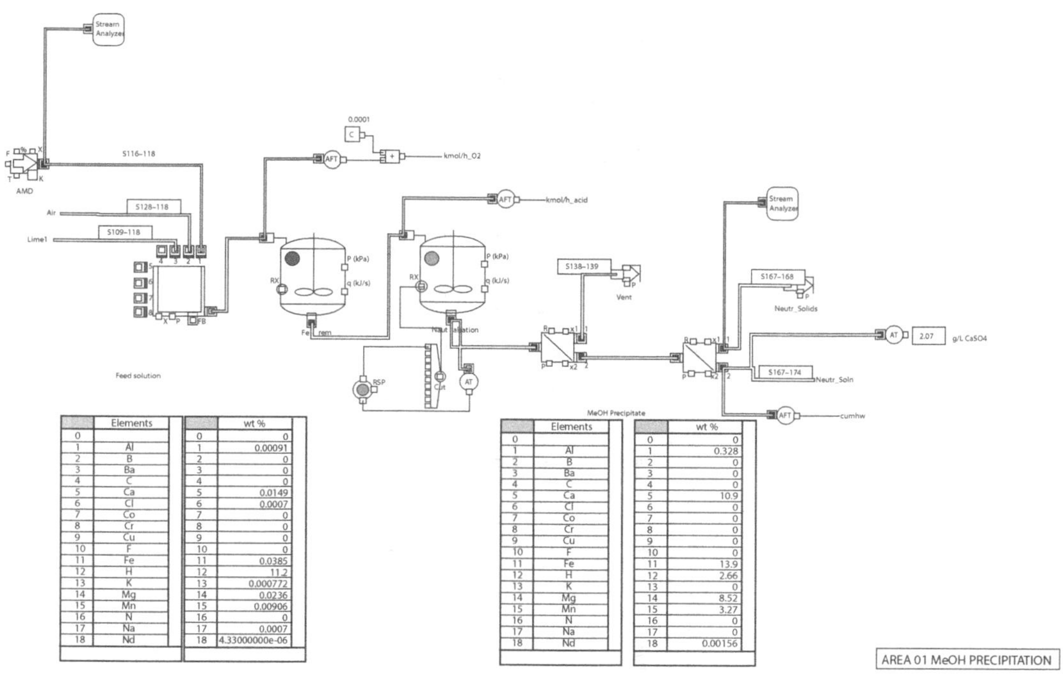

Figure 10.13 Simulation of the upgrade of the REEs in an AMD stream from the Appalachian Coal Basin.

In stage 1 of this process, all the metals including REEs will be precipitated as hydroxides from the AMD stream, thereby concentrating them in a much smaller volume, leading to a much smaller stream to treat for the recovery of these elements. Similarly, any other value metals such as uranium or base metals can then also be recovered from this precipitate.

Running a simulation of such an AMD flowsheet shows that the rare earths can be concentrated from around 40 μg/L to about 15 mg/L, which represents an upgrade ratio of more than 300. Figure 10.13 shows the simulated flowsheet, while Table 10.11 gives the composition of the feed and metal hydroxide precipitate from this simulation of the AMD stream from the Appalachian Coal Basin. Neodymium was used as a proxy for the REEs.

The mixed rare earths can now be recovered selectively from the other metals in the hydroxide precipitate through established hydrometallurgical processes such as the Resin-in-Pulp ion exchange process and purified through a selective precipitation technique. The resultant

Table 10.11 Simulation of the upgrade of REEs by neutralization to pH 12.

Feed solution			MeOH precipitate		
	Elements	**wt%**		**Elements**	**wt%**
0		0	0		0
1	Al	0.00091	1	Al	0.328
2	B	0	2	B	0
3	Ba	0	3	Ba	0
4	C	0	4	C	0
5	Ca	0.0149	5	Ca	10.9
6	Cl	0.0007	6	Cl	0
7	Co	0	7	Co	0
8	Cr	0	8	Cr	0
9	Cu	0	9	Cu	0
10	F	0	10	F	0
11	Fe	0.0385	11	Fe	13.9
12	H	11.2	12	H	2.66
13	K	0.000772	13	K	0
14	Mg	0.0236	14	Mg	8.52
15	Mn	0.00906	15	Mn	3.27
16	N	0	16	N	0
17	Na	0.0007	17	Na	0
18	Nd	0.0000043	18	Nd	0.00156

purified mixed rare earth hydroxide, carbonate, or oxalate can then be sold to an existing separation plant as feed stock and thereby realize about 30% of the contained value of the separated rare earth oxides. A ballpark estimate of the possible value that may be derived from the ±500 t of REO contained in the AMD of such a coal basin is therefore about \$2M – \$4M per year. This is based on 100% recovery and a price of \$16–\$26 per kilogram for a typical REO basket.

10.6 REEs—From AMD to High Technology?

Rare earths have clearly become indispensable in many technological developments in the world today and are likely to become even more sought after in future due to expected supply constraints and increasing demand from new developments using these unique elements. Every source of rare earths will therefore be of interest to the market and this certainly also includes AMD. The question is whether the rare earth content is attractive enough to pursue active recovery from AMD. The case study discussed earlier have shown that there are REEs in AMD and even substantial amounts if all the potential sources of AMD in the world are considered. In addition to this, the technology also exists to recover them from the dilute AMD streams. Unfortunately, all this is not enough and the rare earth content in AMD is likely to remain untapped due to a combination of very low grades and prohibitive production and logistical costs. The same argument will be valid for other value metals such as base metals or uranium. However, even if there is no purely economic driver to recover these value metals, in spite of their potential in new technological developments, it still does not mean that the rare earths will never be recovered as a by-product from AMD. The reason for this is that the true value of AMD lies in a much more mundane and familiar resource, water. As the global population increases, the competition for water resources will become fiercer not only from increased human consumption but also from greater use in mining and industry as well as agriculture. Water is likely to become a critical commodity in many regions and any possible source is likely to become a target for use. AMD will then no longer be regarded as a waste stream that one is forced to treat by law, but rather as a valuable resource worthy of paying a premium to recover.

A by-product is normally regarded as a secondary product from a process that is focused on recovering a viable primary product. When the business case for the recovery of water from AMD becomes compelling, then the recovery of REEs from the AMD as a by-product may become a reality since the processes required to treat AMD will inherently recover and concentrate the rare earth content. So, although the rare earths in AMD is currently likely to remain unexploited, it is probably unwise to disregard the possibility that one day a Nd : YAG laser will shine brightly with rare earths from a lowly waste stream such as AMD.

Acknowledgements

I would like to thank Dr. P.J. van Staden of Mintek for carrying out the simulation for the upgrading of the REE in the AMD.

References

1. Gschneider, K.A. and Eyring, L., *Handbook of the Physics and Chemistry of Rare Earths*, North-Holland Publishing Company, Amsterdam, 1978.
2. Generalić, E., *History of the Rare Earth Elements*, 19 September 2012, [Online]. Available: https://www.periodni.com/history_of_rare_earth_elements.html. [Accessed 19 July 2019].
3. Rowlatt, J., Rare earths: Neither rare, nor earths. *BBC News Mag.*, 2014. Available at: https://www.bbc.com/news/magazine-26687605 (Accessed: 19 July 2019).
4. Levy, S.I., *The Rare Earths: Their Occurence, Chemistry, and Technology*, E. Arnold, London, 1915.
5. Uhlenbeck, G.E. and Goudsmit, S., Spinning Electrons and the Structure of Spectra. *Nature*, 117, 264–265, 1926.
6. Atkins, P.W., *Physical Chemistry*, Oxford University Press, Oxford, 1978.
7. Cotton, F.A. and Wilkinson, G., *Advanced Inorganic Chemistry*, John Wiley & Sons, New York, 1980.
8. Flowers, P., Theopold, K., Langley, R., Robinson, W., *Chemistry - Creative Commons Attribution License 3.0*, Openstax, Houston, 2018.
9. Yost, D.M., Russell, H.R., Garner, C.S., *The Rare-Earth Elements and their Compounds*, John Wiley and Sons, New York, 1947.
10. Hatch, G.P., Dynamics in the Global Market for Rare Earths. *Elements*, 8, 341–346, 2012.
11. Feliczak-Guzik, A., Hierarchical zeolites: Synthesis and catalytic properties. *Microporous Mesoporous Mater.*, 259, 33–45, 2018.
12. Vogt, E.T.C. and Weckhuysen, B.M., Fluid catalytic cracking: Recent developments on the grand old lady of zeolite catalysis. *Chem. Soc. Rev.*, 44, 7342–7370, 2015.
13. C. Plank and E. Rosinski, Catalytic Hydrocarbon Conversion with a Crystalline Zeolite Catalyst, 1964. USA Patent US3140253(A).
14. Sousa-Aguiar, E.F., Trigueiro, F.E., Zotin, F.M.Z., The role of rare earth elements in zeolites and cracking catalysts. *Catal. Today*, 218-219, 115–122, 2013.
15. Weitkamp, J., Zeolites and catalysis. *Solid State Ionics*, 131, 175–188, 2000.
16. Wormsbecher, R., Cheng, W., Wallenstein, D., *Role of the Rare Earth Elements in Fluid Catalytic Cracking*, vol. 108/2010, pp. 19–26, GRACE DAVISON CATALAGRAM®, 2010.
17. Backeberg, N., *Rare Earths: The Engine of Electrification*, Rosskill, London, 2018.
18. Kaspar, J., Fornasiero, P., Graziani, M., Use of CeO_2-based oxides in the three-way catalysis. *Catal. Today*, 50, 285–298, 1999.
19. Li, P., Chen, X., Li, Y., Schwank, J.W., A review on oxygen storage capacity of CeO_2-based materials: Influence factors, measurement techniques and applications in reactions related to catalytic auromotive emissions control. *Catal. Today*, 327, 90–115, 2019.
20. Dutta, P., Pal, S., Seehra, M.S., Shi, Y., Eyring, E.M., Ernst, R.D., Concentration of Ce^{3+} and oxygen vacancies in cerium. *Chem. Mater.*, 18, 21, 5144–5146, 2006.
21. Selvan, V.A.M., Anand, R.B., Udayakumar, M., Effects of cerium oxide nanoparticle addition in diesel and dieselbiodiesel-ethanol blends on the performance and emission characteristics of a CI engine. *J. Eng. Appl. Sci.*, 4, 7, 1–6, 2009.
22. Bafghi, A.A.T., Bakhoda, H., Chegeni, F.K., Effects of Cerium Oxide Nanoparticle Addition in Diesel and Diesel-Biodiesel Blends on the Performance Characteristics of a CI Engine. *Int. J. Mech. Mechatron. Eng.*, 9, 8, 1507–1512, 2015.
23. Venkatesan, H., Sivamani, S., Sampath, S., Gopi, V., Kumar, D.M., A Comprehensive Review on the Effect of Nano Metallic Additives on Fuel Properties, Engine Performance and Emission Characteristics. *Int. J. Renewable Energy Res.*, 7, 2, 825–843, 2017.

24. Gharehghani, A., Asiaei, S., Khalife, E., Najafi, B., Simultaneous reduction of CO and NO_x emissions as well as fuel consumption by using water and nano particles in Diesel-Biodiesel blend. *J. Cleaner Prod.*, 210, 1164–1170, 2019.
25. Sajeevan, A.C. and Sajith, V., Diesel Engine Emission Reduction Using Catalytic Nanoparticles: An Experimental Investigation. *J. Eng.*, 2013, 1–9, 2013.
26. Badwal, S.P.S., Giddey, S., Munnings, C., Kulkarni, A., Review of Progress in High Temperature Solid Oxide Fuel Cell. *J. Aust. Ceram. Soc.*, 50, 1, 23–37, 2014.
27. Omar, S. and Nino, J.C., Consistency in the chemical expansion of fluorites: A thermal revision of the doped ceria. *Acta Mater.*, 61, 5406–5413, 2013.
28. Wang, M., Mu, S., Sun, F., Wang, Y., Influence of Rare Earth Elements on Microstructure and Mechanical Properties of Cast High-speed Steel Rolls. *J. Rare Earths*, 25, 490–494, 2007.
29. Pan, F., Zhang, J., Chen, H., Su, Y., Kuo, C., Su, Y., Chen, S., Lin, K., Hsieh, P., Hwang, W., Effects of Rare Earth Metals on Steel Microstructures. *Materials*, 9, 417, 2016.
30. Lee, J., Evaluation of the nucleation potential of intragranular acicular ferrite in steel weldments. *Acta Metall. Mater.*, 42, 3291–3298, 1994.
31. Goto, H., Miyazawa, K., Yamaguchi, K., Ogibayashi, S., Tanaka, K., Effect of Cooling Rate on Oxide. *ISIJ Int.*, 34, 414–419, 1994.
32. Wang, L., Application prospects and behavior of RE in new generation high strength steels with superior. *J. Chin. Rare Earth Soc.*, 1, 48–54, 2004.
33. Kiełbus, A., Microstructure and Properties of Casting Magnesium Alloys Designed to Work in Elevated Temperature, in: *Magnesium Alloys - Selected Issue*, pp. 53–73, Intechopen, London, 2018.
34. Palanisamy, G., Corrosion Inhibitors, in: *Corrosion Inhibitors*, pp. 1–24, Intechopen, London, 2019.
35. Lucas, J., Lucas, P., Le Mercier, T., Rollat, A., Davenport, W., *Rare Earths*, Elsevier, Amsterdam, 2015.
36. Janoš, P., Ederer, J., Pilařová, V., Henych, J., Tolasz, J., Milde, D., Chemical mechanical glass polishing with cerium oxide: Effect of selected physico-chemical characteristicson polishing efficiency. *Wear*, 262-263, 114–120, 2016.
37. Gin, S., Jollivet, P., Tribet, M., Peuget, S., Schuller, S., Radionuclides containment in nuclear glasses: An overview. *Radiochim. Acta*, 105, 11, 927–959, 2017.
38. Vernaz, E., *Nuclear Waste Conditioning*, CEA Saclay and Groupe Moniteur, Paris, 2009.
39. Donald, I.W., Metcalfe, B.L., Taylor, R.N.J., Review: The immobilization of high level radioactive wastes using ceramics and glasses. *J. Mater. Sci.*, 32, 5851–5887, 1997.
40. Swab, J.J., *Role of Oxide Additives in Stabilizing Zirconia for Coating Applications*, Army Research Laboratory, Aberdeen, 2001.
41. Sergo, V., Schmid, C., Meriani, S., Rare earths in ceramic materials technology. *Mater. Chem. Phys.*, 31, 1/2, 37–43, 1992.
42. Guanming, Q., Xikun, L., Tai, Q., Haitao, Z., Honghao, Y., Ruiting, M., Application of Rare Earths in Advanced Ceramic Materials. *J. Rare Earths*, 25, 281–286, 2007.
43. Yoldjian, G., The use of rare earths in ceramics. *J. Less Common Met.*, 111, 17–22, 1985.
44. Pillai, A.M. and Knapp, F.F., Evolving Important Role of Lutetium-177 for Therapeutic Nuclear Medicine. *Curr. Radiopharm.*, 8, 2, 78–85, 2015.
45. Fricker, S.P., The therapeutic application of lanthanides. *Chem. Soc. Rev.*, 35, 524–533, 2006.
46. Teo, R.D., Termini, J., Gray, H.B., Lanthanides: Applications in Cancer Diagnosis and Therapy. *J. Med. Chem.*, 59, 13, 6012–6024, 2016.
47. Giese, E.C., Rare earth elements: Therapeutic and diagnostic applications in modern medicine. *Clin. Med. Rep.*, 2, 1, 1–2, 2018.

48. Livingstone, J.D., The History of Permanent Magnet Materials. *J. Miner. Met. Mater. Soc.*, 42, 2, 30–34, 1990.
49. Kaufman, J.G., Boardman, B.E., Gerty, C.S.D., Hindin, B., McCormick, S., Sikorsky, P.J., Volk, G.A., *ASM Ready Reference: Properties and Units for Engineering Alloys*, ASM International, Cleveland, 1997.
50. Buschow, K.H.J., New permanent magnet materials. *Mater. Sci. Rep.*, 1, 1, 1–63, 1986.
51. Strnat, K.J., Hoffer, G., Olson, J.C., Ostertag, W., Becker, J.J., A Family of New Cobalt-Base Permanent Magnet Materials. *J. Appl. Phys.*, 38, 3, 1001–1002, 1967.
52. Ray, A.E. and Strnat, K.J., *Research and development of rare earth - transitional metal alloys as permanent magnet materials*, University of Dayton Research Institute, Dayton, 1971.
53. Collocott, S.J., Dunlop, J.B., Gwan, P.B., Kalan, B., Lovatt, H.C., Wu, W., Watterson, W.A., Applications of rare-earth permanent magnets in electrical machines: From motors for niche applications to hybrid electric vehicles, in: *2004 China Magnet Symposium*, Shanghai, China Magnetic Materials and Devices Association, Shanghai, 2004.
54. Legranger, J., *Rare Earth Magnetsfor hybrid and electric cars*, EREAN (European Rare Earth (Magnet) Recycling Network), Leuven, 2014.
55. Boldea, I., A novel, single stator dual PM rotor, synchronous machine: Topology, circuit model, controlled dynamics simulation and 3D FEM analysis of torque production, in: *12th International Conference on Optimization of Electrical and Electronic Equipment*, Brasov, IEEE, 2010.
56. Chau, K.T., Overview of Permanent-Magnet Brushless Drives. *IEEE Trans. Ind. Electron.*, 55, 6, 2246–2256, 2008.
57. Huang, S.Y., Ren, Z.H., Obruchkov, S., Gong, J., Dykstra, R., Yu, W., Portable Low-Cost MRI System Based on Permanent Magnets/Magnet Arrays. *Investig. Magn, Reason. Imaging*, 23, 3, 179–201, 2019.
58. Wang, G., Xie, H., Hou, S., Chen, W., Yang, X., Development of High-Field Permanent Magnetic Circuits for NMRI/MRI and Imaging on Mice. *BioMed Res. Int.*, 2016, 1–11, 2016.
59. Binnemans, K., Jones, P.T., Blanpain, B., Van Gerven, T., Yang, Y., Walton, A., Buchert, M., Recycling of rare earths: A critical review. *J. Cleaner Prod.*, 51, 1–22, 2013.
60. Gutfleisch, O., Willard, M.A., Brück, E., Chen, C.H., Sankar, S.G., Liu, J.P., Magnetic Materials and Devices for the 21st century: Stronger, Lighter and More Energy Efficient. *Mater. Views*, 23, 821–842, 2011.
61. Coey, J.M.D., Topical review. Permanent magnet application. *J. Magn. Magn. Mater.*, 248, 441–456, 2002.
62. Constantinides, S., *The Important Role of Dysprosium in Modern Permanent Magnets - Rev. 150903a*, Arnold Magnetics, Rochester, 2017.
63. Noréus, D., *Substitution of rechargeable NiCd batteries*, Arrhenius Laboratory - Stockholm University, Stockholm, 2000.
64. Omar, N., Firouz, Y., Monem, M.A., Samba, A., Gualous, H., Coosemans, T., van Mierlo, J., Analysis of Nickel-Based Battery Technologies for Hybrid and Electric Vehicles, in: *Reference Module in Chemistry, Molecular Sciences and Chemical Engineering*, Elsevier, Amsterdam, 2014.
65. Van Vucht, J.H.N., Kuijpers, F.A., Burning, H.C.A.M., *Reversible Room temperature absorption of large quantities of hydrogen by intermetallic compounds*, Philips Research Reports, Eindhoven, 1970.
66. Sakintuna, B., Lamari-Darkrim, F., Hirscher, M., Metal hydride materials for solid hydrogen storage: A review. *Int. J. Hydrogen Energy*, 32, 1121–1140, 2007.
67. Young, K. and Nei, J., The Current Status of Hydrogen Storage Alloy Development for Electrochemical Applications. *Materials*, 6, 4574–4608, 2013.
68. Nii, K. and Amano, M., R&D of hydrogen absorbing alloys in Japan. *Acta Metallurgica Sinica (English Letters)*, 10, 3, 249–255, 1997.

69. Cuevas, F., Joubert, J.-M., Latroche, M., Percheron-Guégan, A., Intermetallic compounds as negative electrodes of Ni/MH batteries. *Appl. Physics A*, 72, 2, 225–238, 2001.
70. Lifton, J., *The Future Of The Nickel Metal Hydride Battery And Its Constituent Rare Earth Metals*, Technology Metals Research, Elgin - USA, 2009.
71. Zhao, H., Xia, J., Yin, D., Luo, M., Yan, C., Du, Y., Rare earth incorporated electrode materials for advanced energy storage. *Coord. Chem. Rev.*, 390, 32–49, 2019.
72. Badwal, S.P.S., Giddey, S.S., Munnings, C., Bhatt, A.I., Hollenkamp, A.F., Emerging electrochemical energy conversion and storagetechnologies. *Front. Chem.*, 2, 1–28, 2014.
73. McNaught, A.D. and Wilkinson, A., *IUPAC Compendium of Chemical Terminology ("Gold Book")*, 2nd ed., Blackwell Scientific Publications, Oxford, 2006.
74. Balachandran, G., Extraction of Rare Earths for Advanced Applications, in: *Treatise on Process Metallurgy- Volume 3: Industrial Processes*, pp. 1291–1340, Elsevier, Stockholm, 2014.
75. de Sousa Filho, P.C., Lima, J.F., Serra, O.A., From Lighting to Photoprotection: Fundamentals and Applications of Rare Earth Materials. *J. Braz. Chem. Soc.*, 26, 12, 2471–2495, 2015.
76. Rygel, J.L. and Pantano, C.G., Synthesis and properties of cerium aluminosilicophosphate glasses. *J Non-Cryst. Solids*, 355, 2622–2629, 2009.
77. Alers, P.B., Effects of Gamma Radiation on Cerium-Bearing Phosphate Glasses. *J. Opt. Soc. Am.*, 51, 11, 1251–1254, 1961.
78. Doweidar, H. and Sadeek, Y.B., Effect of La_2O_3 on the structure of lead borate glasses. *J. Non-Cryst. Solid*, 356, 28-30, 1452–1457, 2010.
79. Reisfeld, R., Optical Properties of Lanthanides in Condensed Phase, Theory and Applications. *AIMS Mater. Sci.*, 2, 2, 37–60, 2015.
80. Oliveira, P., *The Elements*, Pedia press, Mainz, 2011.
81. Venable, W.H. and Eckerle, K.L., *Didymium glass filters for calibrating the wavelength scale of spectrophotometers—SRM 2009, 2010, 2013, 2014*, National Bureau of Standards, Washington, 1979.
82. Allen, D.W., Holmium Oxide Glass Wavelength Standards. *J. Res. Natl. Inst. Stand Technol.*, 116, 6, 303–306, 2007.
83. Extavour, M., *Rare Earth Elements: High Demand, Uncertain Supply*, Optics & Photonics News, Washinton, DC, 01 July 2011.
84. Kokavec, J., Wu, Z., Sherwin, J.C., Ang, A.J.S., Ang, G.S., Nd:YAG laser vitreolysis versus pars plana vitrectomy for vitreous floaters. *Cochrane Database Syst. Rev.*, 6, 1–16, 2017.
85. Henderson, A., *Advances in Light Sources: Fiber lasers tune to the visible*, LaserFocusWorld, Endeavor Business Media, Nashville, Tennessee, 9th June 2014.
86. McHale, J., *Future weapons: Solid-state lasers*, vol. 30, Military & Aerospace Electronics, Endeavor Business Media, Nashville, Tennessee, 1 May 2006.
87. Pavel, N., Bärwinkel, M., Heinz, P., Brüggemann, D., Dearden, G., Croitoru, G., Grigore, O.V., Laser ignition—Spark plug development and application in reciprocating engines. *Prog. Quantum Electron.*, 50, 1–32, 2018.
88. Boetti, N.G., Pugliese, D., Ceci-Ginistrelli, E., Lousteau, J., Janner, D., Milanese, D., Highly Doped Phosphate Glass Fibers for Compact Lasers and Amplifiers: A Review. *Appl. Sci.*, 7, 1295–1313, 2017.
89. Dong, L. and Samson, B., *Fiber Lasers - Basics, Technology and Applications*, CRC Press, Boca Raton, 2017.
90. Wandera, C., *Fiber Lasers in Material Processing—Creative Commons Attribution License 3.0*, INTECH, London, 2016.
91. Müller, H.-R., Kirchhof, J., Reichel, V., Unger, S., Fibers for high-power lasers and amplifiers. *C. R. Phys.*, 7, 154–162, 2006.

92. Lin, J.-T., Progress of medical lasers: Fundamentals and applications. *Med. Devices Diagnostic Eng.*, 1, 2, 36–41, 2016.
93. Jelínková, H., *Lasers for Medical Applications—Diagnostics, Theraphy and Surgery*, Woodhead Publishing Limited, Cambridge, 2013.
94. Kaushal, H. and Kaddoum, G., Applications of Lasers for Tactical Military Operations. *IEEE Trans.*, 5, 20736–20752, 2017.
95. Ready, J.F., *Industrial Applications of Lasers*, 2nd Edition, Academic Press, San Diego, 1997.
96. Swift, K., Moore, M.G., Rose-Glowacki, H.R., Sanchez, E., *The Economic Benefits of the North American Rare Earths Industry*, Amercian Chemistry Council, Washington DC, 2014.
97. Petty, T.R., *Final List of Critical Minerals 2018*, Government of the United States of America - Federal Register 83 FR 23295, Washington DC, 2018.
98. European Commission, Report on the Critical Raw materials for the EU—Ares, 2014. (2015)1819503 - 29/04/2015, European Commission, Brussels.
99. Bartekova, E. and Kemp, R., National strategies for securing a stable supply of rare earths in different world regions. *Resour. Policy*, 49, 153–164, 2016. 10.1016/j.resourpol.2016.05.003.
100. Steidle, E. and Forecast, Mineral, *A. D., Volume 2*, State College: School of Mineral Industries, State University of Pennsylvania, 2000, 1952.
101. Wong, P. and Hatch, G., *The IMC Consortium Eliminating the Rare Earth Industry Bottleneck*, Innovation Metals Corporation, Toronto, November, 2012. http://mric.jogmec.go.jp/kouenkai_index/2012/briefing_121109_4.pdf.
102. Papangelakis, G. and Moldoveanu, G., Recovery of rare earth elements from clay minerals, in: *Proceedings of ERES2014*, Milos, ERES2014:1stEuropean Rare Earth Resources Conference.
103. Birk, D. and White, J.C., Rare earth elements in bituminous coals and underclays of the Sydney Basin, Nova Scotia: Element sites, distribution, mineralogy. *Int. J. Coal Geol.*, 19, 1–4, 219–251, 1991.
104. Li, D., Tang, Y., Deng, T., Chen, K., Liu, D., Geochemistry of rare earth elements in coal—A case study from Chongqing, southwestern China. *Energy Explor. Exploit.*, 26, 6, 355–362, 2008.
105. Hower, J.C., Eble, C.F., Dai, S., Belkin, H.E., Distribution of rare earth elements in eastern Kentucky coals: Indicators of multiple modes of enrichment? *Int. J. Coal Geol.*, 160–161, 73–81, 2016.
106. Hedrick, J.B., Haxel, J.B., Orris, G.J., *Rare Earth Elements—Critical Resources for High Technology (Fact Sheet 087-02)*, U.S. Geological Survey, Reston, 2002.
107. Ayora, C., Macías, F., Torres, E., Nieto, J.M., Rare Earth Elements in Acid Mine Drainage, in: *XXXV Reunión de la Sociedad Española de Mineralogía*, Seville, Spanish Mineralogical Society, 2015.
108. Stewart, B.W., Capo, R.C., Hedin, B.C., Hedin, R.S., Rare earth element resources in coal mine drainage and treatment precipitates in the Appalachian Basin, USA. *Int. J. Coal Geol.*, 169, 28–39, 2017.

11

Opportunities and Challenges of Re-Mining Mine Water for Resources

Martin Mkandawire

Department of Chemistry, Cape Breton University, Sydney, Nova Scotia, Canada

Abstract

This chapter discusses the emerging trend of considering resource recovery from mine water as part of a larger strategy for remediation and management of abandoned mining sites. This emerging paradigm is being driven by the increasing need for resources and energy, against a backdrop of growing apathy for opening new mines due to increasingly stringent environmental conditions and emphasis on sustainable practices on one hand, while on another hand, the never-ending socio-economic burden of cleaning abandoned mines. Specifically, the chapter discusses potential resources that can be extracted from mine water and techniques currently used at either research or limited commercial levels. The chapter gives some examples of the potential resources, but it should be noted that the list is not exhaustive. Further, general challenges faced by the approach and the potential solutions to overcome the challenges are discussed.

Keywords: Abandoned mines, green mining, resources recovery, geothermal, ore extraction, remediation, pump-hydrostore electricity

11.1 Introduction

During active mining, the mine is usually at the center of mining communities, being the main source and second order of employment and economic base, which nucleate urbanization. When mines close, the contemporary trend has been, thus, the transformation of the once vibrant mining town into dying or ghost towns, mostly devoid of employment and economic base leading to rising in poverty, deterioration of living standards, and emigration [1, 2]. In addition, most post-mining towns are associated with contamination originating from the mines and mining activities, which are a burden left to be borne by the local communities in terms of high costs of remediation and restoration.

One of the major activities in remediation and restoration of closed or abandoned mining sites is the treatment of mine water. A single mine pool can hold billions of litres of mine water, which often contains corrosive agents, such as acids or alkalis. Thus, left untreated, that mine water might become a substantial environmental and economic risk. Generally, mines are prone to discharging acid mine drainage (AMD) due to exposure of remnant minerals,

Email: martin_mkandawire@cbu.ca [ORCID: 0000-0002-0506-2281]

Elvis Fosso-Kankeu, Christian Wolkersdorfer and Jo Burgess (eds.) Recovery of Byproducts from Acid Mine Drainage Treatment, (315–350)

specifically pyrite, to air, which leads to the oxidation of the mineral and acid generation. The acidic water may further dissolve metals from the surrounding rocks, causing them to leach into streams and groundwater and build up on the banks of waterways. This AMD can have devastating effects on the local environment. For instance, AMD can sometimes have extremely low pH values, low enough to burn human skin and kill aquatic life, and also carry potentially toxic elements, affecting lakes and reservoirs far away from the mining sites.

Consequently, monitoring for AMD generation and treatment of mine water is usually the longest and on-going remediation activity in post-restoration of mine sites. The financial and economic implications are very high and often have been a public responsibility, financed from taxpayers because many mining companies are liquidated immediately after operation. For former mining towns, mine water is a financial and environmental liability as communities already suffering several negative social economic effects of mine and mining activity closure. The total effects of AMD on local economies is difficult to estimate but may be in the billions of dollars annually, ranging from essentially permanent loss of valuable land and resources for farming, fishing, and recreation, as well as local tourism attractions to reduced drinking water sources. Hence, it is often asked: Can mine water be turned into an asset instead of remaining a liability?

Mine water can contain several dissolved non-ferrous metals, including some precious base metals and rare earth elements, depending on factors including site geology, hydrology, ore deposit composition, and mineralogy. AMD can also be a good source of hydrogen and gypsum for energy as well as geothermal and natural gas. Recovering these components of mine water adds an economic value, making mine water treatment a potential commercially viable investment. Recent advances in resource recovery technologies are pushing the prospects of mine water as a new source of commercially viable resources close to reality. Currently, a variety of mine water re-mining techniques, like hydrometallurgy, electrochemical, microbiological, evaporation, precipitation, solvent extraction, ion exchange, and reverse osmosis, are either being tested at research or used at limited commercial levels. However, most of these techniques have usually had limited success either from an economic or technical perspective. For instance, electrochemical techniques offer considerable versatility but can suffer from competing electrode reactions and energy demands depending on the type of cells used. Further, there are challenges to separate the valuable components from each other in a manner that is efficient and economically viable and to obtain resources of high quality and purity of industrial standard. In brief, these and other issues are discussed in detail in this chapter.

11.2 Mine Water and Drainages

11.2.1 Mine Water in Context of This Chapter

The term "mine water" has been used interchangeably with other terms like "mine drainage" and "mine pools" [3, 4]. For this chapter, mine water will broadly mean any water directly affected or influenced by mining activities or mining-related processes. This can be water coming from milling, extraction, or percolation from mine waste, as well as groundwater flooding active mine workings, which is subsequently de-watered [4, 5]. Generally, there is ambiguity between the term "mine water" and "mine wastewater". For clarification and as used in this chapter, the term "mine water" will mean any surface or groundwater present at a mine site, while "mine wastewater" will specifically refer to water that has come into contact with any mine workings.

Thus, mining wastewater in this context will be regarded as just a type of mine water and, in most cases, describes water coming from mine processing [5, 6]. For instance, water used during the crushing and grinding of ore, which are called mill water are a type of mine wastewater. Another example of mining wastewater is process waters, which is water used in the chemical extraction of metals, namely, hydrometallurgy. Both mill and process water as well as water leaching through solid mine wastes or tailing are considered mine wastewater and may contain high burdens of dissolved metals, minerals, process, and other chemicals [3].

11.2.2 General Mine Water Chemistry

The chemistry of mine water highly depends on factors including ore deposit composition and mineralogy, mining methods, climate, site hydrology, and site geological conditions. For example, when deposits of sulfurous coal and pyrite are being worked, acid water is often produced when no buffering minerals exist. In deposits of complex ores, the water is often enriched by Cu, Zn, and Pb. However, the quality of mine water is determined by the ability of the mine water to interact with the mineralogical and geological background in the presence of oxygen. Microbial activities also contribute greatly to the quality of mine water [7, 8], which varies considerably; they may be alkaline, acidic, ferruginous, highly saline, or close to freshwater. Other factors that influence the mine water quality are the hydrological regime and groundwater flow pattern of the site, which affects the mixing of the mine water with other subterranean waters [9]. Nevertheless, these factors may differ within the same mine depending on whether they arise from shallow workings and adits or from deeper levels. Very large areas of interconnected collieries, with multiple seams, worked at various depths, can lead to a large variation of mine water quality.

During active mining, the mines are artificially kept dry through dewatering, which means pumping the water out of the mine. When the mines close, the pumping usually is also stopped. As a result, the groundwater level rises until it reaches the surface or discharges into overlying aquifers. As the groundwater rises, minerals in the rocks, such as sulfide minerals, are exposed to air and release sulfate and soluble metal ions [10]. Flooding of the exposed seams stops the oxidation of the sulfide minerals but dissolves the metal ions and leads sulfur to form sulfuric acid [3, 4, 9]. This is the reason for high concentrations of metals, particularly Fe, Zn, Cu, Pb, Cd, Mn, and Al in the mine water because the highly acidic water dissolves most metal compounds present. Since underground water is low in oxygen and can have low pH, when the mine water first emerges to the surface is often clear as if clean, because the metals are dissolved. As the water is aerated, iron rapidly oxidizes and settles out as an orange deposit of "ocher". However, if the rocks contain carbonate minerals, the acidic mine water can be buffered, and metals may remain immobile. In some deeper mines, the main pollutants are usually sulfate or chloride rather than metals because the water levels usually never reach the surface but may connect with underground aquifers.

11.2.3 Types of Mine Water Sources

11.2.3.1 Overview

As mine water is water that has interacted with the mining environment, including processing activities, it is simple to classify the mine water sources according to the mining and processing procedure and how the interaction happens. Thus, the classification of mine water sources in this chapter is according to the mining technique and how the water interacts with the mining environment. Following this thought, it can be said there are three main classes of mine water sources, namely, the flooded underground mine pool, flooded opencast (pit) lakes, and leachates.

11.2.3.2 *Flooded Underground Mine Pool*

Extensive underground mining results in volumes and hundreds of kilometer-long interconnected mine workings (Figure 11.1). Water in active below-drainage mines is pumped to keep working sections dry. After mine closure, the mine voids can fill with groundwater—a process referred to as mine flooding—creating a vast network of interconnected mine pools [11, 12]. The water in flooded mines comes from adjacent and overlying flooded mines through vertical infiltration from overlying aquifers and from surface water sources. Without pumping, flooded or flooding below-drainage mines form "pools" that may eventually discharge onto the surface or drain into underground aquifers [11].

11.2.3.3 *Flooded Opencast Lakes*

Open-pit mining is a surface mining technique that extracts minerals from an open pit, also known as opencast mining. Upon the completion of mining operations, the pit will fill with water, forming waterbodies known as flooded opencast lakes or pit lakes (Figure 11.2). The opencast mine can be filled artificially or be allowing to fill naturally through precipitation

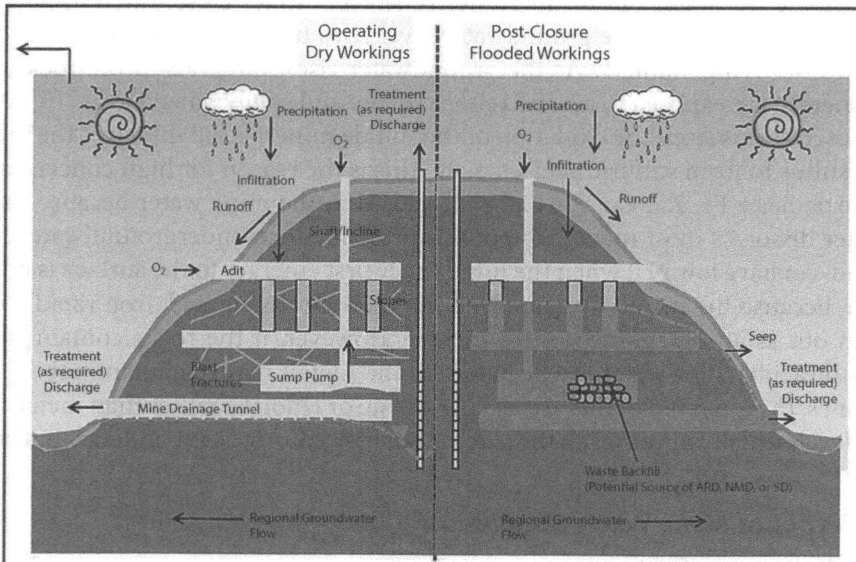

Figure 11.1 Illustration of an active below-drainage mine kept dry by pumping and a post-closure mine with mine pool formation in former mine-working tunnels as water rebound when pumping stops (Reproduced with permission from INAP [13]).

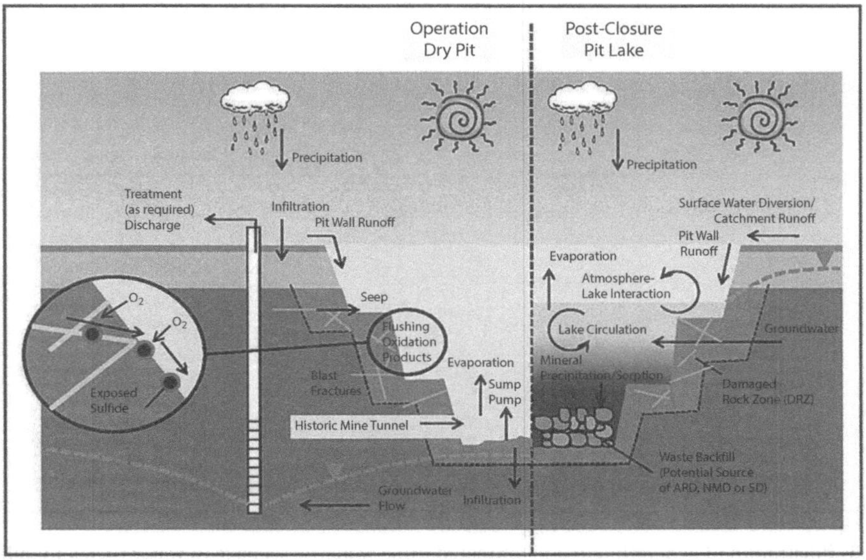

Figure 11.2 Illustration of an active open mine kept dry by pumping and a post-closure mine with water filling in pit mine as water table rebound when pumping stops (Reproduced with permission from INAP [13]).

or groundwater infiltration [14]. These flooded opencast lakes are often complicated sources of mine water. Like natural lakes, they display a huge diversity in mine water quality and content because of being young and therefore being typically in a non-equilibrium state. There are several processes in interplay in the opencast lake, which some are generally common to all lakes, while others are specific to open lakes. Some processes are common to all opencast lakes (e.g., wind forcing and convective mixing), whereas some are distinct to a given mine type (e.g., geochemistry, pit shells and mining methods) [15]. Further, mine pits are geotechnically recommended for storage of mine waste, which also influences the water quality (Figure 11.2).

11.2.3.4 Leachates

In mine water, leachate is water, laden with soluble or suspended solids or any other component of the material, which the water extracts in the course of passing through solid mine waste and tailings. This is one of the major sources of mine water especially those that generate AMD. There, two ways in which leaching leads to contaminated mine water are considered, namely, leachates from mine waste and tailings dams and post-closure leaching in *in situ* leach mining.

In both surface and underground mining operations, large quantities of solid wastes, especially, overburden, waste rock and rejects not valuable to the owners of the mine, are generated (Figure 11.3). Usually, most of the rock deposits contain a very low percentage of extractable mineral material that large quantities of tailings are generated during ore extraction [21]. Mine waste like overburden, rocks, and tailings can contain a lot of geochemical and processing chemical of which some are potentially toxic. Due to exposure to atmospheric conditions, these chemicals may readily leach. Commonly, leachates from mining and mining processing

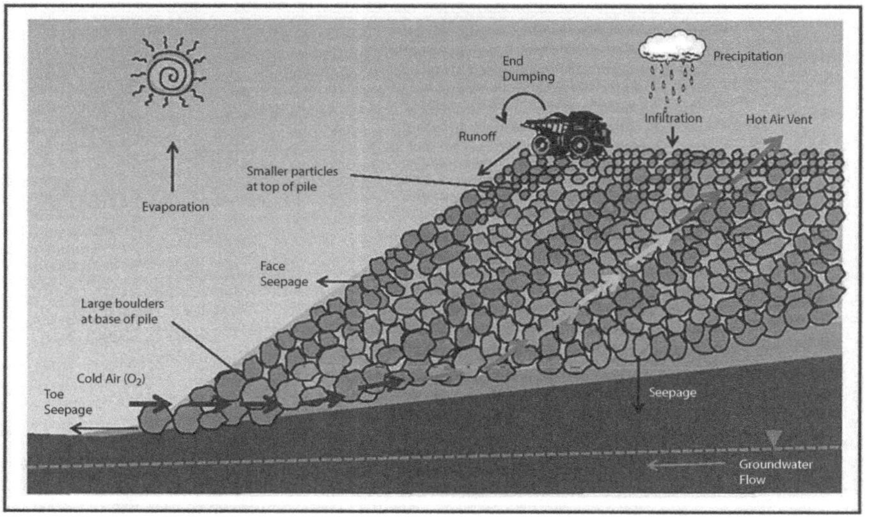

Figure 11.3 Illustration of generation of contaminated mine water by seepage in water rock and tailing piles (Reproduced with permission from INAP [13]).

wastes may contain toxic metals and metalloids like As, Co, Cu, Ca, Pb, Ag, Fe, Cr, Mn, Hg, Mo, Ni, Se, V, and Zn, as well as some radionuclides [6, 16–20]. A special type of leachate comes from *in situ* mining (Figure 11.4). Logically, considerable volumes of leaching agent injected into the mine remain underground for a long time after the mine closes, because it is impossible to remove everything. As a result, they continue to dissolve minerals contained in the rocks, thereby releasing potentially toxic metals and radionuclides into the groundwater.

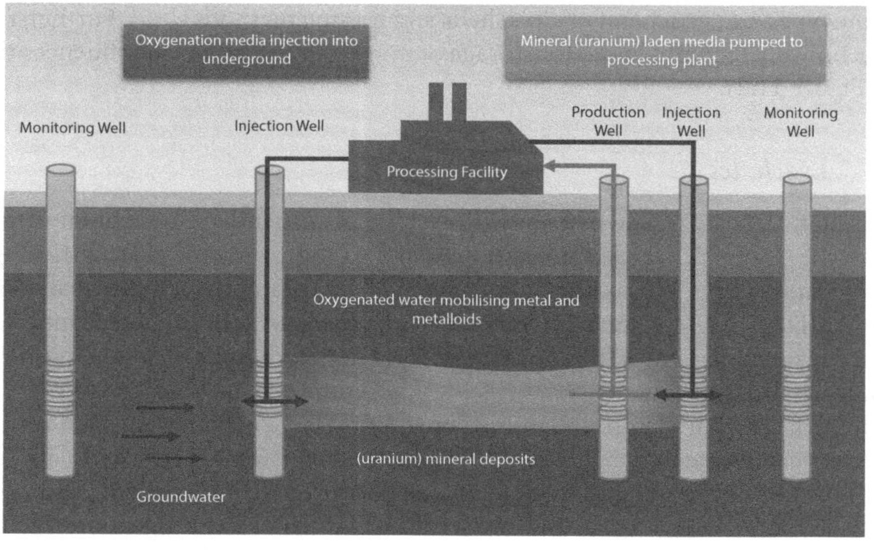

Figure 11.4 Generalized *in situ* leaching (ISL) mining technology, often used in uranium extraction. Depending of the geological background of the deposits, the extract media can be a very strong acid or alkaline fluid (modified from USEPA [21]).

When groundwater enters the leaching zone, mine water forms and can disperse the contaminant depending on the hydrological regime and flow.

11.2.4 Drainages of Mine Water

Mine drainage refers to surface water or groundwater from an active or abandoned mining operation. The drainage can be high-quality like natural waters or contaminated by leftover materials, being either extremely acidic, circum-neutral or alkaline, and is often laden with high concentrations of potentially toxic metals and metalloids. The contamination can be through different processes like interactions with waste rock and tailings or through interactions with geological formations in water-filled abandoned mines. Consequently, the drainages are classified by their pH as either acid ($pH < 5.6$), circum-neutral ($5.6 < pH < \approx 8$), or alkaline ($pH > \approx 8$) mine drainage [3, 4]. However, there are three main types of mine drainage waters, namely: (1) saline formation waters; (2) acidic, metal-containing sulfate waters derived from mineral oxidation, especially pyrite; and (3) alkaline, hydrogen-sulfide-containing, metal-poor waters resulting from buffering reactions or sulfate reduction.

11.2.4.1 Acid Mine Drainage

When mineral-containing rocks are exposed to oxidizing conditions, the sulfur-bearing minerals (i.e., pyrite, Table 11.1) oxidize and upon contact with water results in waters rich in sulfuric acid and dissolved iron [22]. When the exposure process occurs naturally, it is usually known as Acid Rock Drainage (ARD). Exposure of these minerals to atmospheric

Table 11.1 Some important metal sulfide minerals responsible for ARD at mining sites. Adapted from Sheoran *et al.* [10].

Chemical formula	Mineral
FeS_2	Pyrite
FeS_2	Marcasite
Fe_xS_x	Pyrrhotite
Cu_2S	Chalcocite
CuS	Covellite
$CuFeS_2$	Chalcopyrite
MoS_2	Molybdenite
NiS	Millerite
PbS	Galena
ZnS	Sphalerite
FeAsS	Arsenopyrite
Cu_5FeS_4	Bornite

conditions increases during and after mining activities due to shafts, tunnels, and collieries which brings air into the mine. In such a scenario, the high acidic and metal-laden water with a pH below 5.6 discharging from the mines or mine processing sites is called AMD [13, 22]. In most cases, the presence of iron (di-)sulfides such as pyrite and marcasite (FeS_2) cause the AMD formation in coal mining regions, but it may also be formed by chalcopyrite ($CuFeS_2$), covellite (CuS), and arsenopyrite (FeAsS).

Consequently, AMD is initially rich in sulfuric acid and dissolved iron and then further dissolves metals such as Cu, Pb, and Hg into groundwater or surface water. The concentration of metals in AMD depends on the type and amount of sulfide minerals and the presence of alkaline materials that buffer the acidity. This can be understood with revisiting the chemistry of AMD formation.

The initial chemical process involves oxidation of an iron di-sulfide mineral, which releases ferrous iron (Fe^{2+}), sulfate (SO_4^{2-}), and protons (Equation 11.1):

$$FeS_2 + 7/2\, O_2 + H_2O \rightarrow Fe^{2+} + 2\, SO_4^{2-} + 2H^+ \tag{11.1}$$

The resulting Fe^{2+} can further oxidize to ferric iron (Fe^{3+}, Equation 11.2):

$$2\, Fe^{2+} + 1/4\, O_2 + H^+ \rightarrow Fe^{3+} + 1/2\, H_2O \tag{11.2}$$

Ferric iron can undergo two processes. It can either be hydrolyzed to ferric hydroxide, $Fe(OH)_3$, increasing the proton acidity (Equation 11.3), or directly attack pyrite as a catalyst, facilitating generation of more Fe^{2+}, SO_4^{2-}, and acidity (Equation 11.4).

$$Fe^{3+} + 3\, H_2O \rightarrow Fe(OH)_3 + 3\, H^+ \tag{11.3}$$

$$FeS_2 + 14\, Fe^{3+} + 8\, H_2O \rightarrow 15\, Fe^{2+} + 2\, SO_4^{2-} + 16\, H^+ \tag{11.4}$$

The steps in the AMD formation process occur sequentially and interdependently [23, 24]. Therefore, generation of AMD can be inhibited by slowing or halting any of these constituent processes. The rate of pyrite oxidation depends mostly on the physicochemical properties of the minerals and conditions in the milieu. The rate can be influenced by the reactive surface area of pyrite, oxygen availability, pH, catalytic agents, flushing frequencies, and the presence of Fe-oxidizing bacteria (FeOB). Usually, the oxidation of Fe^{2+} to Fe^{3+} (Equation 11.2) is the rate-limiting pyrite oxidation step because the conversion of ferrous iron to ferric iron is slow below pH 5 under abiotic conditions [23, 25]. In presence of FeOB, mostly *Acidithiobacillus* sp., the oxidation of Fe^{2+} to Fe^{3+} can be greatly accelerate the reaction.

11.2.4.2 Alkali Mine Drainage

When the mine bedrock contains little iron sulfide minerals, the originating mine water can be neutralized by local alkalinity or proton buffering capacity. The resulting discharge has a pH of 5.6 or above and is known as circum neutral or Alkaline Mine Drainage (NAMD). Dissolved carbonaceous compounds, existing as hydrogencarbonate (HCO_3^-) or carbonate (CO_3^{2-}) in solution, are the principal sources of alkalinity in mine water. For this reason,

NAMD is said to originate from a reaction of acidic mine drainage with calcite ($CaCO_3$) and other carbonate minerals. This explains why NAMD formation is associated with the presence of calcite or dolomite in the bedrock of the mine. However, the alkalinity or proton buffering capacity is attributable to the local presence of a wide range of anions including OH^-, CO_3^{2-}, Si_4^{4-}, BO_3^{3-}, PO_4^{3-}, NH_3, and organic ligands within the mine drainage [4, 27].

Consequently, the mine conditions that results in NAMD formation include any combination of relatively low concentrations of sulfide minerals, a deoxygenated but sulfide-rich environment, the presence of monosulfides, large grain-sizes that would slow the oxidation rate of pyrite, the presence of highly reducing or alkaline influent water, or naturally highly alkaline groundwater [26, 27]. A few geochemical reactions do contribute to the formation of NAMD. Some of the reactions include the oxidation of other sulfides, dissolution of host rock minerals, precipitation and dissolution of iron hydroxide and hydroxysulfate, co-precipitation and adsorption of dissolved metals, precipitation and dissolution of iron sulfate, gypsum precipitation and dissolution, and CO_2 degassing [28]. NAMD is prone to contamination with divalent metal ions, like ferrous iron and manganese.

11.3 Potential Extractable Resources

11.3.1 Water Supply

11.3.1.1 Opportunities

Most mining activities take place below the water table. Consequently, water inflow into the mine system is inevitable, and the water has to be removed to allow working in the mine (Figures 11.1 and 11.2). In arid and semi-arid zones, mining operations can lower or sometimes deplete the underground aquifers if the recharging amount and rates are lower than the mine dewatering process. Once the mining activities stop, there is no longer a purpose to continue the dewatering process. Thus, the mine gradually floods until it overflows at the surface via a shaft, tunnel, or adit, forming underground water reservoirs or open pitlakes [9]. The efficiency of draining water from the rocks surrounding the mine depends upon the degree of joints and subsidence fractures in the rock that allow the movement of water into mines. The underground water reservoirs and pit lakes can be water sources for municipal, industrial, agricultural, or domestic water supplies, especially in areas with acute water shortage but depending upon the water quantity and quality.

Generally, flooded mines can be a safe source of water because the mine can act as a reservoir where the water is somewhat protected from contamination by humans and animals. Nevertheless, the requirements of the user play a role in what mine water quantity or quality is acceptable for extraction investment.

11.3.1.2 Applicable Extraction Methods

Extracting water from closed or abandoned mines is like any other water extraction process, which involves wells and pumps for underground water, or intake pumps for open pits. For shallow mines, water can be extracted using drainages driven by gravity through soughs or adits to where the water can be directly tapped into pipes ready for further treatment to meet the quality required for a specific use. In deep mines, pumping is usually required [29].

11.3.1.3 Challenges

The major challenges of extracting water from abandoned mines are related to the water quality, quantity, and dynamics resulting from the cone of depression created by the pumping. Changes in water quality may be dynamic and difficult to predict because the quality of mine water is influenced by the mine's position within the groundwater flow system, mineralogy, and enclosing bedrock, the time elapsed since the mine got flooded, and flushing that may have taken place [30]. Furthermore, the water quantity from abandoned deep mines varies greatly, depending on many variables that control flooded volume and recharge rate.

Most mine water have high total dissolved solids (TDSs) due to high degree of mineralization of groundwater facilitated by the relatively slow movement and increased residence times of the water in the distal ends of the groundwater flow system [31]. The challenges are exuberated further due to AMD generation when sulfur-containing minerals are exposed to air and oxidize. Further, there are other stability challenges including geochemical and hydrogeological which may result from dynamics induced by flooding and pumping [32]. Thus, the method used to extract the water from the mines should be flexible and easily adaptable to the geochemical, hydrogeological, and geophysical changes that may take place in the process of water extraction. There are a number of variables that control groundwater discharge from a mine. These variables include the surface area overlying the mine, the position of the mine within the groundwater system, and the structural geology of the site, which determines draining of groundwater into and from the mine. In arid and semi-arid zones, slow recharge rates and sometimes lack of recharge lead to losses in water storage in deep mines.

There is no one-fits-all strategy of mine water extraction. Therefore, extraction of mine water for any use should follow a thorough and informed study that would ascertain the safety of the water and hydrogeochemical and geophysical stability of the decommissioned mines. Consequently, the pre-extraction investments, in terms of studies and investigation, are quite exuberant and expensive as well as time consuming.

11.3.1.4 Counter Options

For a sustainable water extraction from closed or abandoned mines, water quality, and quantity should be analyzed for each potential mine source and should be continuously monitored and analyzed throughout the time of water withdrawal. This should be done to determine any potential changes in quality associated with induced mobility of otherwise slow-moving to stagnant water within a mine. Since water withdrawal creates voids and a hydrogeological gradient, the gas and oxygenated water ingress into the mines should be part of the monitoring process. Further, the voids created by the difference in water withdraw and recharge would lead to change in the geophysical stability of the mines due to several factors including water inflow into the voids.

One important water-quality factor is the length of time after mine closure and subsequent flooding because natural maturing of the mine requires considerable long periods for flushing. Where the need for auxiliary water sources is urgent just after a mine closure, it is advisable to invest strongly into mine water treatment using either active or passive processes depending on the type of discharge. In addition, the mine water treatment process (Figure 11.5) can be augmented with more advanced water treatment methods such as ion-exchange, adsorption, bio-sorption, bio-precipitation, neutralization, coagulation, and precipitation. The extent of application of most of the methods is largely an economic-benefit issue.

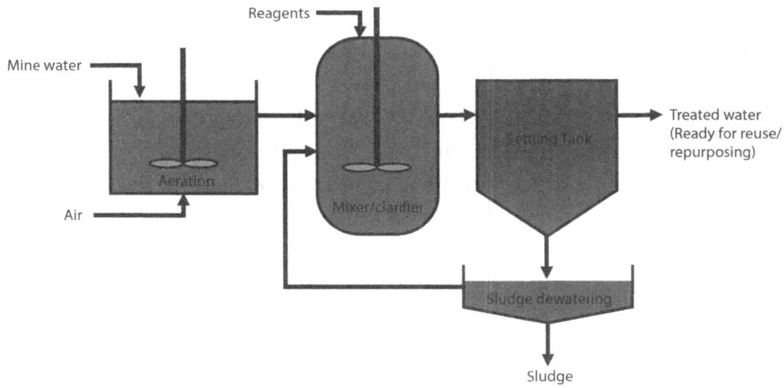

Figure 11.5 Generalized mine water treatment process.

11.3.2 Thermal Resource

11.3.2.1 Opportunities

Basically, the heating of water occurs due to a natural process called the geothermal gradient. As a result, extraction of geothermal energy to provide heating and cooling or to generate electricity is well established [30, 34]. A number of technologies for extracting the geothermal energy already exist, which may involve, among others, exploiting heat directly via drilled wells or buried "ground collectors" from pumped groundwater or from surface water bodies such as lakes [33]. In all these technologies, the initial capital investment for establishing geothermal energy infrastructure is usually very high beyond the cost-benefit value, especially in areas farther away from the "ring of fire". On the other hand, much of the work required in establishing geothermal infrastructures is done in mines to facilitate the mining operation, which are left behind when the mines close or are abandoned and may be an advantage for geothermal energy extraction. Such infrastructure includes major structures that penetrate the subsurface are constructed in open pits and underground roadways and borehole wells for mine dewatering operations. Consequently, the mine water discharges can be a "green" renewable energy asset.

Geothermal energy can be used in two ways: (a) for electricity generation and (b) for direct heating (non-electrical) applications [34]. Rarely, the temperatures of mine water exceed 85 °C, which is required for generating electricity in binary power plants. Thus, mine water is suitable for lower-enthalpy geothermal systems, involving temperatures that are either tepid (i.e., 11.5–13.5 °C) or warm (13.5–25.0 °C) for direct cooling or heating purposes. Nevertheless, the decision to exploit geothermal energy, whether heating or cooling, in closed mines depends on the heat energy potential (H) of the mine water. To extract the geothermal energy, the mine water should flow and transfer its heat to the heat exchanger, leading to a reduction in the water temperature by ΔT [35, 36]. Conversely, if cooling is to be provided, heat must be added to the flowing water, and the temperature of the water will be increased by ΔT. Consequently, the heat energy that can be extracted from a stream of flowing mine water is [35, 36]:

$$H = Q \times \Delta T \times \rho_w \times c_w \tag{11.5}$$

where Q is mine water flow in L s^{-1}, ΔT is temperature change at heat exchanger in °C, ρ_w is density (kg L^{-1}), and c_w is specific heat capacity (J kg^{-1} K^{-1}) of water.

When external energy input is required to extract the geothermal energy from the mine (e.g., a heat pump), the efficiency of operation of the heat pump, also known as coefficient of performance (*COP*), is considered. The *COP* depends on the relative input and output temperatures. Thus, the heat output from the system includes the heat associated with the motor driving the heat pump, which can be calculated using (Equation 11.6):

$$H_{heatpump} = \frac{H_{minewater}}{1-\left[\frac{1}{COP}\right]} \tag{11.6}$$

When the system is used to provide cooling, this heat associated with the heat pump reduces the available cooling output that the heat output of the system becomes (Equation 11.7):

$$H_{heatpump} = \frac{H_{minewater}}{1+\left[\frac{1}{COP}\right]} \tag{11.7}$$

11.3.2.2 *Applicable Extraction Methods*

Since most mines are below the 80 °C area, the most application technologies are low-enthalpy geothermal systems. Conceptually, low enthalpy geothermal energy exploitation involves three interconnected elements, via pumps and heat exchangers in conjunction with either open or closed loops. For mine water, a few modes of operation for heat pump or exchange schemes have been adopted or proposed for different mining systems, depending on the mine water quality and heat quantity as well as place of use. The following are some of the modes in which heat is exchanged with mine water:

11.3.2.2.1 Open-Loop Systems With Disposal of Thermally Spent Water

This mode involves abstracting mine water from a flooded mine via a shaft or borehole, including decommission mine dewatering wells, and passed directly through a heat pump or a heat exchanger. After heat exchange, the mine water is rejected to a surface water reservoir [29]. Such schemes are suitable for abandoned or closed mines where the water quality is relatively good and where no treatment is necessary or where heat exchange takes place prior to mine water treatment. The mode can be applied in both abandoned underground and open pit mines (Figure 11.6a, b).

11.3.2.2.2 Open-Loop Systems With Reinjection of Thermally Spent Water

This mode is similar to the one above, except that after the heat exchange, the water is reinjected back into either the mine workings or a different aquifer unit (Figure 11.7). The advantage of this is that water resources are conserved while treatment and disposal costs

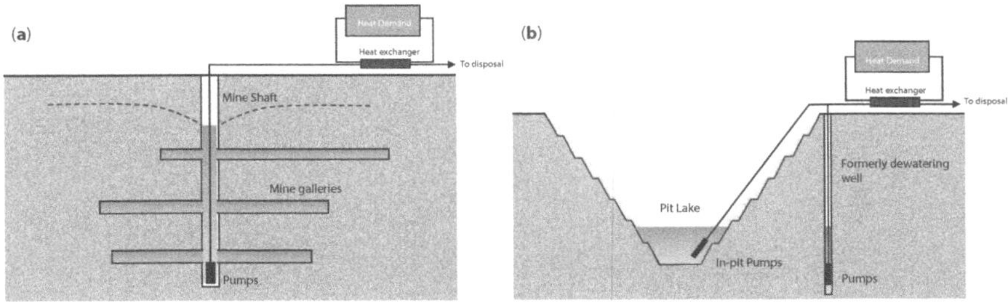

Figure 11.6 Open loop with disposal of water to surface recipient: (a) underground mines; and (b) open pit mines. Modified from Preene and Younger [33] and Banks *et al.* [29].

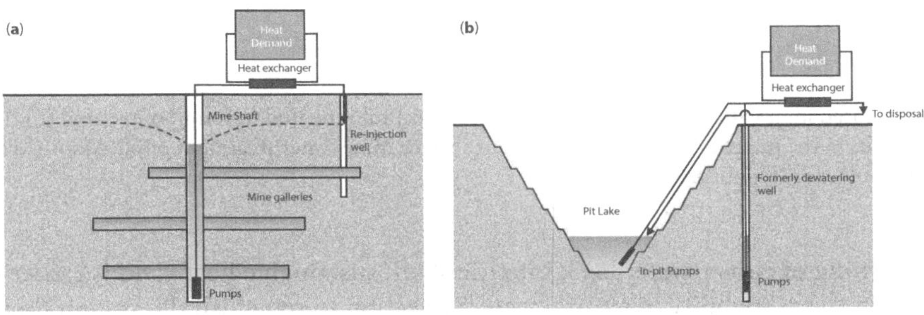

Figure 11.7 Open loop with (a) reinjection into underground mines; and (b) return into the pit lake. Modified from Preene and Younger [33] and Banks *et al.* [29].

are avoided [30]. On the other hand, it requires drilling and maintenance of reinjection boreholes and runs the risk of thermal "feedback" if the connection between the abstraction and injection points is too direct [30].

11.3.2.2.3 Closed-Loop Systems

These systems involve submerging heat exchangers directly into the mine water, and a heat transfer fluid, circulating through the heat exchanger and back to a heat pump servicing a space heating or cooling demand (Figure 11.8). The heat extraction can take place either in the mine itself or within a mine water treatment lagoon after the mine water is pumped to the surface. No mine water is abstracted in closed-loop systems, so that most of the issues relating to the mine water chemistry and treatment are avoided. Replenishment of heat to the heat exchanger takes place by conduction, natural advection, and thermal convection of water in the mine void, which can affect the heat extraction capacity [29].

11.3.2.2.4 Standing Column Systems

This technique involves abstraction of water from a specific depth in a mine shaft. This abstracted water is passed through a heat exchanger, before the water is returned to the same shaft at a different depth and different temperature. Sometimes, a fraction of the water after heat extraction is not returned into the shaft but disposed of at surface. Thus, depending on the fraction disposed on the surface, the system can be either closed- or open-loop systems.

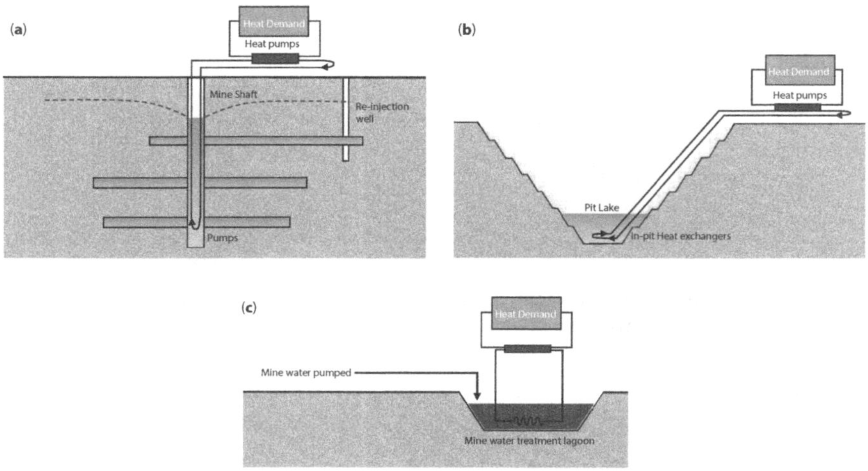

Figure 11.8 Closed loop with (a) heat exchangers place direct into the underground mines and (b) heat exchangers place in the pit lake; and (c) closed loop in surface mine water treatment pond. Modified from Preene and Younger [33] and Banks *et al.* [29].

Due to returning of cooler water back into the shaft, sustainable heat yield is rather limited, dependent on the availability of natural water advection to replenish the system thermally.

11.3.2.3 Challenges

The general challenges of extracting heat from abandoned mines are related to four issues, namely: (a) sustainable rates of geothermal heat extraction of an underground reservoir; (b) ocher clogging of pumps, heat exchangers, pipelines, and reinjection wells; (c) presence of long-term and stable demand; and (d) the proximity of the geothermal sources to the potential energy consumers. Like any other geothermal heat extraction, the total heat resource from the mine water reservoir is the available quantity of heat, which results from natural heat inputs and outputs to the reservoir.

The temperature stability in the geothermal energy reservoir is influenced by inputs and output of energy from sources like solar energy heating the ground, geothermal flux (i.e., for every 100 m depth into the ground), the temperature increases by 3 °C and groundwater flow, which may transfer heat between aquifers as well as strata. Thus, one of the challenges to the geothermal use of mine water is to design the best configuration and the rate at which the heat is extracted or added to a reservoir by the geothermal system [37]. If the geothermal system adds or removes heat at a greater rate than the natural replenishment, the reservoir temperature can change substantially. This can lead to an unstable or fluctuating heat source or may cause the reservoir temperature to cross a cut-off temperature beyond which the geothermal system fails to function efficiently [37, 38].

11.3.2.4 Counter Options

Among the outstanding issues of geothermal energy extraction are the alteration of water temperature of the mine pool due to, *intra alia*, heat extraction or rejection, and nucleation,

Table 11.2 Advantages and disadvantages of various mine water geothermal configurations (modified from Banks *et al.* [29]).

	Configuration			
	Open loop			
Issues	**With discharge**	**With reinjection**	**Closed loop**	**Standing column**
Ocher clogging/fouling	High – depending iron water iron content and exposure to oxygen	High – depending iron water iron content and exposure to oxygen	None	High – depending iron water iron content and exposure to oxygen
Legislative burden	High	High	Low	Depending on regulatory regime
Energy efficiency	High	High	Lower	High or lower, depending on degree of thermal feedback
Parasitic power loss	High – depending on mine water depth	High – depending on mine water depth	Low (circulation pumps only)	Can be high (depending on mine water depth)
Potential thermal capacity	Large	Large	Modest	Modest
Other advantages	Attractive if mine water is already pumped for treatment	No treatment required	Controlled fluid quality Can be installed in surface lagoons and ponds	Single site No water transfer or treatment required
Other disadvantages	Cost of water treatment Effect of temperature change on water treatment	Clogging of reinjection boreholes Requires at least two boreholes or shafts and water transfer pipe between them		

and aggregation of ocher leading to fouling of heat exchangers and pipes. Ocher is the corrosion due to the precipitation of Fe^{3+}-oxides after oxidation of Fe^{2+}. In case of fouling problems, caused by raw unaerated mine water (Table 11.2), the addition of reducing agents helps to hinder the oxidation and precipitation of ferric iron and manganese. This entails that effort should be made to avoid the iron-rich mine water getting into contact with atmospheric oxygen or other oxidizing agents, such that the iron remains in its soluble ferrous (Fe^{2+}) form. The best strategy in this case is using a closed-loop configuration in which the circulating heat transfer fluid, based on a solution of glycol, is passed through the heat exchanger instead of mine water. Otherwise, reducing agents like sodium bisulfite ($NaHSO_3$) and sodium dithionite ($Na_2S_2O_4$) should be added to keep the iron dissolved, just before the water flows into the heat exchanger.

When deciding the construction of geothermal systems, it should be born in mind that the ultimate amount of extractable heat from the system is dictated by the quantity of mine water that can be pumped or discharged and the temperature gradient that can be achieved across a heat exchanger [29, 33].

11.3.3 Electricity Generation Prospects

11.3.3.1 Opportunities

Mines, whether open or underground, intercept the piezometric level, which results into either partial or a total flooding of the mine voids and pits after mine closure. To avoid potential contamination resulting from uncontrolled migration of polluted mine water, large volumes of excess water are pumped from flooded mines and treated before release into the environment, becoming an important post mining remediation cost. These large volumes of water can be used as hydroelectric source (Figure 11.9), which is a relatively underinvestigated potential, in which there are two considered prospects for generating electric energy.

Thus, closed or abandoned mines could be a potential hydroelectric storage reservoir. A pumped hydroelectric energy storage is a commercial utility-scale technology currently used in many locations world-wide, where periods with excess electricity production exist during low demand times [39]. As a result, the excess electricity is used to pump water from a lower reservoir into a reservoir at a higher elevation. When need for electricity arises, the water is released back to the lower reservoir via a hydroelectric turbine to generate electricity. Thus, this technology can be the backbone of a reliable renewable electricity system by supporting bulk electricity storage for intermittent renewable energy, especially solar and wind energy.

Usually, after treatment, the mine water is left to flow with gravity to the receiving water bodies. This has potential for direct gravity hydroelectric generation as well as hydroelectric gravity dams. In a hydroelectric dam, whose reservoir is usually at higher elevation, electricity is generated when water flows downward due to gravity through a narrow tube—a penstock—and turns the blades of the turbine. Thus, the treated mine water can feed into a hydroelectric dam and provide a constant source of uninterrupted power, so long as the mine water treatment takes place.

11.3.3.2 Applicable Extraction Methods

A hydroelectric power system converts potential energy of water from a high elevation to kinetic energy as the water flows and drives a turbine and generator. Thus, the power

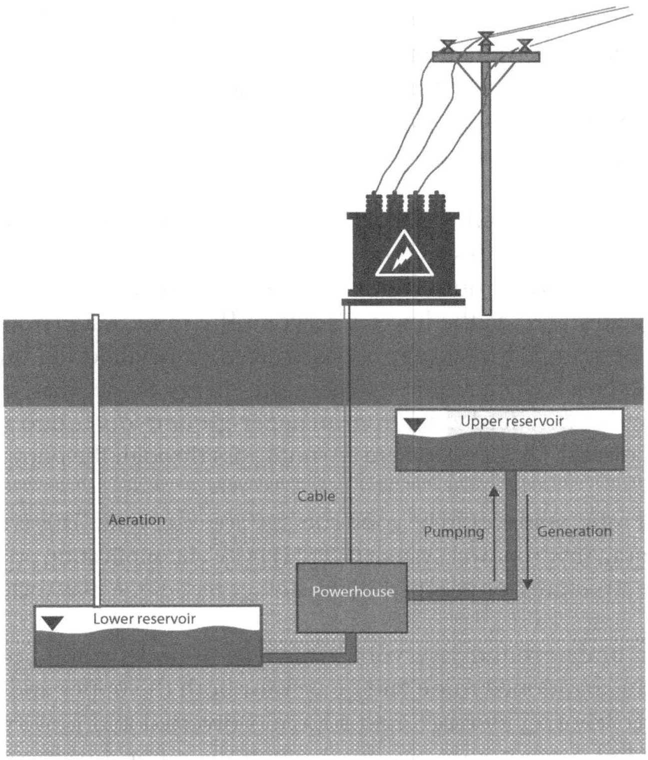

Figure 11.9 Underground pump storage hydropower system applicable in both abandoned underground and open mines (modified from University of Colorado at Boulder).

capacity of a hydroelectric station depends on the volume of water and the elevation difference between the upper and lower reservoir, called the water head. Further, the gradient of the penstock (i.e., the pipe) delivering water from the reservoir to the turbine is laid to contribute to the actual power output. Assuming that the penstock is gradient, the electric power production at a hydroelectric station is approximated using Equation 11.8:

$$P = -\delta(\dot{m}\Delta h) = \delta((\rho\dot{V})g\Delta h) \quad (11.8)$$

where P is power (W), δ is the coefficient of efficiency, ρ is the density of water ($\approx$1000 kg/m^3), $\dot{V}$ is the volumetric flow rate (m^3/s), $\dot{m}$ is the mass flow rate (kg/s), Δh is the change in height (m), and g is acceleration due to gravity.

This principle of hydroelectric power generation is adapted to produce electric energy in abandoned mines using two main methods, namely:

11.3.3.2.1 Pumped-Hydrostorage Technique

The pumped-hydrostorage technique is not an energy source *per se* but an energy storage method, with potential application in balancing and integration of variable and unpredictable renewable energy, especially wind and solar [34, 40]. When wind or solar is unpredictably high and produces electricity higher than the demand, the excess electric energy can be used to pump mine water into a higher reservoir, which can later be released through a turbine to supplement the production from the renewable source.

Mine pools in in pit lakes and underground voids in abandoned mines are a unique potential as reservoir options. Conceptually, open pit lakes are used as the upper or lower reservoir depending on the mine structure, as follows:

i. Open pit lakes used as the upper reservoir: This approach combines two different unconventional reservoirs, namely, the open pit lakes as the upper reservoir and the underground voids as the lower reservoir.
ii. Open pit lakes used as the lower reservoir: The mine water from a lower pit lake may be pumped to an upper constructed or neighboring higher pit lake and released back when required to generate energy. Sometimes, it may involve diverted and gravity-feed water from an upper-ground source into a powerhouse located above the lake level, where it would pass through the pump-turbine.

The geometry of mine voids could be good sub-surface reservoirs. This is a type of an underground pumped storage hydropower system (UPSH) [41]. The application opportunities of these underground mine structures for pumped-hydrostorage technique may include the following:

i. Surface to underground reservoirs: This type of underground pumped storage uses an abandoned mine and quarry for both the upper and lower reservoir. Water from an aboveground source is diverted and gravity-fed into an underground powerhouse, passing through the pump-turbine, thereby generating power. The water is then temporarily stored in the abandoned mine voids until there is excess electric energy production by either wind or solar to be pumped back to the upper reservoir.
ii. Underground to underground reservoirs: This method is employed where both upper and lower reservoirs are below ground, obviously at different elevations. This is particularly applied where the underground mining was done in deposits, at different seams of different elevations. Basically, a shaft connects the two reservoirs, where in most cases the lower reservoir has high capacity. Water is let drop down the shaft flow into a powerhouse containing pump-turbines before being stored in underground voids, and being pumped up using excel renewable energy, or low-cost off-peak power.
iii. In-ground closed-loop system: This a novel and innovative technology developed by Gravity Power, LLC (USA), using underground structures like abandoned mine shafts for a grid-scale electricity storage system. Briefly, the technology is similar to the pumped storage, except that a large and heavy piston suspended in a sealed shaft, which is fitted with a side-loop return pipe that passes through the turbine (Figure 11.10). When the piston is let to drop, it pushes water down the storage shaft. The water escapes up via the return pipe and spins the generator turbine, before filling the part of the same shaft but above the piston. To store energy for instance, excess solar or wind energy is used to pump the water in reverse, down the return pipe and up into the shaft to lift the piston to a higher position. None of their announced projects has been able to show the viability of the method so far.

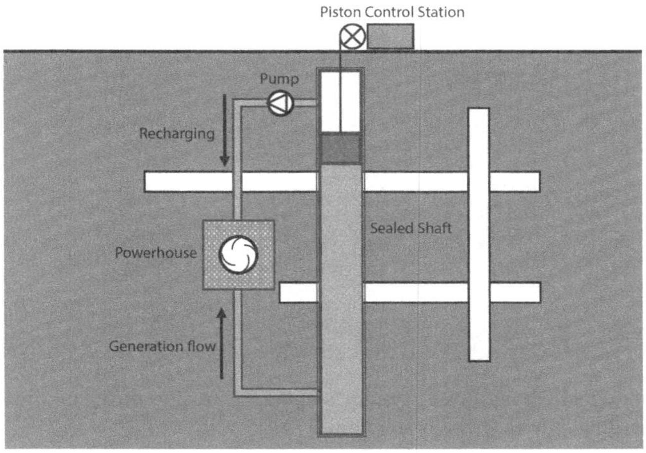

Figure 11.10 An in ground closed loop technique of generating electricity using abandoned underground mine shaft, a technology developed and potentially commercialized by Gravity Power, LLC (Developed and modified after concept by Gravity Power, LLC [42]).

11.3.3.2.2 Post Mine Water-Treatment Hydroelectricity Technique

The common practice in mine site remediation is to treat AMD in either passive or active treatment plants to prevent pollution of the receiving environment. Once the mine water is treated to ambient quality, it is released into receiving water bodies. In most cases, high volumes of water are treated and released daily, and the treatment activity can take years to centuries. Thus, there is an assurance of constant water flow from the treatment facility. Instead of just releasing the treated mine water, the water can be tapped for hydroelectric generation, as the water flows to the water bodies. There are many ways of harnessing the kinetic energy of the treated mine for hydroelectricity, but two potential examples are presented and summarized in Figure 11.11:

i. Run-of-the-treatment effluent stream hydroelectricity: Hydroelectric stations are planted using only the treated water, directly coming from a mine water treatment plant for generation of electricity at that moment, and any oversupply pass unused. This requires a constant supply of water from an abandoned mine, which is an important determining factor to use a site for this hydroelectric generation. Electricity is generated when water from the effluent stream is gravity fed through the penstock to spin the turbine; and
ii. Conventional (dams) hydroelectricity: This approach is slightly different from the above, that the treated mine water is led into a small reservoir. The reservoir can be constructed as a mini hydroelectric dam on the mine treatment effluent stream with a relatively high volume of water. Electricity is generated by forcing the water by gravity through a penstock. The conventional hydroelectricity can provide a constant source of uninterrupted power, as long as the mine water treatment continues.

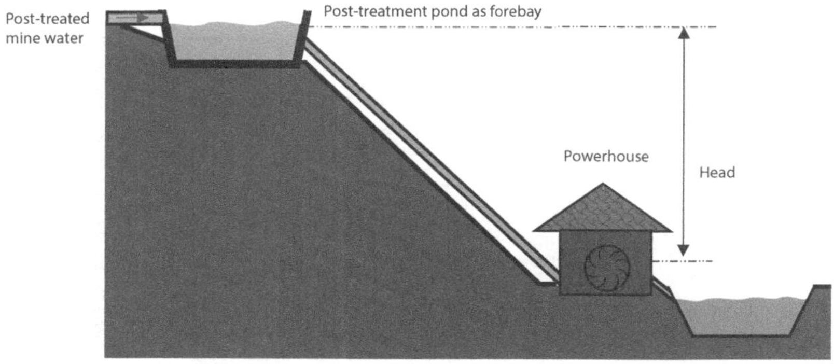

Figure 11.11 A conventional hydroelectricity system adapted for use post-treated mine water to generate electricity before discharging in to receiving water bodies.

11.3.3.3 Challenges

Implementation of hydroelectric generation from mine water or abandoned mines is challenged from economic, geotechnical, and geochemical perspectives, among others, except in a post mine water treatment plant. The use of mine water hydroelectricity is mainly for energy storage and not energy source. Therefore, it is only applicable where there is need to store either excess energy at low demand or unpredictable renewable sources like solar or wind.

One relevant technical aspect in pumped-hydrostorage technique using existing mine infrastructure is related to the storage structure for the lower reservoir. The construction of an underground storage reservoir is technically feasible, but it heavily depends on availability of competent rock, especially at reservoir depths [43]. Technically, using mine shafts and tunnels for lower reservoirs in the underground pumped-hydrostorage system can affect the mine stability, the same way water inrush can threaten the mine. As a result, there is always need for re-enforcement of the tunnel or shaft walls and pillars or extract new cavities. However, the extraction of large cavities is as technically demanding and financially expensive as the construction of a new mine, which seems not worth the effort and investment or reasonable [42].

In underground pumped-hydrostorage systems, the plants interact with the surrounding porous medium through exchanges of groundwater. This a common challenge for consideration because the exchanges may affect the surrounding aquifers, affecting the head difference between the reservoirs and thus influencing the efficiency of the pumps and turbines. This can also affect the geotechnical stability of the mines. As a natural physical process, the lower reservoir is connected to a ventilation for smooth flow of water. Consequently, the mine water is constantly supplied with oxygen rich air which eases the development of AMD in the underground mine pool. Thus, with time, the AMD becomes an issue that may threaten the life of the pumped-hydrostorage system including pipes, turbines and pumps.

For post-mine water treatment, the major challenge is to find the best terrane with a good gradient to supply enough gravity generated water. Further, techno-economic and environmental issue can be limitation, especially where a reservoir involving dam construction is involved.

11.3.3.4 Counter Options

For the pumped-hydrostorage system using the mine structure, it is important to understand the groundwater exchange with the surrounding porous medium in designing the system, because each site has its own specific complexity. Generally, systems development in high groundwater exchange has high-energy generation efficiency. Thus, the system is most efficient when the underground reservoirs are located in a transmissive porous medium, allowing free groundwater exchange [39]. Further, the knowledge of the mine water geochemistry should always be determined and monitoring through the implementation of the hydroelectric system. The main criteria for the use of mine water and abandoned mines for energy storage or production should consider both the economic benefits and the environmental impact. Further, it is advisable to indulge into such a project only when there is a necessity to store excess or renewable energy.

11.3.4 Mineral Resource Extraction

11.3.4.1 Opportunities

The growing need for low-cost resources, the increasingly environmental consciousness and stringent regulation for opening new mines, as well as the search for new economy bases for rejuvenating ghost town created by closure of mining activities has led a fresh consideration of mine water for mineral resource recovery and production [44, 45]. Ironically, most of these dying ghost towns have a huge burden of mine water remediation. Thus, the approach to resource recovery from mine water can be part of a larger water treatment remediation strategy. However, recovering resources from mine water is not a new phenomenon—it has been there for a long time.

As mentioned before, the earth is the source of all chemical elements, extracted through either mining or farming. It has also been repeated that deposits of sulfide minerals, such as pyrite (FeS_2), are a substantial part in the earth's crust and they are associated with most mining operations, including petroleum. By virtue of their deposition in deep earth, they are shielded from the atmospheric conditions. However, mining and excavation exposed these minerals to atmospheric conditions rich in oxygen and water, oxidizing to form either metal-rich solutions of sulfuric acid, or secondary acidic metal sulfate minerals like jarosite $(K,Na,H)Fe^{III}_3(OH)_6(SO_4)_2$, melanterite ($FeSO_4{\bullet}7H_2O$), römerite ($Fe^{II}Fe^{III}_2(SO_4)_4{\bullet}14H_2O$), or copiapite ($Fe^{II}Fe^{III}_4(SO_4)_6(OH)_2{\bullet}20H_2O$) [44, 46].

Thus, the flooded abandoned mines typically contain water rich in dissolved metals, sulfate and can be acidic. Obviously, economic value base metals or rare earth elements may be recovered from the mine water, such as Fe, Al, Cu, Pb, Zn, Mg, Mn, Ca, and Cd, high concentrations of SO_4^{2-} and rare earth elements like lanthanides plus scandium and yttrium, just to mention a few. Generally, mine water from abandoned coal mines is rich in rare earth elements because coal contains humic acids, which chelate metals and rare earths. Over time, humic acids tend to naturally degrade, resulting in drop of their concentration in coal and releasing the rare earth elements, which eventually get incorporated into aluminosilicate clays [44, 47]. Typically, metal concentrations for most mine water from precious metal mines occur in decreasing concentrations in the order Fe > Al > Zn, Cu > Cd > Pb. Therefore, Fe and Al dominate in the schemes of resource recovery from mine water.

The overall reaction of oxidizing (di-)sulfide minerals forming AMD has been presented in admittedly simplified form in Equation 11.1. The acid produced in the reaction may be neutralized by reaction with carbonate or silicate minerals, if the host rocks contains iron-bearing carbonates such as siderite ($FeCO_3$) and ankerite ($Ca(Fe,Mg,Mn)(CO_3)_2$) [44, 46, 47]. The neutralization reaction is presented in Equation 11.9:

$$2\ Fe^{2+} + 4\ SO_4^{2-} + 4\ H^{+}(aq) + 4\ CaCO_3 \rightleftharpoons 2\ Fe^{2+} + 4\ SO_4^{2-} + 4\ Ca^{2+} + 4\ HCO_3 \qquad (11.9)$$

Thus, many neutralized mine waters are relatively rich in base cations and hydrogen-carbonate alkalinity. Neutralization reactions can release yet more ferrous iron to solution.

11.3.4.2 Applicable Extraction Methods

Methods for extracting valuable elements from mine water follow into two main broad categories, namely: chemical and biological. Though the methods are not only applicable to metal extraction, the processes can be thought of as a form of hydrometallurgy. Chemical extraction methods stand for some of the more attractive options for resource recovery from mine water. The techniques used in chemical extraction methods can be subdivided into pure chemical, physicochemical, and electrochemical (Table 11.3) [43, 45, 48, 49] and the biological methods (Table 11.4) may further be divided into microbiological and phytomining techniques [50]. Most of the techniques in the table can be combined, used as stand-alone but also in combination with other microbiological methods as well as chemical methods.

11.3.5 Re-Mining Mine Water Treatment Sludge

11.3.5.1 Opportunities

One of the major inevitable activities in mine site remediation is the treatment of mine drainages. Mine water treatment systems fall into two categories: (a) Active treatment systems involve application of treatment chemicals (for neutralization and flocculation) to remove iron from the water and use of pumps and other energy-consuming equipment and (v) Passive technologies, in strict sense, harnessing natural processes, with little human intervention, to treat mine water. In most cases, passive treatment systems use a combination of natural materials, geophysical, and geochemical processes and biotic processes and use a gravity-fed water flow system [53]. A typical design of a passive treatment system allows the mine water to flow down an elevated cascade—for maximum aeration to oxidize Fe^{2+} to Fe^{3+}—into settling ponds, designed to collect precipitating iron and other contaminants. The settling pond has water residence times designed based on the sedimentation rate of the precipitates. Then, the water flows into a constructed wetland where treatment continues through natural chemical reactions and biological processes [52, 53].

An unavoidable consequence of mine drainage treatment is the production of large amounts of metal-rich sludge. Apart from iron, the sludge is also laden with various metals like Mg, Mn, Ca, and Al. Consequently, management and disposal of the sludge becomes an environmental and economic burden of the operation in both active and passive treatment

Table 11.3 Chemical methods for extraction of mineral and rare elements from mine water (Nordstrom *et al.* [44], El-Nadi *et al.* [45], Mohammadesmaeili *et al.* [48], Islam *et al.* [49], and Nariyan *et al.* [51]).

Technique	Subdivision	Description	Applicability
Pure chemical	Oxidative precipitation	Involves use of an oxidant agent to generate a colloidal precipitate, which can then be separated by filtration or sedimentation. Common oxidants in use include air, Cl_2, O_3, SO_2, O_2, $KMnO_4$	Extraction of ferric oxides, Manganese
	Metal reduction	Involves reduction of metals either with carbon or by electrolysis, or with a more reactive metal: (a) Using carbon – metal oxide is heated with carbon, losing electron to carbon, mostly producing intermediates metal chlorides because most metals react with carbon to form the carbide instead (b) Electrolysis – breaking down ionic substances into their precursor elements by electricity (c) more reactive metal – usually applied instead of carbon reduction	Extraction of titanium (mainly carbon) Cu cementation
	Solvent extraction	Involves separating metal complexes based on their relative solubilities in two different immiscible liquids, a polar (e.g., water) and non-polar (e.g., an organic solvent)	Recovery of U, Th, most rare earth, Cu, Ni, Au, Pd, Zn
	Sulfide precipitation	Involves reacting sulfur containing reagents (e.g., $H_2S(g)$, NaHS and Na_2S) to form highly insoluble metal sulfides: $HS^- + Me^{2+} \rightarrow MeS + H^+$	Recovery of valuable metals like Cu, Zi, Ni, and Co

(*Continued*)

Table 11.3 Chemical methods for extraction of mineral and rare elements from mine water (Nordstrom *et al.* [44], El-Nadi *et al.* [45], Mohammadesmaeili *et al.* [48], Islam *et al.* [49], and Nariyan *et al.* [51]). *(Continued)*

Technique	Subdivision	Description	Applicability
Physicochemical	Reverse osmosis	Involves removing water out of AMS by forcing water under pressure through a semi-permeable membrane, recovering suspended solids and ca. 99% of soluble cations and anions	Extraction of Cu, Ni, Co, Mg, Ca, Si
	Evaporation	Involves evaporating the water out of high concentrated mine water to recover solute metals	Fe oxides, Ca, U
	Ion exchange	Involves extraction of metallic ion from mine water with solid polymeric or mineralic ion exchangers, typical resins like functionalized porous or gel polymer, zeolites, montmorillonite, clay, and soil humus.	Extraction of Cu, Cd, Sb, Ni, Zn, Co, lanthanides, and actinides
Electrochemical	Electrowinning/ electroplating	An electrolytic process, entailing application of an electrical potential between inert Pb-alloy anodes and stainless steel (occasionally copper) cathodes to extract non-ferrous metals from mine water	Most common electrowon metals are Pb, Cu, Au, Ag, Zi, Al, Cr, Co, Mn, and rare-earth and alkali metals
	Electrocoagulation	Involves applying direct current to metal electrode plates, which dissolve and their metal ions form wide range of collectable coagulated species and metal hydroxides at an appropriate pH	Extraction of a number of ions including Al, B, Ca, Fe, K, Li, Mg, Na, SiO_2
	Electrodialysis	Involves separating and concentrating ions of interest using an ion-selective membrane under the influence of an electric field	Extraction of a number of ions including Al, B, Ca, Fe, K, Li, Mg, Na, SiO_2
	Electrokinetics	Involves application of an electrical current to induce the flow of water from the anode to the cathode, thereby simultaneously inducing the migration of ions. Eventually, the metal cations precipitate at the cathode where they are collected and dried.	Applicable to most major metallic ions

Table 11.4 Biological methods for extraction of mineral and rare elements from mine water (Nordstrom et al. [44], El-Nadi et al. [45], Mohammadesmaeili et al. [48], Islam et al. [49], and Nariyan et al. [51]).

Technique	Subdivision	Description	Applicability
Microbiological	Sulfate bioreduction	Involves sulfate-reducing bacteria (SRB) utilize alternative electron acceptors such as metal oxides in respiration, thereby reducing of metal oxides. Examples of SRB for application in mine water includes species of *Desulfobacteraceae*, *Desulfuromonadaceae*, *Rhodobacteraceae*, and *Desulfobulbaceae*	Extraction of ferric oxides, Manganese
	Iron bioreduction	Involves enzymatic dissimilatory oxidation of ferrous iron or reduction of ferric iron by acidophilic prokaryotic species widely distributed within the domains Bacteria and Archaea	Extraction of Ti, Cu cementation
	Metal bioreduction	Bioleaching process involving two steps: (1) Production of Fe^{3+} ions by ferrous iron oxidizing bacteria (e.g., *Thiobacillus ferrooxidans, Leptospirillum ferrooxidans, Gallionella ferruginea* and *Mariprofundis ferrooxydans*); (2) Fe^{3+} ions acting as a catalyst in oxidize the ore to release other metals.	Recovery of U, Th, most rare earth, Cu, Ni, Au, Pd, Zn
	Sulfide bioprecipitation	Bioleaching process involving two steps: (1) Production of Fe^{3+} ions by sulfur oxidizing bacteria (e.g., *Acidithiobacillus ferrooxidans* and *Acidithiobacillus thiooxidans*); (2) Fe^{3+} ions acting as a catalyst in oxidize the ore to release other metals.	Recovery of valuable metals like Cu, Zi, Ni, and Co
Phytomining	–	Involves bio-harvesting of metals from biomass of hyperaccumulating plant species grown, particularly, in mine water receiving and tailing ponds. It has recently advanced to produce low volume, sulfide-free "bio-ore"	Extraction of most metal capable to accumulate in hyperaccumulating pants

systems. These burdens can be lessened or eliminated by extracting potentially marketable components of the sludge, particularly iron oxide constituents for a variety of industrial and manufacturing applications. Further, with a few modifications, the active treatment system can simultaneously be used for treatment of water and synthesis of commercially valuable Fe-hydroxides, gypsum, and limestone.

11.3.5.1.1 Iron Oxide From Sludge

On average, iron oxide makes up about 60% of sludge from the mine water treatment process. Iron oxide has potential application such as in iron and steel manufacturing, many alloys production, in metallic jewellery and lenses polishing, in cosmetics, in pigment, in magnetic storage and recording media, in photocatalysis, and in medicine. It is used during copper mining for the extraction and purification of copper. Furthermore, the physical and chemical characteristics of the sludge facilitate their use as soil amendments or cementing agents. As a result, the demand for iron oxides has been ever increasing. Thus, mine water treatment sludge could prove a good resource.

11.3.5.1.2 Gypsum and Limestone Synthesis

AMD sulfate laden and pH neutralization of mine water in treatment systems are mostly done though addition of lime (CaO). If right thermodynamic conditions would prevail, CaO can transform to anhydrite ($CaSO_4$), which turns into gypsum ($CaSO_4 \bullet 2H_2O$) in water. Therefore, mine water treatment processes can be easily modified to produce gypsum. Gypsum is used in manufacturing of wallboard, cement, plaster of Paris, soil conditioning, a hardening retarder in Portland cement. It has also many applications in art and pottery and in medicine as casts for broken bones or as dental moulds for making artificial teeth. Its many other uses include the "paste" part of toothpaste, modern chalk used in classrooms, and as filler for paper and paints.

11.3.5.2 Applicable Extraction Methods

11.3.5.2.1 Iron Oxide From Sludge

The iron exists in one of two oxidation states, either ferrous having a +2 charge soluble in water at any pH, or ferric having a +3 charge, soluble at a pH less of than about 3.5 and precipitates if the pH is higher than 3.5, forming an orange/yellow compound called yellow boy. Since mine water can have high acidity, it may appear crystal clear, just because both ferrous and ferric iron are completely dissolved. Thus, the principal treatment strategy for mine water is neutralizing the acidity by adding alkalinity and, virtually, has all the iron in the ferric state *vis* increasing oxygen availability in the water for all ferrous to become ferric iron. This leads to the formation of mostly ferric (oxy)hydroxide (yellow boy), which eventually separates from the water into sludge and naturally sinks. Sludge has usually low density.

During the process, ferric iron may co-precipitate with other elements. The mine water sludge is commonly known as ocher, due to an orange to red coloration imparted by the ferric hydroxide, and is a mixture of hydrated iron oxides, ranging from amorphous ferric hydroxide ($Fe(OH)_3$), ferrihydrite (metastable nanocrystalline $Fe_2O_3 \bullet (H_2O)_n$, where $n = 1.8 - 0.5$), goethite (FeOOH) to haematite (Fe_2O_3) [29, 54].

The salient chemical reactions involved are mainly two: one that describes the reaction in which ferrous iron is converted to ferric iron (Equation 11.10); and the other that describes the actual hydrolysis and precipitation of ferric hydroxide (Equation 11.11).

$$4\ Fe^{2+}(aq) + O_2(g) + 4\ H^+\ (aq) \rightarrow 4\ Fe^{3+}\ (aq) + 2\ H_2O\ (l) \tag{11.10}$$

$$4\ Fe^{3+}(aq) + 12\ H_2O(l) \rightarrow 4\ Fe\ (OH)_3\ (s) + 12\ H^+\ (aq) \tag{11.11}$$

For extraction for economic use, high density sludge is needed. Hence, the process is enhanced by mixing the sludge with the lime slurry, which alters the physical structure of the sludge particles, resulting in larger and more rounded and smooth particles. The particles are hydrophobic and tend to attract metals or metallic compounds while repel water. As a result, they are denser and drain more readily to achieve high solids content.

11.3.5.2.2 Gypsum and Limestone Synthesis

In a classical active treatment system, the first step is usually to raise the mine water pH though addition of lime (CaO), in conjunction with aeration and clarification. Magnesite ($MgCO_3$) equally neutralizes and removes metals from the mine water, due to the interaction of magnesite with sulfate species available in the mine water lead to the precipitation of metals as hydroxides [55]. From the process, magnesium is freed and forms complexes of $MgSO_4$ on contact with sulfate rich mine water. When CaO is simultaneously added, interaction of CaO and $MgSO_4$ leads to the formation of gypsum ($CaSO_4{\bullet}2H_2O$). In brief, the reactions involved in the process are:

$$CaO + H_2O \rightarrow Ca\ (OH)_2 \tag{11.12}$$

$$MgSO_4 + Ca(OH)_2 + H_2O \rightarrow Mg(OH)_2 + CaSO_4 \bullet 2\ H_2O \tag{11.13}$$

At this point, bubbling the magnesite-lime treated water with CO_2 leads to synthesis of limestone because the residual Ca in reactions 11.12 and 11.13 reacts further with CO_2, and also stabilizes the pH, as shown in Equation 11.14.

$$CO_2 + Ca^+ + H_2O \rightarrow CaCO_3 + 2\ H^+ \tag{11.14}$$

The reaction above is a simplification of the process because CO_2 dissolves in water, forming carbonic acid (i.e., H_2CO_3)—the protons ($2H^+$) and initially hydrogencarbonate (HCO_3^- ions. The hydrogencarbonate ions donate protons and generate a carbonate ion. This process can be used as a double sided sword for gypsum manufacturing and sequestration of carbon dioxide.

11.3.5.3 Challenges

The market value of an iron oxide product is largely determined by its purity (i.e., FeOOH, Fe_2O_3, or Fe_3O_4 content), the presence of metal contaminants and the particle size [52]. Unfortunately, iron oxides extracted from mine sludge are extremely inconsistent in size and have a high number of other metals that it requires a lot of purification to compete with iron oxide obtained from direct ore mining, synthetic, and even by products from steel manufacturing. Iron oxides from direct mining are present as goethite, hematite, and magnetite deposits. Synthetic iron oxides are produced from oxidized and neutralized ferrous sulfate solutions. They are usually of high quality, pure, and it is easy to manipulate their grain size according to market requirements. This makes them the most valuable on the market.

Similarly, breaking into the market with gypsum synthesized from mine water is severely challenged by massive gypsum deposits in the world. Quality is usually not an issue, because no gypsum deposits are 100% pure. They are usually found with deposits of a combination of limestone, sand, shale, anhydrite and sometimes rock salt. To be a commercial deposit, gypsum content should be at least 75%, which the gypsum synthesized from the mine water surpasses.

11.3.5.4 Counter Options

Despite the challenges faced with mining of iron oxides from mine water treatment sludge, extraction iron oxide and development of products that do not require high purity still has very high potential of commercialization. The strategy would be changing the iron oxides into forms and quality required by target industries or application. Therefore, the quality and content of the sludge as well as the chemical, physical, and mechanical properties of the sludge should be known before the extraction starts. This helps to target the specifics of the iron oxides demanding industries and to decide on the products to be made.

For example, the raw iron oxide recovered from mine water treatment sludge can be processed into value-added products such as fertilizer additive. It has become customary in most golf courses to apply 2% iron-containing fertiliser a few days before a tournament to obtain a deep green color in specific areas of the course. Since the sludge contains a very high percentage of ferric oxides, cheap procedures are required for reducing the Fe^{3+} to the soluble, labile, and mostly bioavailable form, Fe^{2+} species. In brief, there are many iron oxide products that can be produced from mine water treatment sludge extracts.

11.3.6 Mine Methane Gas Extraction

11.3.6.1 Opportunities

Mines gases, especially related to coal mining, are responsible for underground death and mine explosions. These mine gases are dominantly composed of methane, counting for about 50%, followed by almost equal amounts of nitrogen and carbon dioxide, and several low quantities of other gases (Table 11.5). Methane and carbon dioxide are a product of geochemical transformation—carbonification—of organic substances that formed coal, starting hundreds of million years ago. Therefore, methane dominates among gases entrapped and adsorbed in cracks, crevices, and pores within coal seams and adjoining rocks [56]. The methane gas is responsible for most explosions in underground coal mining. Thus, methane and other mine gases are extracted and vented out of the mine for the safety of miners. Further, methane gas can be released naturally from the ground into the atmosphere by diffuse migration through cracks, crevices, and pores of overlying rock [55–59]. After the closure of a coal mine, the mine gases continue being released due to cracks and crevices created in the ground by the mining activity.

Table 11.5 Generalized composition of most mine gas. Modified from Robinson [59] and Osunmakinde [60].

Fraction	Composition
Methane, CH_4	50.0%
Carbon dioxide, CO_2	24.0%
Nitrogen, N_2	24.3%
Ethane, C_2H_6	1.0%
Propanol, C_3H_8	0.2%
Oxygen, O_2	0.5%

Methane is used primarily as fuel to make heat and electricity as well as to manufacture organic chemicals. The composition of mine gas is reasonably stable because the gases change very slowly due to very little oxygen, and the methane content ranges from 60–80% [61]. As a result, the composition of abandoned mine methane (AMM) gas has no technical difficulties for combustion in gas engines [61]. Thus, sealed abandoned mines offer an excellent opportunity for the production and extraction of high-quality methane. Further, recovery of AMM from abandoned coal mines for energy (thermal and electrical) production adds advantage of reducing atmospheric emissions of methane, a potent greenhouse gas.

11.3.6.2 Applicable Extraction Methods

11.3.6.2.1 Methane Drainage System

This method uses infrastructure from active mining to extract methane from underground coal workings, where the coal is extracted. When coal is removed, it leaves gobs, which usually fill with methane and other mine gases. Generally, extraction of AMM using the drainage system consists of a borehole detector, a wellbore, and a surface methane drainage system (Figure 11.12). In an abandoned mine, water influx from aquifers into the abandoned gobs is one of the most relevant problems, substantially affecting the gob methane flow to the wells. Thus, the borehole detector is used to detect aquifers above the abandoned gobs, so that they can be sealed before drilling begins, to avoid groundwater leakages, pouring into the bottom of the well. The surface well is used to drain the abandoned gob methane and is drilled through the topsoil overburden and enters the deep bedrock aquifers, into the gob.

The methane in the gobs is extracted using the negative pressure at the well bottom, which is created by vacuum pumping. Generally, wells located in the fractured zone near the caved zone are most effective for gob methane drainage because such locations are enriched by gob stress-relief methane [62–64]. The methane flow in abandoned mine gobs can be described by Darcy's law [61]:

$$u = \frac{k_p}{\mu} \nabla P \tag{11.15}$$

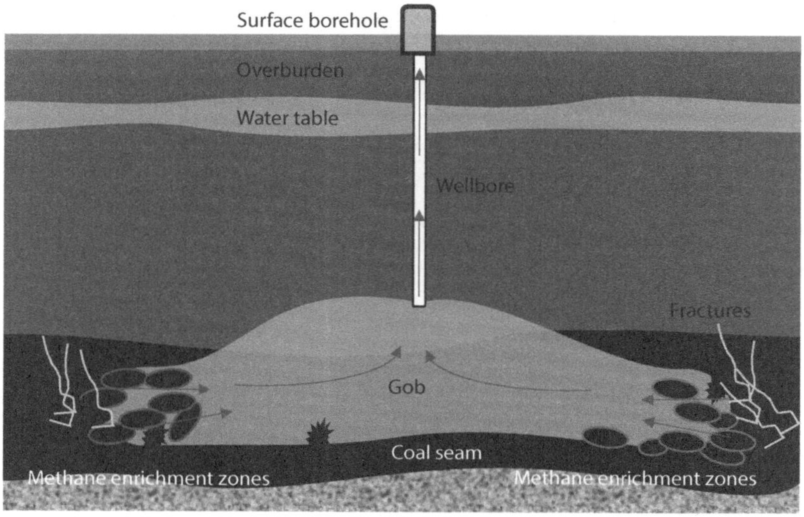

Figure 11.12 Methane drainage from a methane rich zone in an unmined coal seam of gobs of closed mines.

where u is the relative gas flow velocity between the fluid and the solid structure, μ is the dynamic viscosity of the fluid, ∇P is the pressure gradient that forces the fluid to flow through the porous media, and k_p is the permeability. Under normal state, the volume of methane (V_g) in the gob is proportional to residual pore volume (φ_r) the pressure exerted by methane (p) in relation to the pressure at normal state (p_0) and absolute temperature (T_0). Their relationship is described by Mariotte's law [62, 63]:

$$V_g = \frac{\varphi_r p T_0}{p_0 T Z} \tag{11.16}$$

where T is the reservoir temperature and Z is the methane compressibility factor.

11.3.6.2.2 Hydraulic Fracturing (Fracking)

This technique of extracting AMM is similar in many aspects to the drainage system, except for injection of fluids into the underground to enhance gas flow. The fluids are water mixed with particulate material (mostly sand) and chemical additives. The injection pressure of the fluids creates fractures and while the particulate coarse material holds the fractures open and the water displaces the gas [65]. Traditional hydraulic fracturing is quite an effective method for the extraction of coalbed methane and can be similarly applied in the AMM extraction. However, gobs have specific issues related to the mining activity such as water pressure, resulting from halting of dewatering when the mine closes, sometimes, there are large volumes of fractures, requiring special sealing. Consequently, there are several technology innovations, mostly modifications to improve methane extraction using hydraulic fracturing being developed. One such innovation is the newly developed technology of pulse hydraulic fracturing, which involves exciting oscillation from pulsating water pressure to weaken the rock strength. The gas pressure weakens the integrity of the coal seam, inducing micro-cracks generation, creating spots for facture formation even under a lower water pressure [66].

11.3.6.2.3 *In Situ* Gasification

Though not directly involving mine water, *in situ* gasification is applied when the gobs of the abandoned coal mines still contain a lot of un-mineable coal. The process involves drilling injection and production boreholes apart from each other into the coal despite remnants and the gob, respectively. Then, a permeable channel is created between two boreholes which enables a sufficient flow of gases [65, 66]. The injection borehole is used to feed the gas and oxygen mixture, while the production borehole is where the gas is captured and pumped to the surface for further processing. To initiate the gasification process, the coal is ignited to raise the coal temperature while the mixture of air/oxygen and steam are injected into the remnant deposits to gasify the coal (Figure 11.13) [67]. Generally, all types of coal can be gasified through this technique, although coal with lower ash content is preferable. The advantage of *in situ* gasification in abandoned mines is that there are minimal major issues of emissions of S, NO_x, and Hg.

11.3.6.2.4 Underground Bio-Gasification

In circumstance that methane in constant quantities and pressure exists, the production can be augmented or stimulated biologically. Generally, methane is produced biologically through microbial activity, mostly living anaerobically and using coal as the electron donor carbon (i.e., energy source). The growth and activities of aboriginal microorganism in the abandoned mine can be stimulated by addition of nutrients. Another approach is to augment the activities to break down *in situ* coal into simpler compounds, methane by introducing microorganisms.

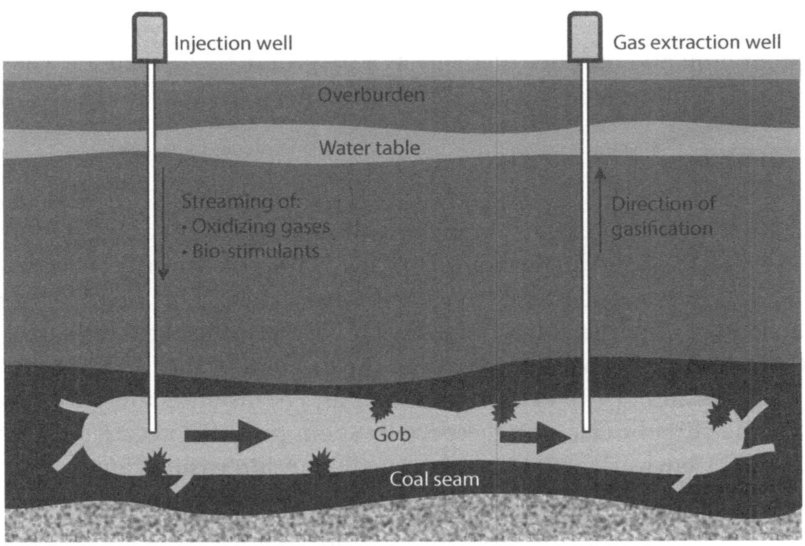

Figure 11.13 Schematic illustration of *in situ* biogasification of coal in an abandoned coal mine.

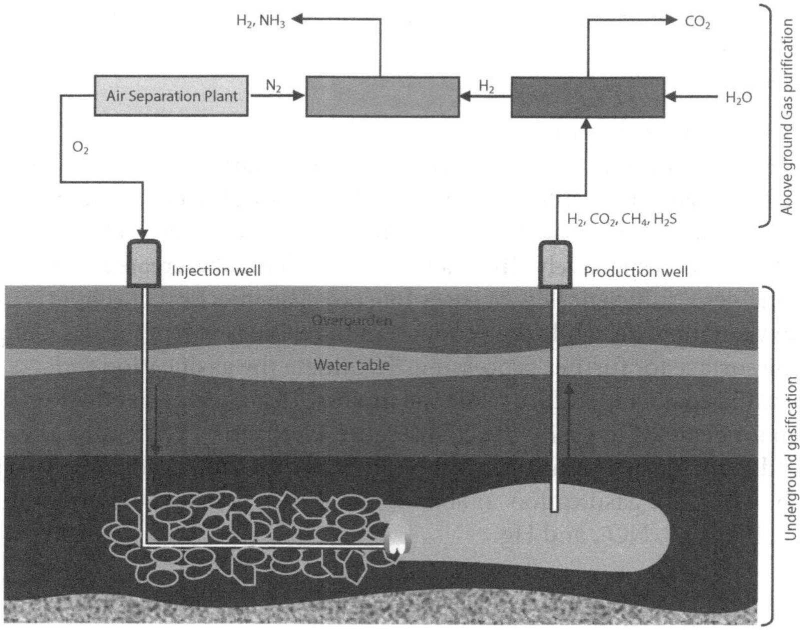

Figure 11.14 Principle of underground coal biogasification using the gob of an abandoned coal mine as bioreactor.

To carry out the process of *in situ* biogasification of coal in gobs, the system requires two boreholes via an injection and a production borehole through the cracked coal seam and in gob, respectively (Figure 11.14). The gasification takes place in the gob between the two boreholes, and the gas is collected through the production borehole. To stimulate or enhance the biogasification, nutrients or microorganisms as well as oxidizing agents are injected into the gob through the injection well. Most mine water establishes conditions for growth of aboriginal anaerobic microbe species, which are known to use coal as carbon source for respiration and produce methane as a by-product. The methane gas is captured in producing wells, pumped, processed, and compressed into tanks for further use.

11.3.6.3 Challenges

The major challenge with gas extraction in mines is mostly related to the age of the concept. Though some concerns are real, the concept of extracting gas from abandoned mines and mine water still sounds hypothetical to many. Consequently, there are a lot of myths that surround abandoned mine gas extraction, which has resulted in challenges to obtain social licence as well as obtaining authorization for the projects. Other challenges are directly related to specific techniques for extraction of the gas. For instance, there are generally public concerns with hydraulic fracturing, especially about potential negative effects on drinking water and other environmental damage. Further, *in situ* gasification in the abandoned mines has issues related to the nongaseous products of the gasification reaction, which can leach and contaminate groundwater. Other challenges are the possibility of developing surface subsidence due to consumption of the coal in deep seed during gasification.

11.3.6.4 Counter Options

Principally, a solution to the challenges faced with extraction of methane from abandoned mines lies in further research to generate information, to help differentiate the real from hypothetical issues. Further research will help to develop mitigation of the risks related to the problems. Some literature suggests careful analysis of the geological setting and selective gasification of seam areas would help in reducing both subsidence and production of a chemical product that may leach into ground water [66]. Further, stability mentoring of geophysical and geochemical conditions during and after the gasification should be mandatory to any *in situ* biogasification projects.

11.4 Conclusion

This chapter has examined a few examples of resource recovery from mine water, which can form part of a larger strategy for remediation and management of abandoned mine sites as well as rejuvenation of ghost post-mining towns, born out of loss of the main economic drive—the mining activity. Some of the potential resources and techniques discussed in the chapter are still at research levels, while others have already reached pilot stages as well as commercialization. There is very high potential in extracting a number of dissolved non-ferrous minerals from AMD, including some precious base metals and rare earth elements, depending on factors including site geology, hydrology, ore deposit composition, and mineralogy. Further, mine water can be a source of hydrogen and gypsum for energy as well as geothermal energy and natural gas. Recovery of these components of mine water adds an economic value, making mine water treatment a potential investment. Advances in resource recovery technology are pushing the extraction of these components of mine water close to reality. Like every innovation, however, there are several technical and socio-political challenges for this approach and its techniques to overcome to attain both full commercialization and social licences. For example, there are challenges to separate the valuable components from each other in a manner that is efficient and economically practical, and to obtain resources of high quality and purity of industrial-standard. Nevertheless, it is highly optimistic that mine water, even AMD, may once fuel new economy growth for most ghost post-mine towns. Finally, decisions on extracting resources and the methods to use should be grounded on well informed knowledge of the types of mine water that might be met.

It would be impossible to adequately represent all types of mine water compositions as well as abandoned mines. Hence, a few generalizations as well as conditions that may apply universally have been made and discussed.

References

1. Johnson, D.B. and Hallberg, K.B., Acid mine drainage remediation options: A review. *Sci. Tot. Environ.*, 338, 3–14, 2005.
2. Siyongwana, P.Q. and Shabalala, A., The socio-economic impacts of mine closure on local communities: Evidence from Mpumalanga Province in South Africa. *GeoJournal*, 84, 367–380, 2019.
3. Younger, P.L. and Robins, N.S. (Eds.), *Mine Water Hydrogeology and Geochemistry*, The Geological Society, London, 2002.

4. Younger, P.L., Banwart, S.A., Hedin, R.S., *Mine Water: Hydrology, Pollution, Remediation*, Springer, Dordrecht, The Netherland, 2002.
5. Bowen, G.G., Dussek, C., Hamilton, R.M., Pollution resulting from the abandonment and subsequent flooding of Wheal Jane Mine in Cornwall, UK, in: *Groundwater Contaminants and their Migration*, vol. 128, J. Mather, D. Banks (Eds.), pp. 93–99, Special Publications, Geological Society, London, 1998.
6. Carvalho, I.G., Cidu, R., Fanfani, L., Pitsch, H., Beaucaire, C., Zuddas, P., Environmental Impact of Uranium Mining and Ore Processing in the Lagoa Real District, Bahia, Brazil. *Environ. Sci. Technol.*, 39, 8646–8652, 2005.
7. Rawat, N.S. and Singh, G., The role of micro-organisms in the formation of acid mine drainage in the North Eastern Coalfield of India. *Int. J. Mine Water*, 1, 29–36, 1982.
8. Colmer, A.R. and Hinkle, M.E., The Role of Microorganisms in Acid Mine Drainage – A Preliminary Report. *Science*, 106, 253–256, 1947.
9. Banks, D., Younger, P.L., Arnesen, R.T., Iversen, E.R., Banks, S.B., Mine-water chemistry: The good, the bad and the ugly. *Environ. Geol.*, 32, 157–174, 1997.
10. Sheoran, A.S. and Sheoran, V., Heavy metal removal mechanism of acid mine drainage in wetlands: A critical review. *Miner. Eng.*, 19, 105–116, 2006.
11. Donovan, J.J. and Perry, E.F., Mine Flooding History of a Regional Below-Drainage Coalfield Dominated by Barrier Leakage. *Geofluids*, 2019, 16, 2019.
12. Hawkins, J.W. and Dunn, M., Hydrologic Characteristics of a 35-Year-Old Underground Mine Pool. *Mine Water Environ.*, 26, 150–159, 2007.
13. INAP, *Global Acid Rock Drainage Guide (GARD Guide)*, INAP: The International Network for Acid Prevention, Melbourne, Victoria, Australia, 2009, https://www.gardguide.com.
14. Castro, J.M. and Moore, J.N., Pit lakes: Their characteristics and the potential for their remediation. *Environ. Geol.*, 39, 1254–1260, 2000.
15. Guest, R. (Ed.), *Literature Review of Global Pit Lakes*, Suncor Energy Inc. and Golder Associates Ltd, Toronto, 2017, Report No. 1777450.
16. Shaoping, H., Xincai, C., Jiyan, S., Yingxu, C., Qi, L., Particle-facilitated lead and arsenic transport in abandoned mine sites soil influenced by simulated acid rain. *Chemosphere*, 71, 2091–2097, 2008.
17. Leopold, K., Michalik, B., Wiegand, J., Availability of radium isotopes and heavy metals from scales and tailings of Polish hard coal mining. *J. Environ. Radioact.*, 94, 137–150, 2007.
18. Banks, M.K., Schwab, A.P., Fleming, G.R., Hetrick, B.A., Effects of plants and soil microflora on leaching of zinc from mine tailings. *Chemosphere*, 29, 1691–1699, 1994.
19. Landa, E.R., Phillips, E.J.P., Lovley, D.R., Release of Ra-226 from Uranium Mill Tailings by Microbial Fe(III) Reduction. *Appl. Geochem.*, 6, 647–652, 1991.
20. Mkandawire, M., Biogeochemical behaviour and bioremediation of uranium in waters of abandoned mines. *Environ. Sci. Pollut. Res.*, 20, 7740–7767, 2013.
21. EPA, *2017 Notice of Proposed Rulemaking*, vol. 40, p. CFR Part 192 2, United States Environmental Protection Agency, Washington, D. C., 2017.
22. Johnson, D.B. and Hallberg, K.B., Acid mine drainage remediation options: A review. *Sci. Tot. Environ.*, 338, 3–14, 2005.
23. Mielke, R.E., Pace, D.L., Porter, T., Southam, G., A critical stage in the formation of acid mine drainage: Colonization of pyrite by *Acidithiobacillus ferrooxidans* under pH-neutral conditions. *Geobiology*, 1, 81–90, 2003.
24. Elsetinow, A.R., Borda, M.J., Schoonen, M.A.A., Strongin, D.R., Suppression of pyrite oxidation in acidic aqueous environments using lipids having two hydrophobic tails. *Adv. Environ. Res.*, 7, 969–974, 2003.
25. Benzerara, K., Morin, G., Tyliszczak, T., Casiot, C., Bruneeld, O., Farges, F., Brown, G.E., Jr., Nanoscale study of As biomineralization in an acid mine drainage system. *Geochim. Cosmochim. Acta.*, 72, 3949–3963, 2008.

26. Hedin, R.S., Iron Removal by a Passive System Treating Alkaline Coal Mine Drainage. *Mine Water Environ.*, 27, 200–209, 2008.
27. Shikazono, N. and Utada, M., Stable isotope geochemistry and diagenetic mineralization associated with the Tono sandstone-type uranium deposit in Japan. *Miner. Deposita.*, 32, 596–606, 1997.
28. Woulds, C. and Ngwenya, B.T., Geochemical processes governing the performance of a constructed wetland treating acid mine drainage, Central Scotland. *Appl. Geochem.*, 19, 1773–1783, 2004.
29. Banks, D., Athresh, A., Al-Habaibeh, A., Burnside, N., Water from abandoned mines as a heat source: Practical experiences of open- and closed-loop strategies, United Kingdom. *Sustainable Water Resour. Manage.*, 5, 29–50, 2019.
30. Wolkersdorfer, C., *Water Management at Abandoned Flooded Underground Mines: Fundamentals, Tracer Tests, Modelling, Water Treatment*, Springer, Berlin, 2008.
31. Masindi, V., Madzivire, G., Tekere, M., Reclamation of water and the synthesis of gypsum and limestone from acid mine drainage treatment process using a combination of pre-treated magnesite nanosheets, lime, and CO_2 bubbling. *Water Resour. Ind.*, 20, 1–14, 2018.
32. Tsibaev, S., Renev, A.A., Kalinin, S.I., Zorkov, D.V., Filimonov, K.A., Evaluation of The Effect of Flooding the Mine Workings with Water on Bolts Stability and on Deformations Coal and Rock Strata. *Vestn. Kuzbass State Tech. Uni.*, 18, 37–45, 2018.
33. Preene, M. and Younger, P.L., Can you take the heat? – Geothermal energy in mining. *Min. Technol.*, 123, 107–118, 2014.
34. Ismail, B.I., Power Generation Using Geothermal Low-Enthalpy Resources and ORC Technology, in: *Renewable Geothermal Energy Explorations*, B.I. Ismail (Ed.), pp. 1–8, IntechOpen, London, UK, 2019.
35. Dickson, M.H. and Fanelli, M., Geothermal background, in: *Geothermal Energy, Utilization and Technology*, UNESCO, *Paris*, 2003.
36. La Touche, G.D. and Preene, M., The potential use of ground energy in the mining industry–exploration to closure, in: *Mine Water – Managing the Challenges*, T.R. Rüde, A. Freund, C. Wolkersdorfer (Eds.), pp. 161–166, IMWA, Aachen, Germany, 2011.
37. Sui, D., Wiktorski, E., Røksland, M., Basmoen, T.A., Review and investigations on geothermal energy extraction from abandoned petroleum wells. *J. Pet. Explor. Prod. Technol.*, 9, 1135–1147, 2019.
38. Golabi, K., Scherer, C.R., Tsang, C.F., Mozumder, S., Optimal energy extraction from a hot water geothermal reservoir. *Water Resour. Res.*, 17, 1–10, 1981.
39. Pujades, E. *et al.*, Underground pumped storage hydropower plants using open pit mines: How do groundwater exchanges influence the efficiency? *Appl. Energy*, 190, 135–146, 2017.
40. de Oliveira e Silva, G. and Hendrick, P., Pumped hydro energy storage in buildings. *Appl. Energy*, 179, 1242–1250, 2016.
41. Menendez, J., Oro, J., Loredo, J., Fernandez, J.M., Velda, M.G., Underground pumped-storage hydro power plants with mine water in abandoned coal mines, in: *Mine Water and Circular Economy*, C. Wolkersdorfer, L. Sartz, M. Sillanpää, A. Häkkinen (Eds.), pp. 1–12, IMWA, Lappeenranta, Finland, 2017.
42. Gravity Power, *In-ground closed loop modular pumped storage*, http://www.gravitypower.net, 2017.
43. Menéndez, J. and Loredo, J., Use of closured open pit and underground coal mines for energy generation: Application to the Asturias Central Coal Basin (Spain). *E3S Web Conf.*, 80, 01005, 2019.
44. Nordstrom, D.K., Bowell, R.J., Campbell, K.M., Alpers, C.N., Challenges in recovering resources from Acid Mine Drainage, in: *Mine Water and Circular Economy*, C. Wolkersdorfer, L. Sartz, M. Sillanpää, A. Häkkinen (Eds.), pp. 1138–1146, IMWA, Lappeenranta, Finland, 2017.
45. El-Nadi, Y., Solvent extraction and its applications on ore processing and recovery of Metals: Classical Approach. *Sep. Purif. Rev.*, 46, 3, 195–215, 2017.

46. Freitas, R.M., Perilli, T.A.G., Ladeira, A.C.Q., Oxidative precipitation of Manganese from Acid Mine Drainage by Potassium Permanganate. *J. Chem.*, 2013, 287257, 2013.
47. Balaram, V., Rare earth elements: A review of applications, occurrence, exploration, analysis, recycling, and environmental impact. *Geosci. Front.*, 10, 1285–1303, 2019.
48. Mohammadesmaeili, F., Badr, M.K., Abbaszadegan, M., Fox, P., Mineral recovery from inland reverse osmosis concentrate using isothermal evaporation. *Water Res.*, 44, 6021–6030, 2010. https://doi.org/10.1016/j.watres.2010.07.070.
49. Islam, S.M.D.-U., Electrocoagulation (EC) technology for wastewater treatment and pollutants removal. *Sustainable Water Resour. Manag.*, 5, 359–380, 2019.
50. Sheoran, V., Sheoran, A.S., Poonia, P., Phytomining: A review. *Miner. Eng.*, 22, 1007–1019, 2009.
51. Nariyan, E., Sillanpää, M., Wolkersdorfer, C., Electrocoagulation treatment of mine water from the deepest working European metal mine – Performance, isotherm and kinetic studies. *Sep. Purif. Technol.*, 177, 363–373, 2017.
52. Hedin, R., Recovery of marketable iron oxide from mine drainage in the USA. *Land Contam. Reclam.*, 11, 93–97, 2003.
53. Branan, N., Mining for iron oxides in coal mine sludge. *Earth*, 54, 2009. https://www.earthmagazine.org/article/mining-iron-oxides-coal-mine-sludge.
54. Masindi, V., Gitari, M.W., Tutu, H., DeBeer, M., Synthesis of cryptocrystalline magnesite–bentonite clay composite and its application for neutralization and attenuation of inorganic contaminants in acidic and metalliferous mine drainage. *J. Water Process Eng.*, 15, 2–17, 2017.
55. Pashin, J.C., Variable gas saturation in coalbed methane reservoirs of the Black Warrior Basin: Implications for exploration and production. *Int. J. Coal Geol.*, 82, 135–146, 2010.
56. Pini, R., Ottiger, S., Burlini, L., Storti, G., Mazzotti, M., Sorption of carbon dioxide, methane and nitrogen in dry coals at high pressure and moderate temperature. *Int. J. Greenhouse Gas Control*, 4, 90–101, 2010.
57. Pophare, A.M., Mendhe, V.A., Varade, A., Evaluation of coal bed methane potential of coal seams of Sawang Colliery, Jharkhand, India. *J. Earth Syst. Sci.*, 117, 121–132, 2008.
58. Wierzbicki, M. and Dutka, B., The influence of temperature changes of the structurally deformed coal-methane system on the total methane content. *Arch. Min. Sci.*, 55, 547–560, 2010.
59. Robinson, R., Mine gas hazards in the surface environment. *Min. Technol.*, 109, 228–236, 2000.
60. Osunmakinde, I.O., Towards safety from toxic gases in underground mines using wireless sensor networks and ambient intelligence. *Int. J. Distrib. Sens. Netw.*, 9, 159273, 2013.
61. Hu, S. *et al.*, Methane extraction from abandoned mines by surface vertical wells: A case study in China. *Geofluids*, 2018, 9, 8043157, 2018.
62. Karacan, C.Ö., Analysis of gob gas venthole production performances for strata gas control in longwall mining. *Int. J. Rock Mech. Min. Sci.*, 79, 9–18, 2015.
63. Sang, S., Xu, H., Fang, L., Li, G., Huang, H., Stress relief coalbed methane drainage by surface vertical wells in China. *Int. J. Coal Geol.*, 82, 196–203, 2010.
64. Liu, A., Fu, X., Wang, K., An, H., Wang, G., Investigation of coalbed methane potential in low-rank coal reservoirs – Free and soluble gas contents. *Fuel*, 112, 14–22, 2013.
65. He, C., Zhang, T., Vidic, R.D., Use of Abandoned Mine Drainage for the Development of Unconventional Gas Resources. *Disruptive Sci. Technol.*, 1, 169–176, 2013.
66. Xu, J., Zhai, C., Qin, L., Mechanism and application of pulse hydraulic fracturing in improving drainage of coalbed methane. *J. Nat. Gas Sci. Eng.*, 40, 79–90, 2017.
67. Mao, F., Underground coal gasification (UCG): A new trend of supply-side economics of fossil fuels. *Nat. Gas Ind. B*, 3, 312–322, 2016.

Index

Printed in the USA/Agawam, MA
September 17, 2026

910661.035